Universalgenie
Helmholtz

Rückblick nach 100 Jahren

Universalgenie Helmholtz

Rückblick nach 100 Jahren

Herausgegeben
von Lorenz Krüger

Akademie Verlag

Gedruckt mit Unterstützung des Forschungsschwerpunktes
Wissenschaftsgeschichte und -theorie der Förderungsgesellschaft
Wissenschaftliche Neuvorhaben mbH

Die Deutsche Bibliothek – CIP-Einheitsaufnahme
Universalgenie Helmholtz : Rückblick nach 100 Jahren / hrsg.
von Lorenz Krüger. – Berlin : Akad. Verl., 1994
ISBN 3-05-002667-7
NE: Krüger, Lorenz [Hrsg.]

Der Akademie Verlag ist ein Unternehmen der VCH-Verlagsgruppe.

Gedruckt auf chlorfrei gebleichtem Papier.
Das eingesetzte Papier entspricht der amerikanischen Norm ANSI Z.39.48 – 1984 bzw. der europäischen Norm ISO TC 46.

Druck und Bindung: Druckhaus „Thomas Müntzer“ GmbH, Bad Langensalza

Printed in the Federal Republic of Germany

Vorwort

Am 8. September 1994 jährt sich zum 100. Male der Todestag von Hermann von Helmholtz - ein besonderer Anlaß für würdigende Rückblicke auf Leben und Werk des großen Wissenschaftlers. Helmholtz hat in einer Zeit sich rasch verzweigender Entwicklungen in Wissenschaft und Technik die grenzüberschreitende Geisteskraft eines Universalgenies bewiesen und uns damit auf seine Art die Einheit kultureller Aufgaben sichtbar gemacht, wie sie sich in der industriellen Lebenswelt stellen. Das Studium einer solchen Leistung, ihrer fachwissenschaftlichen, geistesgeschichtlichen, wirtschaftlichen und politischen Bedingungen und Auswirkungen ist noch längst nicht abgeschlossen; neue Einsichten in den komplexen Prozeß des Werdens der wissenschaftlich-technischen Zivilisation sind zu gewinnen.

Vom 3. bis 8. Januar 1994 fand sich eine internationale Gruppe von etwa 50 Fachwissenschaftlern, Wissenschaftshistorikern und Philosophen auf Burg Ringberg, der Tagungsstätte der Max-Planck-Gesellschaft, zum Austausch und zur kritischen Diskussion neuer Ergebnisse der Helmholtz-Forschung zusammen. Die Initiative und die Trägerschaft der Tagung sowie der Publikation lag beim Forschungsschwerpunkt Wissenschaftsgeschichte und Wissenschaftstheorie (Berlin) der Förderungsgesellschaft Wissenschaftliche Neuvorhaben mbH (München).

Der Schwerpunkt fand Rat und Ermutigung von seiten seines Wissenschaftlichen Beirats. Der großzügige internationale Zuschnitt der Tagung wurde dank der Unterstützung durch die Deutsche Forschungsgemeinschaft möglich. Die vorzügliche Betreuung durch die Tagungsstätte sorgte für ein optimales Gesprächsklima. Vielfache organisatorische Hilfen durch die Verwaltung der Förderungsgesellschaft und Angehörige des Schwerpunkts werden dankbar anerkannt. Unabhängige Gutachter haben mit ihren Ratschlägen bei Auswahl und Ausführung der Drucklegung entscheidend geholfen. Die Verantwortung für das Endprodukt liegt dessen ungeachtet natürlich bei den Autorinnen und Autoren und dem Herausgeber. Die Mühe, einen Index zu erarbeiten, nahmen Horst Kant und Angelika Irmscher (Berlin) auf sich. Besonderer Dank gebührt Sven Rosenkranz (Göttingen), der mich in allen Phasen der Tagung und Drucklegung unermüdlich beraten und unterstützt, sowie Tatjana Tarkian

(Göttingen), die das Manuskript korrekturgelesen und redigiert, und Ute Boldt (Göttingen), die den gesamten Text des Buches in kamera-fertige Form gebracht hat. Dem Akademie-Verlag, insbesondere den Herren Egel und Dr. Giesler, bin ich für die erfreuliche Zusammenarbeit verbunden.

Göttingen, Juli 1994 Lorenz Krüger

Inhalt

I
Helmholtz: Akademische Wege und Wirkungen

The Role of Johannes Müller in the Formation of Helmholtz's Physiological Career

Frederic L. Holmes

Johannes Müller is renowned not only for his own scientific achievements, but for the distinguished scientists counted as his students. Historical treatment of the relations between Müller and his students has, however, been ambivalent. They are said, on the one hand, to have been united by bonds of strong affection to their teacher, and on the other hand to have been separated from him by sharp discontinuities. His vitalism has been contrasted with their reductionism, his anatomical approach with the quantitative, instrumental directions they took, especially after 1840. Timothy Lenoir has described the three most prominent of these students, Helmholtz, Ernst Brücke, and Emil du Bois-Reymond, as "rebellious students" who "rejected Müller's view of the subject".[1]

1 Karl E. Rothschuh, *History of Physiology*, tr. Guenter B. Risse (Huntington, NY: Krieger, 1973), 212; Karl E. Rothschuh, *Physiologie: Der Wandel ihrer Konzepte, Probleme und Methoden vom 16. bis 20. Jahrhundert*, (Freiburg: Karl Alber, 1968), 253; Peter W. Ruff, *Emil du Bois-Reymond* (Leipzig: Teubner, 1981), 16; Timothy Lenoir, *The Strategy of Life: Teleology and Mechanics in Nineteenth Century German Biology* (Dordrecht: D. Reidel, 1982), 195; Timothy Lenoir, Laboratories, medicine and public life in Germany 1830-1849, *The Laboratory Revolution in Medicine*, ed. Andrew Cunningham and Perry Williams (Cambridge: Cambridge University Press, 1992), 14-71.

That Brücke, du Bois-Reymond, and Helmholtz each openly dismissed vitalistic views that included implicitly those of their own mentor, is clear enough from their published writings. A picture of the three younger physiologists at odds with Müller himself is, however, difficult to reconcile with the remarkable degree of personal solidarity that all three maintained with Müller throughout the formative stages of their scientific careers. A superficial explanation for this apparent paradox might be that they were dependent upon Müller's support to launch them into their academic trajectories: support which Müller provided in abundance for each of them at crucial junctures. I believe, however, that the bonds holding them together were far deeper. However much they differed with Müller in principle over the philosophical foundations, boundaries, or ultimate explanatory categories of physiology, they were in full harmony with their mentor about how to conduct concrete physiological investigations.

In an address given in 1877 at the Institute in which he had completed his medical education 35 years earlier, Helmholtz himself reminisced eloquently about this relationship:

> "There was one man in particular who aroused our enthusiasm for work in the right direction - the physiologist Johannes Müller. On theoretical issues he still favored the vitalistic hypothesis, but on the most essential points he was a natural philosopher, firm and immovable; to him, all theories were only hypotheses which had to be tested by facts [...]. Although he relied most heavily upon techniques of anatomical investigation, which were most familiar to him, he familiarized himself also with the alien methods of chemistry and physics".

After mentioning some of Müller's work that involved chemistry and physics, Helmholtz summarized briefly his achievements in the physiology of the nervous system. The principle of the specific energies of the nerves, and the distinction between motor and sensory nerves, "emerged from Müller's hands", Helmholtz declared, "in a state of classical perfection." Müller's "scientific spirit and his example had a strong influence upon his students", among whom Helmholtz included himself.[2]

Like all retrospective accounts this was not a transparent memory of how Helmholtz had perceived the situation in the 1840s. It was a reconstruction that fitted an early episode in his own life into the trajectory of his prior and subsequent experiences. There is, however, ample contemporary evidence, not

---

2 Russell Kahl, ed., *Selected Writings of Hermann von Helmholtz* (Middletown, CT: Wesleyan University Press, 1971), 352-353.

only to substantiate the general view he outlined in this succinct verbal portrait, but to specify in more concrete ways how Müller's "scientific spirit and example" inspired Helmholtz and his fellow students.

Historians have characterized Müller's approach to physiology in various ways, but have generally summarized it as qualitative, based in comparative anatomy, more morphological than experimental. He has been situated historically in an intermediate position between a speculative *Naturphilosophical* physiology and the new directions of the 1840s that are supposed to have left him behind.[3] Such characterizations fail to encompass the astonishing range and diversity of Müller's own physiological investigations, still less of the grand critical synthesis he achieved in his *Handbuch der Physiologie*.

Over the long course of his career Müller did carry out far more extensive investigations in comparative, developmental, or pathological anatomy than in functional physiology; but he also performed an impressive number of vivisection experiments, especially in his exploration of the nervous system. Those by which he confirmed in frogs the sensory and motor roots of the spinal nerves were models of experimental rigor. He was among the first to apply the improved achromatic microscopes that began to be available in the early 1830s to physiological questions, as well as to what was at first called "finer anatomy". He mastered contemporary chemical methods sufficiently to carry out important analyses of the blood and lymph and to initiate studies of digestion that were brilliantly continued by his assistant, Theodor Schwann. He incorporated new physical instruments, such as the voltaic pile and the galvanometer, into his physiological experiments. He cultivated contacts not only with comparative anatomists such as Cuvier, but with chemists such as Berzelius, and physically oriented physiologists such as the Webers.[4] Alert to all contemporary trends in physiology, he appreciated fully that anatomical ob-

3 Rothschuh, *History of Physiology*, 202; Gottfried Koller, *Johannes Müller: Das Leben des Biologen, 1801-1858* (Stuttgart: Wissenschaftliche Verlagsgesellschaft, 1958), 219-220; Lenoir, Laboratories, 45; Ruff, *Du Bois-Reymond*, 17. Brigitte Lohff has given an important critical assessment of these characterizations and shown that Müller had a deep understanding of the nature of physiological experiment. See Brigitte Lohff, Johannes Müller und das physiologische Experiment, in: *Johannes Müller und die Philosophie*, ed. G. Hogner and B. Wahny-Schmidt (Berlin: Akademie Verlag, 1993), 105-123.

4 Wilhelm Haberling, *Johannes Müller: Das Leben des Rheinischen Naturforschers* (Leipzig, Akademische Verlagsgesellschaft, 1924), 100, 463-464.

servation, animal experimentation, physical and chemical methods and reasoning were all essential to progress in this multifaceted field.

These aspects of Müller's work are essential to understand his relation to the directions taken by his later students. Space does not permit a detailed discussion of them here, but I would like to mention briefly one strategic example. In his lengthy discussion of the nervous system in his *Handbuch der Physiologie*, Müller's dominant orientation was expressed in the title of this section *Physik der Nerven*, and particularly in that of the third chapter, "The mechanics of the nervous principle". By the latter he meant "the same thing that is understood by the mechanics of light in physics; namely, the laws according to which the conduction of the effects in nerves takes place". Typical of the "laws" that Müller established was "the motor force acts in the nerves only in a direction toward the primitive fibers entering the muscles, or in the direction of the branching of the nerves, and never backwards". By "physics of the nerves", therefore, Müller did not mean a reduction of the processes in the nervous system to laws of physics, but an analysis of physiological laws analogous to the way in which physical phenomena were analysed. Although he did not report the details of the experiments underlying his analysis, it is evident from his discussion that he had performed a large number of them to arrive at these laws.[5]

Despite his view that the mechanics of the motions of the nerve principle did not depend on knowledge of its nature, Müller also took up that question. Ever since the discovery of galvanism, he noted, some people had identified the "active principle of the nerves" with electric currents. His own view was that nerves are conductors of electricity, but that "electricity and the nerve force are entirely distinct", a position for which he offered substantial evidence. Among the arguments against their identity were that those who had attempted to measure electric currents in nerves by means of the recently invented galvanometer had failed to detect them.[6] Müller testified that he had confirmed that a magnetized "needle hung from a silk thread showed [no] trace

5 Johannes Müller, *Handbuch der Physiologie des Menschen*, 3d. ed. (Coblenz: J. Hölscher, 1838), 1: 597-866, esp. 685, 688. For a fuller discussion of Müller's investigations of the nervous system, see Brigitte Lohff, Johannes Müller: Von der Nervenwissenschaft zur Nervenphysiologie, in: *Das Gehirn - Organ der Seele*, ed. Ernst Florey and Olaf Breidbach (Berlin: Akademie Verlag, 1993), 39-54.

6 See Prévost and Dumans, Sur les phénomènes qui accompagnent la contradiction de la fibre musculaire, *Journal de Physiologie*, 3 (1823): 301-338.

of declination when brought into the vicinity of a muscle and nerve in action".[7] In another place he reported that he had applied to such experiments two multiplicators (galvanometers) sensitive enough so that the weak galvanic current produced by only "two small zinc and copper plates" was sufficient to cause its needle to deviate by about 100 degrees on its compass, but "with this instrument I have never observed any trace of a reaction in nerves".[8] Du Bois-Reymond later wrote that he knew that Müller "had made many unsuccessful attempts to induce electrical effects in nerves". In the anatomical museum du Bois-Reymond found some multiplicator windings around a glass tube, which indicated to him that Müller had considered "whether the nerve principle might perhaps be conducted only through fluids, and could in that way be brought to have an effect on the magnetic needle".[9]

Müller did not take these negative results to be decisive. "Although it is certain that the experiments conducted with the galvanometer to test the electricity of the nerves yield no proof for their electricity", he wrote, "they can prove no more rigorously that no electricity is developed in the nerves, because these instruments are too imperfect". Briefly mentioning the limitations of a galvanometer even for measuring "true electricity developed by metal plates", he concluded "one can see clearly enough from this, that even if electricity also acts in the nerves, it will not easily be demonstrated through the galvanometer".[10]

When Müller arrived at the University of Berlin in 1833 to become Professor of Anatomy and Physiology in the Medical Faculty, he intended to make physiology the central thrust of his activity there. The field as he envisioned it would maintain a careful balance between gross anatomy, "the new aid of microscopic anatomy, experimentation, physics, and chemistry".[11] The demands of his position, particularly as Director of the Royal Anatomical Museum, reshaped his personal research program, directing his most sustained research efforts toward problems in comparative anatomy originating in his examination

7 Müller, *Handbuch*, 3rd. ed., 1: 645.

8 Karl Friedrich Burdach, with Johannes Müller, *Die Physiologie als Erfahrungswissenschaft*, 4 (Leipzig: Leopold Boß, 1832), 103-116.

9 Emil du Bois-Reymond, Gedächtnisrede auf Johannes Müller, in: *Reden von Emil du Bois-Reymond*, ed. Estelle du Bois-Reymond (Leipzig: Von Veit, 1912), 1: 191-192.

10 Müller, *Handbuch*, 3rd ed., 1: 647.

11 Haberling, *Müller*, 147-150.

of specimens in the museum collections.[12] During the rest of the decade, however, he found time also to carry out important microscopic investigations of the fine structure of tumor tissue,[13] and to publish in 1839 a remarkable monograph "On the Compensation of Physical Forces in the Human Vocal Organs". For the latter he performed delicate quantitative experiments to "determine the relation between the decreasing tension [of the vocal cords] and the growing strength of blowing with certainty and in numbers". Commenting later on this work of his teacher, du Bois-Reymond wrote, "One saw Müller, heretofore known only as an anatomist and an experimental physiologist [...] suddenly appear with great assurance in the field of physical investigation".[14]

After 1840 Müller devoted himself almost exclusively to morphological studies, particularly in invertebrate comparative and developmental anatomy. In the absence of reliable knowledge about why he made this shift, historians have speculated that he abandoned physiology because he was unsympathetic to its new directions. There is no contemporary evidence, however, that he made such a decision.[15] He continued to direct his students as strongly toward physiological as toward anatomical subjects. Like other heads of research schools, he undoubtedly associated himself with work carried out with his personal and institutional support, even when he was no longer in a position to perform it with his own hands.

In Berlin Müller had three positions for assistants - two prosectors, who made anatomical preparations for his lectures, and one general "Gehülfen". He used these posts "not merely for the needs of the museum, but to take into account the needs of science" by providing "talented young scholars a desired opportunity to prepare themselves for a scientific career". The general assistant when Müller came was Jacob Henle. In 1834, when one of the prosectorships became open, Müller promoted Henle to it and offered the assistantship to Theodor Schwann, who had just completed his medical education at Berlin.[16]

---

12 Koller, *Johannes Müller*, 113-120.

13 Johannes Müller, *Ueber den feinern Bau und die Formen der krankhaften Geschwülste* (Berlin: G. Reimer, 1838), 2-3.

14 Johannes Müller, *Über die Compensation der physischen Kräfte am menschlichen Stimmorgan* (Berlin: Hirschwald, 1839), passim, esp. 3, 8, 29; Koller, *Johannes Müller*, 231; du Bois-Reymond, Gedächtnisrede, p. 225.

15 Lenoir, Laboratories, 43-44; Haberling, *Johannes Müller*, 231; du Bois-Reymond, Gedächtnisrede, 225.

16 J. Müller to Ministry of Culture, February 1840, Rep. 76, Va, Sekt. 2, Tit. X, Bd. IV, 54-55, Geheimes Staatsarchiv Preussischer Kulturbesitz (Henceforth: GSPK);

For four years Müller, Henle, and Schwann worked together so closely that it is difficult entirely to separate their respective contributions to the fruitful investigations that took place during those years in the anatomical institute. Others who soon joined the group included Robert Remak and Karl Reichert.

That Müller possessed special qualities that drew talented young scientists to him and that helped launch them on auspicious "*wissenschaftliche Laufbahnen*" is clear from the testimony of several of those who worked in the two small rooms where they conducted their research, or in the anatomical museum. Friedrich Bidder, who spent several months there in 1834, wrote afterward of the precious advantages he had gained from this experience. Among them was the opportunity

> "to witness all the scientific discussions that Müller carried on with his two trusted associates [Henle and Schwann], and to learn about all the questions with which he and his two friends busied themselves. I could also take from him the example of the tireless enthusiasm and the enduring perseverance with which Müller devoted himself to the investigations necessary to resolve the questions posed".[17]

The qualities that initially attracted Theodor Schwann were Müller's scientific achievement, his "open loyal character", and the clarity and intensity of his lectures. When he began to work with Müller, Schwann received strong encouragement "to devote myself to an academic career". "Association with Müller", Schwann recalled later, "was extraordinarily stimulating". He encouraged one "to pursue through investigation and experiment, every new idea that one expressed to him".[18]

Perhaps because of his own broadly ranging scientific interests, Müller could guide his young students and assistants into fruitful lines of investigation in widely diverse areas of anatomy and physiology. Henle pursued problems in both comparative and microscopic anatomy. Reichert continued to expand his work in comparative embryology.[19] Schwann, himself the most versatile of Müller's students, carried out brilliant investigations on four different fundamental problems: on digestion, in which he characterized the digestive ferment

---

J. Müller to Ministry of Culture, October 18, 1834, Rep. 76, Va, Sekt. 2, Tit. X, No. 11, Bd. V, 164-165, GSPK.

17 Friedrich Bidder, Vor hundert Jahren im Laboratorium Johannes Müllers, *Münchener Medizinische Wochenschrift*, 12 January 1934, 61-62.

18 Schwann to du Bois-Reymond, Dec. 22, 1858, Gedächtnisrede, 287-288.

19 Müller to Ministry of Culture, February 1840; Vladislav Kruta: Reichert, Karl Bogislaus, *Dictionary of Scientific Biography*, ed. C.C. Gillispie (New York: Charles Scribner's 1970-1980) 10: 360-361.

that he named "pepsin";[20] on fermentation, where he concluded long before Pasteur that a growing microorganism is essential to the process;[21] on muscle contraction, in which, as he later claimed, "for the first time, so far as I know, a vital phenomenon was subjected to mathematical laws expressed in numbers".[22] Climaxing these endeavors was the microscopic examination of animal tissues that led Schwann in 1838 to his celebrated cell theory.[23]

As is well known, it was Schwann, not the students who entered Müller's Institute after 1840, who first challenged Müller's vitalistic philosophy of physiology. Schwann later recollected that "already as a student in Bonn my intellectual direction was very different from his".[24] In the *Microscopic Investigations* that he published at the end of 1838, just before leaving Berlin to take up a chair in physiology in Belgium, Schwann drew a contrast between the "teleological" and the "physical" view of life[25] which opened the confrontation between vitalism and antivitalism that became so prominent during the next decade.

It was characteristic of Müller as a mentor that he did not mind this attack by his own student on views with which he had identified himself in his *Handbuch* and in other writings. Recognizing that Schwann was not only a "richly gifted talent" but that his "precious independent discoveries" had already assured him "an outstanding place among the physiological investigators and observers of the first rank", Müller did his utmost to secure for Schwann an appropriate academic position in Prussia. He did not succeed, but the description he gave of his former assistant reveals how highly Müller valued both Schwann's experimental achievements and his theoretical independence. "In fact", Müller wrote the Minister of Culture, "through his work the material has completely unexpectedly been delivered for a theory of the organism, especially of the nature of animals, and he himself has largely carried it out in the course of his work on cells". Schwann was "so significant as an observational,

20 Theodor Schwann, Ueber das Wesen des Verdauungsprozesses, *Archiv für Anatomie, Physiologie*, 1836, 90-138.

21 Theodor Schwann, Vorläufige Mittheilung, betreffend Versuche über die Weingährung und Fäulniss, *Annalen der Physik und Chemie*, 2d. ser., 11 (1837), 184-193.

22 Schwann to du Bois-Reymond, Gedächtnisrede, 288-289; Johannes Müller, *Handbuch der Physiologie des Menschen*, vol. 2 (Coblenz: J. Hölscher, 1840), 59-62.

23 Theodor Schwann, *Mikroskopische Untersuchungen über die Uebereinstimmung in der Struktur und dem Wachstum der Thiere und Pflanzen* (Berlin: Sander, 1839).

24 Schwann to du Bois-Reymond, Gedächtnisrede, 288.

25 Schwann, *Mikroskopische Untersuchungen*, 220-230.

as well as a speculative investigator", that the investigations to which he had devoted himself now defined the "dominant direction of research".[26] Müller himself adopted Schwann's cell theory and applied it to his ongoing investigations of pathological tumors. He never accepted Schwann's anti-teleological views,[27] but there is no evidence that he ever felt threatened by them.

Of the trio of Brücke, du Bois-Reymond and Helmholtz, we can follow in most detail the interactions between du Bois-Reymond and their common mentor. During his medical student years, and the time that he found his way into the experimental investigation that defined his life-work, du Bois-Reymond wrote to his youthful friend Eduard Hallmann long letters in which he mentioned Müller so frequently that we can construct a full picture of the role that Müller played in the formation of du Bois-Reymond's career. Although their relationship was shaped as much by du Bois-Reymond's intense personality as by that of Müller, du Bois-Reymond's experiences can nevertheless help to illuminate the more sparsely documented role of Müller in Helmholtz's early scientific life.

Du Bois-Reymond took all of Müller's lecture courses and anatomical *repertoria*, and studied his *Handbuch der Physiologie* thoroughly. Because Hallmann had been at odds with Müller and described him to du Bois-Reymond as untrustworthy, du Bois-Reymond was wary in his first encounters with Müller, but very quickly began to feel the "power" of his personality. Within six months he had formed a strong admiration for Müller, but believed that the best way to get what he wanted from him was to treat him with "firmness, if not to say rudeness". Far from putting Müller off, this aggressive approach succeeded so well that du Bois-Reymond soon entered Müller's inner circle, with full access to the research facilities of the anatomical museum.[28]

While Müller was on an extended travel to Italy in the fall of 1840, du Bois-Reymond came under the strong influence of Müller's assistant and former student Karl Reichert, who had just published a book on the embryology

26 Johannes Müller to Ministry of Culture, December 7, 1838, Rep. 76-Va, Sekt. 2, Th. X, Xte Abtheilung, No. 11, Vol. VI, 47-52, GSPK.

27 For an excellent discussion of the way in which Müller adapted Schwann's cell theory to his own teleological viewpoint, see François Duschesneau, *Genèse de la théorie cellulaire* (Montréal: Bellarmin, 1987), 212-231.

28 K.E. Rothschuh: Du Bois-Reymond, Emil Heinrich, *DSB* 4: 200-205; Estelle du Bois-Reymond, ed., *Jugendbriefe von Emil du Bois-Reymond an Eduard Hallmann* (Berlin: Reimer, 1918), 1-7, 27, 32-33, 35, 39-42, 51, 56-57.

of vertebrates oriented around Schwann's cell theory. In contrast to Schwann, however, Reichert maintained an organismic philosophy more uncompromising than that of Müller, to which du Bois-Reymond became strongly attracted. Under Reichert's tutelage, du Bois-Reymond undertook in the spring of 1841 a study of the cleavage of frog eggs from the standpoint of the cell theory. Within a few weeks, however, he became somewhat disillusioned with Reichert and transferred his allegiance to another member of the Müller circle, Ernst Brücke, a fellow medical student. Under Brücke's influence du Bois-Reymond began to move toward the view of physiology maintained by Schwann, that the phenomena of life are not essentially different from physical phenomena.[29]

It is conspicuous that du Bois-Reymond did not identify Johannes Müller with the opposition to the physicalist direction of Schwann. Far from it. It was, he wrote to Hallmann, "Entirely through Müller that I have again been led back into the field of physics". That had happened in March, when as du Bois-Reymond then wrote, "Müller pressed on me most urgently (entirely on his own, because he believed that the task is designed for me, as I am created for the task) the repetition, extension, and testing of the earlier and most recent experiments of Matteucci on the frog current and on the relation of the nerve principle to electricity".[30]

In 1840 Carlo Matteucci had reported, as Müller summarized it in a later edition of his *Handbuch*, that when a preparation consisting of a frog spinal cord connected only through a nerve to a hind leg was placed in a vessel filled with a salt solution, and the leg in a second such vessel, "and the ends of the leads of a galvanometer are brought into contact with the solutions, there follows a deviation of several degrees of the magnet needle, which always shows a current from the foot toward the head".[31]

It is common knowledge that Müller suggested the investigation that du Bois-Reymond turned into the foundation for his scientific career. What the

29 Vladislav Kruta: Reichert, Karl Bogislaus, *DSB*, 10: 360-361; Karl Bogislaus Reichert, *Das Entwicklungsleben im Wirbelthier-Reich* (Berlin: Hirschwald, 1840); Estelle du Bois-Reymond, *Jugendbriefe*, 72, 75, 79, 81, 84, 89, 97-99. See K.B. Reichert, Ueber den Furchungs-Process der Batrachier-Eier, *Archiv für Anatomie, Physiologie*, 1841, 523-541.

30 Estelle du Bois-Reymond, *Jugendbriefe*, 85-99.

31 Johannes Müller, *Handbuch der Physiologie des Menschen*, 4th ed., vol. 1 (Coblenz: J. Hölscher, 1844), 556.

foregoing account should make clearer, however, is that in doing so Müller was not reaching to find a topic suitable to someone whose talents and outlook were fundamentally different from his own. He was urging du Bois-Reymond to take up a problem that had been of deep personal interest to him, that he had formerly investigated himself but no longer had time to pursue, and that had been placed in a new light by Matteucci's claim to have discovered what had previously eluded both Müller and others. He had already before then suggested in his *Handbuch* that, if these currents existed, the failure to detect them was due to the imperfections of the galvanometers. When du Bois-Reymond accepted Müller's charge to him, his first task was to construct a galvanometer far more sensitive than any that had previously been available. As he immersed himself in what proved to be a difficult and prolonged project, he continued to enjoy Müller's full support. The bonds between the impetuous young scientist and his mentor became far too strong to be disturbed by philosophical or other differences of opinion peripheral to their shared commitment to rigorous, critical scientific investigation.[32]

Fewer details concerning the personal interactions of Ernst Brücke with Müller are available than for those of du Bois-Reymond, but there is every reason to infer that Brücke and Müller also formed strong bonds of mutual esteem. After completing his MD degree in November 1842, Brücke planned to become a surgeon, but was instead drawn deeper into Müller's orbit. In recommending his appointment as *Gehülfen* in 1843, Müller wrote that "Dr. Brücke is a decisive and outstanding talent, of whom it can be definitely predicted that he will have further significant achievements in science". Three years later, in recommending him for a professorship at Königsberg, Müller said that Brücke's investigations had "fully justified the expectations that I held for him". After enumerating his research publications and praising his knowledge of physiology, physics, and chemistry, Müller added that "Physiology as the experimental physics of the living body is in all of this work the dominating direction and, given Brücke's intellectual foundation, it will remain so". Müller made it clear that he regarded this direction, as well as the physical and mathematical methods Brücke applied to his work, as rare but important assets for the future growth of physiology.[33]

---

32 Estelle du Bois-Reymond, *Jugendbriefe*, 89-90.

33 Hans Brücke, Einleitung, in: *Ernst Wilhelm von Brücke - Briefe an Emil du Bois-Reymond*, ed. Hans Brücke et al. (Graz: Akademischer Druck, 1978), XIV; Johannes Müller to Ministry of Culture, January 1841, September 25, 1843; Rep. 76-Va, Sekt. 2,

We can now try to fit Hermann Helmholtz into this group portrait of Müller and his physiological students in the early 1840s. At first sight it appears unproblematic that Helmholtz must have experienced as fully as du Bois-Reymond and Brücke the research culture surrounding Müller at the anatomical institute, as well as the "spirit and example" of the leader himself that Helmholtz invoked so strongly 35 years later. According to his first major biographer, Leo Koenigsberger, after Helmholtz returned to Berlin in the winter of 1841,

> "he attacked anew the investigation that his teacher Johannes Müller had, at least in several general suggestions, indicated for him, and he now lived with his thoughts and aspirations entirely in the circle of Müller's youngsters, now already befriended by the two-year older young physiologists Brücke and du Bois-Reymond, who, like him, were attached to their teacher with inspiration and admiration".[34]

If we examine critically this glowing vision of Helmholtz's participation in Müller's circle, however, we find little direct contemporary evidence to validate it. With the publication by David Cahan of the full texts of Helmholtz's letters to his parents, it is easier to see that these letters provided the only direct documentation available to Koenigsberger to reconstruct this phase of Helmholtz's life. The letters contain few entries mentioning Müller. The first two, in a letter of May 5, 1839, merely list Müller's physiology among the lectures he attended regularly during the spring semester. Ten days later he added the brief evaluation "Müller's physiology is excellent". Only once, on August 8, 1842, did he describe a personal contact with Müller:

> "Today I went to Professor Müller with my dissertation. He received me in a very friendly way, and after he had had the main results and the proofs for them stated, he explained that it was, to be sure, of great interest, because it demonstrated an origin for the nerve fibers which had been suspected for the higher animals, but could not be proven. He advised me, however, to investigate the matter first in a more complete series of animals than I had previously done, so as to provide a more stringent proof than can be obtained from the investigation of 3 or 4. He identified several for me, in which one could expect to find the best results, and even invited me, in case my instrument [i.e. microscope] was not adequate, to use his in the anatomical museum. If I were not in a hurry for my promotion, he advised me to use the holiday for further work in order to bring into the world a perfect child that would have no further attacks to fear. Since I knew

Tit. X, Bd. IV, 79-82, 102-103, GSPK; Müller to Ministry of Culture, August 30, 1847, Rep 76 - Va, Sekt 11, IVte Abt., No. 13, 7-8, GSPK.

34 Leo Koenigsberger, *Hermann von Helmholtz* (Braunschweig: Vieweg, 1902-1903), 1: 44.

> nothing sensible to say against that, and would probably have said most of it myself",

Helmholtz asked his parents if they would mind the delay that this extended investigation entailed.[35]

It appears evident that Koenigsberger projected onto this sole contemporary testimony by Helmholtz, which he quoted in his biography, retrospective statements by Helmholtz such as the one quoted earlier in the present paper, together with some imaginative embroidery, to arrive at his picture of a Helmholtz fully integrated into Müller's group of young investigators. If we restrict ourself to the letter itself, however, Helmholtz can be viewed as approaching Müller for the first time about an investigation on which he had already done the groundwork by himself. Moreover, although the date of completion of his dissertation (November 2, 1842) indicates that he probably did extend his observations through the holiday, there is no confirmation that, having previously worked with his own microscope, and probably in his own quarters, he accepted the invitation to move his work to the Museum and to use Müller's instrument.

What additional indirect evidence can we bring to bear on the extent and nature of Helmholtz's interactions with Müller or his students? There exists in the Helmholtz *Nachlaß* a notebook containing 78 pages of notes that he took during Müller's lectures on pathological anatomy in the summer semester of 1840.[36] As David Cahan's introductory essay for the Helmholtz letters shows, we can assume also that, like all medical students at Berlin, he attended all four of the lecture courses Müller offered, and dissected cadavers under his nominal supervision.[37] Beyond this, we can infer from the experiences of the other students of Müller that we have followed, some things that we would expect also to have happened with Helmholtz. From the imperative advice that du Bois-Reymond received to read through Müller's *Handbuch*, we can surmise that Helmholtz, too, probably studied that work carefully. From the prominent place that Müller gave Schwann and the cell theory in his lectures, Helmholtz could hardly have escaped learning how important both appeared from the vantage point of the anatomical museum. Finally, from the pattern of

35 David Cahan, ed., *Letters of Hermann von Helmholtz to his Parents* (Stuttgart: Steiner, 1993), 58, 59, 63, 91.

36 H. Helmholtz, Collegien Heft der Vorlesungen von Johannes Müller, Helmholtz-Nachlaß no. 538, Archives of the Akademie der Wissenschaften, Berlin.

37 Cahan, *Letters*, 17.

support that Müller had consistently offered to "talented young scholars", we would expect that he was not merely being polite when he suggested that Helmholtz could work at the museum. Surely, he would quickly have recognized that this young medical student was extraordinarily gifted, and would have made every effort, as he had done with Schwann and others before him, to bring Helmholtz into his circle.

The subject of Helmholtz's doctoral dissertation, which he dedicated to Müller, constitutes in itself suggestive evidence that Müller had helped to shape it from its inception, rather than only in its late stages. In general, it fits within the domain of microscopic studies carried out within the framework of Schwann's cell theory that was characteristic of the group working in this period around Müller. More specifically, the question that Helmholtz explored was one that Müller had pointedly defined in his *Handbuch* in 1838:

> "An important question is whether the large globules of the gray substance in the brain and in the ganglia are without broader connections. Certain prongs that one can see under favorable conditions extending here and there out from these globules make a connection of the globules with one another or with the fibers probable".

Müller mentioned that he himself had first observed such prongs in the medulla oblongata of the lamprey eel, and that Remak had soon after found in the globules of the gray matter and ganglions extensions several times as long as the globules themselves. According to Müller, Remak's observations made it "to a certain degree probable, or at least likely, that the gray fibers of organic nerves originate" from these ganglionic globules.[38]

To generalize from such isolated and uncertain observations was risky, however, so there was clear need to study further situations favorable to the detection of connections between nerve fibers and globules. His familiarity with invertebrate comparative anatomy would have made it evident to Müller that the simplicity of their nervous systems and the organization of their nervous tissue into a series of small ganglions connected by nerve cords provided especially favorable conditions for studying the relation between fibers and globules.

We can readily imagine that, if Helmholtz did come to Müller from the start for advice concerning his dissertation research, Müller would have been able to suggest that a microscopic examination of the invertebrate nervous system was an excellent topic. He could anticipate that even a fledgling investigator

---

38 Müller, *Handbuch*, 3d ed., 1: 612.

might be able to discover the connections between fibers and globules that he believed must exist.

Helmholtz was quickly able to provide the desired evidence. By the end of the winter term of 1841-42, he had observed in three or four types of invertebrates that some of the fibers in the nerve cords passed into or extended outward from the globular bodies contained in the ganglia. After his conference with Müller in August, he extended his observations to include several species each of crustaceans, leeches, molluscs, insects, and arachnoids.[39]

As plausible as it appears that Helmholtz must have received assistance from the beginning of his project, rather than to have conceived it on his own, taught himself the necessary skills, and first approached Müller with a successful result already in hand, the direct evidence does not rule out the latter. Even if he did not make personal contact with Müller at the start, Helmholtz must have owed the idea for his investigation, at least in part, to reading Müller's *Handbuch* or hearing his lectures. Otherwise it is hard to see where he could have learned that it was an important problem.

After receiving his medical degree in November 1842, Helmholtz began his required military service at the Charité hospital in Berlin. Having some time to spare, he undertook an experimental study of fermentation. According to Koenigsberger, the inspiration of "his revered master Johannes Müller" led Helmholtz to take up this "methodically pursued investigation", and he carried it out in "the laboratory of Müller". The subject of the experiments again lends plausibility to the view that they would be conducted in Müller's anatomical institute. They were direct refinements and extensions of the fermentation experiments that Schwann had performed there six years earlier. In the letters to his parents during that period, however, in which he described in detail his clinical activities at the Charité, Helmholtz never mentioned Müller or the anatomical institute. (He wrote, in fact, only one short passage about his progress on the fermentation experiments.) Koenigsberger provides no other documentary support for his assertion about the location in which Helmholtz worked.[40]

If we put together the direct and indirect evidence for his interactions with Müller, and add to Helmholtz's recollection of 1877 cited earlier, the sentence

39 Hermann Helmholtz, De Fabrica Systematis nervosi, Evertebratorum, *Wissenschaftliche Abhandlungen*, (Leipzig: J.A. Barth, 1882-95), 2: 663-679.

40 Koenigsberger, *Helmholtz*, 1: 50-51; Cahan, *Letters*, 92-106; Helmholtz, Ueber das Wesen der Fäulniss und Gährung, *Archiv für Anatomie, Physiologie*, 1843, 453-462.

"as fellow students [of Müller] I met E. du Bois-Reymond, Virchow, Brücke, Ludwig, Traube, J. Meyer, Lieberkühn, and Hallmann",[41] we may appear to have a strong circumstantial case that Helmholtz did become a full member of the group surrounding Müller sometime between 1841 and his departure for Potsdam in October 1843. There is, however, an obstacle to our acceptance of this picture. According to a letter of du Bois-Reymond to Hallmann, he first made "Helmholtz's acquaintance" in Berlin in October 1845.[42] Reported soon after the encounter, this account seems more reliable than Helmholtz's distant recollection that he had known du Bois-Reymond as a fellow student of Müller (especially in view of the glaring error Helmholtz made in including Carl Ludwig as a student of Müller). How would it have been possible for Helmholtz to frequent Müller's circle for any extended period between 1841 and 1843 without ever running into one of its most prominent and most assertive members?

I have not searched exhaustively for other sources that may resolve this discrepancy, and must for the present leave it as an enigma. It is as difficult to imagine Helmholtz carrying out all his observations on the invertebrate nervous system and his experiments on fermentation in isolation from Müller's group as it is to imagine him conducting them in Müller's vicinity without contact with du Bois-Reymond. Tentatively we might cover the situation by saying that, even if Helmholtz did join Müller's circle, he probably did not enter its inner orbit.

That Helmholtz may have been less fully or literally a member of the "Müller school" than he has customarily been portrayed to be does not make Müller less important to the early orientation of his career. In his later recollections Helmholtz consistently maintained a central place for Müller among his formative experiences. In his "autobiographical sketch" he stated in 1891: "In my studies I came immediately under the influence of a profound teacher, the physiologist Johannes Müller".[43] Distant memories can often be incorrect in details, yet accurate in their central meaning. I believe that we can trust that, no matter how skimpy our reliable knowledge about his role in Helmholtz's student days may be, Müller was certainly a powerful presence for him. To have attended his lectures alone could have been sufficient to inspire Helm-

41 Kahl, ed., *Selected Writings*, 352.
42 Estelle du Bois-Reymond, *Jugendbriefe*, 122.
43 Kahl, ed., *Selected Writings*, 470.

holtz to take up the anatomical and physiological investigations that initiated him into a "wissenschaftliche Laufbahn". It is hard to imagine any other comparable source, within the setting in which Helmholtz studied, for the idea that these were promising directions in which to begin his scientific travels.

According to Koenigsberger, when Helmholtz became one of the initial members of the *Physikalische Gesellschaft* in 1845, he entered "a broader circle of scientifically prominent men".[44] His gradual reorientation toward physics is often considered to have begun with this move; but Müller and his circle did not suddenly recede from Helmholtz's scientific horizon. For the next three years Helmholtz's contacts with Müller seem to have been limited to correspondence concerning the publication in Müller's *Archiv* of the two elegant papers on muscle contraction that Helmholtz produced during that period.[45] When the opportunity came for Helmholtz to leave military service and begin an academic career, however, Müller gave him the same strong, warm support that he had given Schwann, Brücke, and others who had worked and taught side-by-side with him in the anatomical institute.

The occasion for Helmholtz's shift was a position as teacher of anatomy at the *Kunstschule* in Berlin that became available in 1848 by Brücke's departure for Königsberg. At the request of the Ministry of Culture, Müller wrote an advisory opinion about the candidate for the position:

> "In his inaugural dissertation of 1842, Dr. Helmholtz had already shown himself to be gifted and full of talent. In various writings and publications since this time [...] he has further documented his capacity. He shows himself in them as an anatomical-physiological observer of great skill and many-sided education, of whom science can expect great further achievements. Among the talented men who have received their education in the field of anatomy and physiology here, some of whom have already filled academic chairs here and abroad, Helmholtz is one of the rare great talents that I would pick out as exceptional. His education and his strengths are simultaneously outstanding in several directions. For what can be said in recognition of his anatomical-physiological work can be reiterated in the same way for his physical studies and his deep mathematical knowledge.

44 Koenigsberger, *Helmholtz*, 1: 58.

45 Christa Kirsten, ed., *Dokumente einer Freundschaft: Briefwechsel zwischen Hermann von Helmholtz und Emil du Bois-Reymond 1846-1894* (Berlin: Akademie-Verlag, 1986), 74, 84.

> Under these conditions I would seize every opportunity that will permit Dr. Helmholtz to dedicate himself completely to the scientific life [*wissenschaftliche Laufbahn*], as I have always made it my responsibility to support young men of such capacity in every way".[46]

The language is familiar, consistent with the qualities Müller had admired in the other "talented young scholars" drawn to him ever since he had come to Berlin, and with the actions he had taken on their behalf. In his tone, however, we sense that even by his usual standards for judging talent, Müller recognized that Helmholtz was an ascendent star.

Müller did more than to write a resplendent letter for Helmholtz. He also arranged, as he had for others in the past, to use his assistant posts to help further Helmholtz's career. In requesting, in June 1848, that Helmholtz also be appointed as assistant at the anatomical museum, Müller pointed out that it was his policy that

> "this place be used preferably for a young man of physiological orientation and decided talent for experimental physiology, who will gain the opportunity, through the budgetary means available to the anatomical museum for physiological apparatus, to further his training in this field, and to support instruction in physiology by instituting practical exercises".[47]

Helmholtz remained as assistant to Müller for only one year, before he too departed for Königsberg in the wake of Brücke's further call to Vienna.

In closing, I would like to suggest that Müller and his circle may have played a deeper role in the orientation of Helmholtz's professional life than is implied in the foregoing, or in previous accounts of his career. Late in his life, when he reminisced about its beginnings, Helmholtz particularly remembered his early interest in physics. His father explained to him that because of his limited means, "he knew of no way I could study physics other than by taking up the study of medicine in the bargain".[48] Elaborating on this cue, historians have generally tended to treat his career as a predestined trajectory, in which medical training, military service, and the years he spent as a physiologist are seen as the long way around he was forced to traverse in order to reach the physics that remained throughout his firm goal. That Helmholtz could in this way give coherent meaning to his own life after his career had come to a cli-

46 Koenigsberger, *Helmholtz*, 1: 94. The original of this letter, listed in the Findbuch for the Kunstschule in the GSPK, was apparently destroyed in World War II.

47 Johannes Müller to Ministry of Culture, [June 1848], Rep. 76-Va, Sekt 2, Tit. X, No. 11, Bd. VI, 144, GSPK.

48 Kahl, *Selected Writings*, 470.

max with his achievements in physics does not mean, however, that he could foresee his future course from the beginning. There is very little in his early letters, or other contemporary documents, to support such clairvoyance on his part. His letters to his parents during his medical school years are, in fact, remarkably devoid of thoughts about his future. That may be because he wrote mainly about the immediate events he thought would most interest his family; but it might also be because he did not yet know what his aspirations would become. If that is plausible, then the examples of Müller and of Müller's prominent students may have offered Helmholtz, not only his initial anatomical and physiological bearings, but the very idea that a medical student could become a scientific investigator rather than a physician or surgeon. That is what they themselves had done. Müller, Schwann, Reichert, du Bois-Reymond, and Brücke were all medical students who had chosen academic careers in place of the medical practice toward which they had once been headed. Reichert and du Bois-Reymond, like Helmholtz, faced interludes of military service, though they freed themselves from it more quickly than he did. In the case of Schwann and du Bois-Reymond, Müller's example and encouragement were clearly motivating factors in these choices. We can at least speculate, even if we cannot prove, that whether Helmholtz mingled closely with this group, or only viewed them from the distance of a seat in the anatomical lecture hall, their life courses could easily have served him as a model for the early steps in his own.[49]

49 This paragraph owes much to conversations with Kathryn M. Olesko about the approach we intend to pursue in our projected collaborative volume on Helmholtz's early scientific career.

Civic Culture and Calling in the Königsberg Period

Kathryn M. Olesko[1]

I. Helmholtz and Königsberg

The six years that Helmholtz spent in Königsberg, 1849-1855, constituted a crucial phase in the formation of his career as well as his identity, both professional and personal. Here in the isolated easternmost corners of Prussia, he began his private life as husband, father, and head of household. He also inaugurated his public roles as university professor, administrator, and citizen. He had taught earlier in metropolitan Berlin, but we hear of no students; within a short time at the provincial university in Königsberg he found himself surrounded by many from which sprang a small but distinctive and successful school, the direct result of the coherence he brought to physiological instruction. In this, his first academic position, he quickly climbed the rungs of the academic ladder, eventually accepting, at the age of thirty-three, responsibility for leading his senior colleagues as dean of the medical faculty. His activities, scientific and civic, secured his reputation locally, nationally, and internationally in the brief span of less than two years after his arrival.

1 I would like to thank: the participants of the conference on Helmholtz in Tegernsee, Bavaria for their comments; an anonymous referee whose perceptive comments I am still pondering; and Dr. Helmut Rohlfing for his bibliographic assistance. This essay is drawn from my work-in-progress on a biography Frederic L. Holmes and I are writing on Helmholtz's early career.

Helmholtz's intellectual activity at Königsberg defied the scarce resources made available to him. Exposed to a different local scientific culture, his experimental style altered decisively, but not permanently, within months after his arrival. In a city notorious for its lack of skilled mechanics and machinists, he constructed or designed new instruments for measuring small intervals of time and for viewing the eye, most notably the *opthalmoscope*. With some instruments he integrated exactitude into physiology. At first devoted to continuing his nerve velocity experiments, he widened the scope of his research considerably thereafter, embracing human sense perception, muscle physiology, debates on central forces, electric currents, color theory, and vision studies. In addition, for his peers he wrote path-breaking review essays on animal electricity, theoretical acoustics, and physiological heat; for the Königsberg public, he delivered popular lectures on the measurement of small intervals of time, Goethe, sense perception, and the interaction of forces. Several of his intellectual efforts from this period spawned research traditions lasting until well into the twentieth century. This intellectual variety alone marks the Königsberg period as one of the most creative of Helmholtz's life.

It has been largely through his research and the particulars of his fortunate academic appointment, including his official travel abroad, that we know the Helmholtz from this period, as if his career were defined by the institutional setting of the university or encapsulated in the life of the mind or in the activities of the discipline he helped to create. In surveying his life in Königsberg, then, Königsberg the city - as a cultural resource, political environment, and social context - has largely disappeared from view. Sometimes the omission is justified. Helmholtz himself remarked to his father in March 1850 that Königsberg was "splendid" for working because the city did "not tempt one to do much else".[2] It is ironic, though, that in a city of little art and few artists[3], Helmholtz entered deeply into the study of color and visual perception. In this one sense, at least, the issues he took up transcended his immediate local environment.

2 Hermann von Helmholtz to F. Helmholtz, 29 March 1850 (Koenigsberger 1965: 67).

3 The poor artistic environment of Königsberg was legendary. The anatomist Karl Ernst von Baer suffered there for lack of a competent draftsman. Von Baer finally realized that no artist would voluntarily come to Königsberg (von Baer 1986: 252-253). Helmholtz himself called one Königsberg exhibition from 1851 "horrendously shabby" and used it as the bottom line for evaluating other art exhibits (Hermann Helmholtz to Olga Helmholtz, 14 September 1851, Kremer 1990: 90).

Helmholtz's silence about the cultural resources of Königsberg suggests a distance from that city's upper class culture. Königsberg was, for instance, a city of music and, to a lesser extent, theater. Yet in contrast to his Berlin period, we hear of no performances, only of a review of literature in theoretical acoustics. The literature to date has accentuated that distance by focussing on Helmholtz's activity outside Königsberg. Much emphasis has justifiably been placed upon Helmholtz's travels abroad in 1851 and 1853 in securing his international reputation, including by Helmholtz himself; for on these travels he demonstrated with great success both his opthalmoscope and his so-called "frog curves". Upon closer inspection however, his distance from Königsberg's upper class culture proves illusory. It is precisely the episodic nature of his travel that compels us to examine the continuities of his daily life, for only against that background can the significance of his travels, as well as of his silence, be assessed.

How did Helmholtz acknowledge and participate in Königsberg's civic culture? What did his choices mean? Several scholars - Arleen Tuchman, Richard Kremer, David Cahan, and Frederic L. Holmes - have suggested that Helmholtz's public and professional activities during the earlier Berlin and Potsdam years contributed to his career advancement (Tuchman 1993; Kremer 1990; Cahan 1993; Holmes 1994). They have demonstrated how Helmholtz made use of the "cultural capital" of those metropolitan centers, both formal (in the form of official contacts and credentialling) and informal (his social and cultural life). The contrast with Königsberg is striking. Most of his Berlin social and cultural worlds were shaped for him. In Königsberg, by contrast, Helmholtz arrived having few contacts and almost no friends. Most of the Berlin/Potsdam cultural capital Helmholtz had inherited, either from his family or from educational institutions. At Königsberg cultural capital was not a matter of inheritance but of selectively choosing between assimiliating into ongoing cultural practices or instituting new ones.

Helmholtz appropriated three elements of Königsberg's civic culture: the city's ubiquitous and unavoidable commercial culture, which was tied to a landed aristocracy; its local cultural societies; and its medical community, including the university's medical faculty and physicians involved in public health. Helmholtz's public life bears on issues concerning the social role of the sciences in mid-nineteenth century Germany; how he stood vis-a-vis the medical profession at this crucial stage of his career; and most importantly, his conception of his professional identity or calling, and the ethical commitments that calling entailed.

II. Commercial Culture and the Idiom of Measurement

Königsberg - heart of agricultural East Prussia, center of a liberal movement, government seat, military outpost, home to Jews but very few Catholics - was a city of trade and commerce whose merchants felt their fortunes rise and fall with the tonnage that entered and left the harbor. Grain trade, the product of the large traditional estates surrounding Königsberg, dominated when Helmholtz arrived there, just as it always had. Recently the city's smaller but economically significant wood trade had diminished with the disappearance of Königsberg's great forests in the 1840s when brown coal began to replace wood as an energy source. Grainery owners expectedly played a dominant role in Königsberg society, forging links between local merchants in town and the owners of old agricultural estates found in the region around Königsberg, the Masur, and Lithuania. Industry was precarious and for the most part tied to classical eighteenth century areas of production: the manufacture of the colonial goods sugar and tobacco; paper, dyes, and textiles; and a few breweries. State regulations protected many of these firms, and so when officials lifted trade barriers and in 1834 established the *Zollverein*, many of these early manufactories disappeared. Machine-related industries appeared sporadically from 1830 onward, but they were never significant in the economy of the city or the region. Königsberg's economic profile remained closely tied to its traditional social base. The working class in evidence in Helmholtz's Berlin of the 1840s was scarcely to be found in Königsberg where more pressing social concerns were population growth, immigration from the east, and the poverty both had created.

At an earlier time, university professors had preached a free market economy. Reality now spoke otherwise. The city's heavy dependence upon trade, especially grain trade, meant that its economy reacted like a sensitive barometer to bureaucratic social and economic reforms, changes in customs and tolls, the vagaries of international affairs, as well as periodic agricultural crises. Throughout the period of Helmholtz's residence, agriculture dominated economic cycles, investment, and employment opportunities. Bureaucratic economic strategies designed to buffer the region's economy from these disturbances may not have been clear on the social goals of periodic protective measures, but their unstated intention was beyond dispute. Noble large landowners were to survive intact, poised to increase their financial well-being through agrarian reform, technical improvements, and capitalist productive relations in agriculture, survival strategies they first began to learn in the days of mercantilism. So although Königsberg's face changed while Helmholtz was in residence - eastern and western rails were completed, gas street lighting was introduced, and larger streets and new homes with more windows for greater light and ventilated air (in part a health reform) replaced the protective character of the city's medieval architecture - the economic and social profile of the city differed in degree but not kind from what it had been in the late eighteenth century.

By 1850, however, rational techniques regulated the economy to a far greater extent without significantly changing the traditional social base of the area. In part the extension of eighteenth century mercantilist practices, these techniques refined and intensified the quantitative element in commercial transactions. Particularly during economic crises such as those following the Napoleonic Wars, bureaucrats, merchants, Junkers, and citizens turned to refined forms of quantification to achieve a measure of economic stability and control. Hence, Prussia's weights and measures reform, instituted after the Napoleonic Wars but only completed in 1839, assumed a central role in the rational regulation of trade, the economy, and, insofar as the reform concerned the recalibration of standards, even the regulation of human behavior in buying and selling. For an agricultural region like East Prussia, quantitative techniques recast the significance of traditional economic factors and united the state and the academy. So, for instance, Friedrich Julius Richelot - Königsberg's mathematician, Helmholtz's good friend and godfather to Helmholtz's first child - improved techniques for computing a grain ship's tonnage more accurately. Reforms that broke up large agricultural estates in East Prussia and redistributed land required new, more accurate property maps, especially cadastral ones. Prussian officials eventually adopted for the purpose a surface measuring instrument, the polarplanimeter, the invention of Jakob Amsler-Laffon, erstwhile student in the Königsberg mathematico-physical seminar, directed by the physicist Franz Ernst Neumann, where Amsler had learned the techniques of precision measurement before Helmholtz's arrival (Olesko forthcoming).

Königsberg's astronomer, Friedrich Wilhelm Bessel, played a central role in crafting both several techniques of quantification useful to Prussian economic concerns and the idiom used to convey their significance in social and other contexts. It was Bessel who, in 1839, recalibrated the Prussian foot on the basis of measurements of the simple seconds pendulum made in Königsberg and then in Berlin. The emphasis he placed in his investigation upon error analysis, especially the computation of accidental errors by the method of least squares, redefined standards construction, standards replication, and the certification of copies in daily commercial transactions. At the simplest level, the specification of error established the bounds within which a standard could be trusted. Error thus became essential in a legal sense; for it specified the boundary between improper and proper behavior in the marketplace of trade. Proper use of a standard thus meant that the customs official, trader, or salesperson was trustworthy; error literally defined the onset of criminal behavior. Bessel's method of standards calibration, with its emphasis upon error, itself became a standard in Prussia and later in the united German nation where it prevailed in the determination of units of electrical resistance at the end of the century (Olesko 1991; Olesko 1993).

Besides possessing economic and legal dimensions, the idiom of Bessel's type of standards determination, with its emphasis upon error analysis, linked intellectual creativity and individual character traits, thus giving social meaning to the act of measuring. In a state that was still in the process of accommodating socially to recognizing talent, and that still was not quite sure of what to do with large groups of educated, trained men, precision measurement became a way to channel and to recognize certain kinds of intellectual creativity and originality in socially acceptable ways. Neumann's seminar offered no finer example of this. Neumann had based his measuring physics on the work of Bessel, who had died three years before Helmholtz joined the faculty. Bessel's influence lingered in the seminar's curriculum and spread to other locations through the seminar's students who, like Amsler, applied Bessel's techniques to new situations. Like Bessel, they viewed error as specifying an epistemological limit on what could be known with certainty about the world. Claims about measurement had to specify this limit, which became a sign of the investigator's integrity and humility as well as of the trustworthiness of quantitative claims. The traditional sources of honesty and trust in commercial transactions thus became those of scientific investigations as well (Olesko 1991; Olesko 1994; Olesko forthcoming).

When Helmholtz arrived in Königsberg in 1849, techniques of error analysis, especially least squares, were virtually unknown in physiology. Shortly after his arrival and in the context of his nerve propagation experiments, he adopted Bessel's exacting techniques of error analysis, especially the method of least squares, to secure the "reality" of his findings and to persuade his readers of the trustworthiness of his claims (Olesko and Holmes 1993). We might be tempted to view this alteration in his scientific style and mode of argumentation as a purely intellectual decision, an expedient change dictated by the complexity of his experiment and of his data, were it not for Helmholtz's subsequent elaboration in public on the function of error. On several occasions during his Königsberg years, Helmholtz cited or alluded to the importance of error in quantitative determinations. As he cast his findings in popular form at the end of 1850 (Helmholtz 1850) and continued his nerve velocity experiments on humans, especially on the time measurement of reflex and sense perceptions in 1852, Helmholtz, others thought, provided a physiological foundation for another type of error discovered by Bessel, the personal equation (Helmholtz 1852; Hermann and Volkmann 1895: 69). By quantifying the effect of sense experience in the acquisition of knowledge in this way, Helmholtz continued the Besselian tradition of specifying the *exact* limits of the unknown. Thus, although we do not find Helmholtz to be engaged as directly as Richelot or Bessel were in investigations of direct relevance to the commercial concerns of the region, we do find that his most important quantitative work during the Königsberg period deployed the idiom stemming from Prussia's standards reform and based on Bessel's work. It is noteworthy that in his popular speech on the measurement of small intervals of time, delivered in 1850 to Königsberg's cultural elite, Helmholtz said more about error than he had in his essay on nerve velocity of five months before (Helmholtz 1850). As he later explained, "the intelligent portion of the scientific public"[4] in Königsberg could be convinced by only certain kinds of arguments, and Helmholtz deployed the idiom others seem to have been accustomed to hearing.

III. Cultural Societies and the Political Economy of Culture

Karl Rosenkranz's description of Königsberg's various *Stände* - commercial and business, bureaucratic and military, and academic - coexisting in a "beautiful equilibrium" is probably overstated, but the harmony of these *Stände* is indisputable (Rosenkranz 1857). The professoriate did not dominate the social life of Königsberg the way they did the university towns of Halle and Göttingen; yet they were an independent lot, giving the educational ministry

4 Helmholtz to Carl Ludwig, 1855 (Koenigsberger 1965: 138).

pause to think in 1837 when they supported the Göttingen Seven, in 1844 when they cast the university's tricentennial celebration in terms of liberal causes, and in 1848 when some among them vigorously participated in the events of that March. With few social contacts when he arrived in Königsberg, Helmholtz's first instinct was to seek out members of the professoriate whose interests were closest to his own. Besides members of the medical faculty, Helmholtz befriended the physicist Neumann, the mathematican Richelot, and the astronomer Alexander Busch. Although he soon had a devoted coterie of students of his own, we hear mostly about his interactions with Neumann's students (or explorations into their notebooks from Neumann's courses), especially with Gustav Kirchhoff, Emil Schinz, and Emil du Bois-Reymond's brother, Paul. Helmholtz's university friendships became very strong; he later described parts of his official farewell party in 1855 as "tearful".[5]

Cultural societies were important forms of social organization and expression in nineteenth century Germany. By Helmholtz's time special interest societies - having statutes, officers, regular meetings, and often, but not always, stated purposes - provided occasions for Königsberg's social elite to gather together, rarely for political purposes, sometimes to work for public causes, most often to stake their claims to and exchange certain forms of knowledge. Through the social cohesiveness they provided, these societies in effect defined and sustained the city's intelligentsia. The isolation of Königsberg made literary societies and reader's circles especially popular. The most well known was the *Deutsche Gesellschaft*, which focused on history and biography under the direction of the Königsberg historian and friend of Helmholtz, Friedrich Wilhelm Schubert. Several societies popularized science or promulgated its practical uses. Some, like the *Physikalisch-ökonomische Gesellschaft* (1789) and the *Physikalisch-medizinische Gesellschaft* (1808) were originally entirely utilitarian but became less so without entirely losing a practical function. Others, like the newer *Verein zur Förderung der Landwirtschaft* (1838), the *Gewerbeverein* (1845), the *Polytechnische Gesellschaft* (1845), and the *Verein für wissenschaftliche Heilkunde* (1851) reinstated more strongly connections to useful knowledge.

Helmholtz's previous association with the *Berliner Physikalische Gesellschaft* and the progressive 1847 group might have suggested, if present interpretations of those groups are correct, that he would have found the newer and

5 Helmholtz to Olga Helmholtz, 19 July 1855 (Kremer 1990: 149).

more economically oriented *Polytechnische Gesellschaft* or *Gewerbeverein* more appealing. But following a tradition established by university professors before him, Helmholtz became instead an active member of the *Physikalisch-ökonomische Gesellschaft*. His choice, more conservative on several counts, is thus revealing of the social position he sought in Königsberg civic culture. Established in 1789 and among Königsberg's oldest, the Society was originally dedicated to a social levelling of sorts, preaching the "complete equality" of members, the meaninglessness of rank and social differences, and even the acceptance of women as ordinary members (Stieda 1890; Schiefferdecker 1864). Be those statutes as they may, the original stated function of the Society - to make known techniques and knowledge that would be useful in agricultural production - was designed to further the interests of a traditional social elite.

By the time Helmholtz joined, the Society's membership profile had changed somewhat, but was still drawn from several different occupations and professions: the bureaucracy, military, businessmen, merchants, physicians, secondary school teachers, book manufacturers and sellers, large landowners, engineers, factory owners, and of course, professors (Anon. 1869). What had not changed was the fact that the Society remained a stronghold for the economic elite of Königsberg, an elite that still upheld a traditional social hierarchy with large land owners dominating. Society lectures included more scientific topics, but they did not exclude discussions of useful knowledge. Bessel in the 1820s lectured on the practical uses of probability calculus and the uses of astronomy in ship travel; others discussed wheat trade, poverty, how to sharpen razors for shaving, prisons, and brown coal. Lectures treating aspects of industry and manufacturing were rare; the few there were included Karl Gottfried Hagen's lecture on the steam engine and Moritz Jacobi's guest lecture on how machines made human labor dispensable (Baer 1834). Offering topics of relevance to public welfare, the Society's public lectures - Bessel's innovation - proved immensely popular, disseminating scientific and technical knowledge to an audience still largely bound to traditional social identities, albeit supported by novel and sometimes capitalistic pursuits, especially in agriculture.

In his discussion of the political economy of culture, the sociologist Randall Collins emphasizes the importance of dominant individuals who, in formal and informal gatherings, by their spoken words create the ties and bonds that define social groups and the self-images that represent aspects of reality (Collins 1979: 49-72). In societies like Prussia's where an educated elite was so highly valued for its ability to promulgate and create reality-defining symbols and self-definitions, we find numerous examples of individuals in control of the political economy of culture: one need only think of the powerful symbols of classical culture sustained by a professoriate that defined and maintained their social mission through Prussia's educational system. University-based natural scientists were both a part of this educated elite and a challenge to it. Although themselves classically educated, their arguments for the replacement of classical philology by the natural sciences as the defining form of knowledge introduced a new set of cultural symbols that focused on the future rather than the past. That challenge was never sufficiently compelling to replace entirely earlier classical symbols and to dislodge former elite groups. The social role of a scientific and technological culture in Prussia, and then in Germany, was itself continually circumscribed by the stronger social role of classical culture.

Scientific societies like the *Physikalisch-ökonomische Gesellschaft*, strewn all over Prussia, were safe havens for the articulation of a scientific culture. Sadly we do not yet know their impact on the transformation of German culture as a whole. At a local level, however, these societies were powerful venues for the integration of scientific thinking and technological know-how into the highest levels of public life. In Königsberg, Karl Gottfried Hagen, Karl Ernst von Baer, and Friedrich Wilhelm Bessel nicely fit Collins's description of "dominant individuals" whose public pronouncements and activities held the potential for transforming the local political economy of culture. Each was strongly connected to the highest levels of Königsberg society or politics; each made novel and fundamental statements about the role of science in daily life. Through his vocal contempt of classical culture, Bessel in particular sought to transcend the reality-defining symbols then in place (Olesko 1991: 53). When we compare Helmholtz to these three scientists active earlier in the *Physikalisch-ökonomische Gesellschaft*, though, we do not find him to be as strongly positioned to redefine cultural symbols.

Never as strongly tied to Königsberg's political culture as were Bessel and Hagen, Helmholtz in his public speeches remained remarkably conservative in terms of the images he projected of the sciences and cautious in terms of how he delineated the role of science in practical and public life. For instance, much as his discovery of the finite velocity of the nerve impulse held out the possibility of reframing the philosophical interpretation of human sense perception, Helmholtz emphasized in December 1850 that insofar as "practical interests" were concerned, our actions in the world were not in any way noticeably affected by delays in the transmission of sense impressions that were due to the finite velocity of the nerve impulse (Helmholtz 1850: 186). That tendency to cordon off the role of scientific knowledge in daily life was manifest in Helmholtz's own tendency to speak in terms of two views of reality, the aesthetic and the material/mechanistic. In his 1853 speech on Goethe he tried to reconcile the aesthetic and the material/mechanistic views of nature by appeal to a moral imperative not to be blindly controlled by the mechanical. Goethe, he admitted, had been hostile to machinery. Yet in Helmholtz's view it was the responsibility of the natural philosopher to examine the machines at work behind the aesthetic appearances of reality. "We cannot triumph over the machinery of matter by ignoring it," he argued. "We can triumph over it only by subordinating it to the aims of our moral intelligence. We must familiarize ourselves with its levers and pulleys, fatal though it be to poetic contemplation, in order to be able to govern them after our own will. Therein lies the complete justification of physical investigation and its vast importance for the advance of human civilization" (Helmholtz 1853: 73).

A year later, after learning of Victor Regnault's measurement of specific heats and following his own celebrated trip to England where he met William Thomson, Helmholtz addressed more fully the features of a machine culture, this time by way of the measurement of the expenditure of force and its relationship to the amount of work performed. The matrix of considerations he brought to bear upon that machine culture had expanded considerably. But despite the obvious impact that Regnault, Thomson, and in all likelihood Franz Neumann too had had on how Helmholtz framed the problem of how to measure the expenditure of force, Helmholtz cast reality in terms of a machine culture only to a point. So despite provocative assertions like *"Arbeit ist Geld"* (Helmholtz 1854: 53), which seemed to cast all forms of work in terms of profit relations, Helmholtz distinguished human labor from mechanical labor by the skills acquired either through talent or training. Retaining traditional views of the meaning of human labor, Helmholtz argued that skills made the value of human work both qualitatively different from, and of considerably higher value than, strictly mechanical work. Hence he claimed that in contrast to human work, "the idea of the amount of work in machines is therefore restricted to considering the expenditure of force" (Helmholtz 1854:54).

Yet Helmholtz admitted that this mechanical way of looking at the world was limited in scope. He explained that "the view into the confined laboratory of the physicist with its minute ratios and entangled abstractions will not be as attractive as the view of the wide sky above us, of the clouds, rivers, forests, and living creatures" (Helmholtz 1854:67). As he had done in his Goethe speech, Helmholtz grappled with the integration of a machine culture with the more aesthetic one he saw plainly visible in the natural world. Once again he compared the mechanical with the aesthetic, but he gave the latter a higher valuation. The symbols he projected thus created two images of reality, not yet converged. Although he addressed a machine culture, he did not project a new image of reality for his listeners. Speaking to an audience still very traditional at heart, Helmholtz neither created the symbols of a new political economy of culture nor was overly desirous of doing so.

IV. Medicine and Public Health

Helmholtz as well as his later biographers have looked upon his appointment in the medical faculty of Königsberg as the beginning of his *"wissenschaftliche Laufbahn"*. Historians have delineated how his medical training was important for his subsequent scientific career; how his scientific methods influenced the medical community; and how he created an "autonomous physiological science" (Tuchman 1993; Kremer 1990). The literature has tended to view *Medizin* as a deviation or departure from his original intentions. His connections to the medical community in Königsberg have been viewed almost entirely in terms of his invention of the opthalmoscope, which also raised his profile in the scientific community.

Yet Helmholtz's greatest participation in Königsberg's civic culture was in the realm of medicine and public health. In this academic appointment, his first, Helmholtz's primary responsibility was the training of medical students. Although it is difficult to tell just how much he increased student enrollment in subjects he taught because there was a general increase in the student population at Königsberg during the years of his watch, Helmholtz did succeed in creating a small school whose members dedicated themselves to issues raised in his courses.[6] On his travels outside Königsberg in 1851, he demonstrated his frog curves or his opthalmoscope, depending on his audience; the most significant audience for his opthalmoscope was the medical community. Throughout his Königsberg years he maintained especially strong ties to the

---

6 Two years before Helmholtz's arrival there were 53 medical students (SS 1847); three years after his departure that number had almost doubled, totalling 100 in the SS 1858. Among the students who belonged to his school were Ernst Christian Neumann, the son of Franz Ernst Neumann. I use the term "school" here in a loose sense (Olesko 1993: 21-24).

medical community. And finally, in accepting the "scarlet mantel" as dean of the medical faculty at Königsberg, Helmholtz placed himself in the position of gatekeeper, judging the quality of both students and new faculty members. He policed examinations that certified physicians and upheld strict compliance to the course requirements of the pre-clinical years (Olsztyn IV).

When he assumed the appointment at Königsberg in 1849, Helmholtz became a part of an ongoing local tradition in medical education which, even prior to the arrival of his colleague Ernst Brücke a year earlier, had already valued the role of the natural sciences in medical education and to a certain degree had already integrated practical exercises into instruction. As was the case in other medical faculties across Prussia, at Königsberg the natural sciences were required prerequisites to the clinical years of instruction. Yet despite the prior stated emphasis on the natural sciences in medical instruction, very little can be said with certainty about their actual role in training physicians. Appeals made to the natural sciences, here as elsewhere, were often rhetorical in nature. At Königsberg the medical faculty viewed instruction in the natural sciences as a means to expose future physicians to methodological considerations: as a method for achieving greater certainty in the evaluation of the sick, as a method for improving judgment; and as a method for conducting research in applied areas such as epidemiology (Olsztyn III).

Before Helmholtz's arrival, faculty members had already identified physiology as the single most important natural science in the pre-clinical curriculum. Physiology was the science that held all others together: "the theory of human beings [physiology] is the nucleus around which is grouped the entire conception of nature" (Olsztyn III). Physiology was thus essential to considering human beings in their relationship to the totality of nature; subjects like epidemiology could not be understood without it. Helmholtz's teaching of physiology centered on the senses, the nervous system, and the muscles, for which he drew upon his considerable research in each of these areas (Göttingen I). His novelty and original contribution to the medical curriculum was, however, less his further articulation of physiology proper - although his accomplishments were considerable here - than his offering of subjects that the faculty had identified as lacking in the curriculum: physiological chemistry, or chemistry in its application to medicine, and especially pathological chemistry, used to study disease and illness (Göttingen I: allgemeine Pathologie, 51-56).

In this vein, the laboratory experiences invoked for learning the basic sciences were of two sorts.

Like other medical faculties, the Königsberg faculty had integrated laboratory exercises into its teaching, especially in physiology. Helmholtz certainly expanded the practical element in medical instruction at Königsberg, especially through his courses on physiological and pathological chemistry. These exercises historians have traditionally identified as novel in the German medical curriculum of this period. But here in Königsberg (and undoubtedly also elsewhere) there was another laboratory experience, equally important in medical instruction and in the *Physikalisch-medizinische Gesellschaft*, the city's main organization for physicians: the laboratory culture of epidemiology and public health. The local environment, the domain of epidemiology, was no less a laboratory than the practica affixed to courses in physiology because in it one could find and analyze specimens (the sick) that represented deviations from the "normal" natural state of affairs. Helmholtz encouraged this two-pronged approach to the scientific study of medicine, in part through his own example when he tested his opthalmoscope on patients with eye diseases. Especially while he was dean he promoted the laboratory of epidemiology. He allowed local practicing doctors to teach courses, especially on applied subjects such as forensic medicine, rabies, and chemical testing of local water resources. He worked to endow prize competitions for students, especially on applied topics related to epidemiological problems (Olsztyn IV).

Helmholtz's commitment to medical issues in the community is perhaps best exemplified in his supporting role in the establishment, on 6 November 1851, of a new medical society in Königsberg, the *Verein für wissenschaftliche Heilkunde*, for which he served as the first director. The social and political reasons for creation of the *Verein* deserve closer study. On the surface, the *Verein* appeared to be a break with past tradition, for it so clearly was an alternative to the older *Physikalisch-medizinische Gesellschaft*. Like the latter, the *Verein* was both a public expression of the medical faculty's support of the sciences and the central role of physiology among them and of the public health tradition with its emphasis upon epidemiology. Like other societies to which Helmholtz belonged, this one became a public forum for his self-expression. At the first meeting of the *Verein*, on 11 November 1851, Helmholtz made his first formal public announcement of the opthalmoscope, noting its usefulness to the practitioner. In subsequent gatherings he continued to address issues concerning opthalmology with practicing physicians (Hilbert 1901: 3-11, 16-17; Anon. 1859; Anon. 1860).

Scientific medicine, in the sense of a laboratory-based medicine, was not the central thrust of the *Verein's* activities. The *Verein* became a center for physicians to gather to discuss problems arising in the treatment of disease, to learn about other areas of medicine, and to have, in general, a public forum for keeping up to date in medicine. Consonant with the emphasis upon physiology, especially physiological chemistry, as a focal point for understanding human beings in their environment, the first project in the *Verein* undertaken during Helmholtz's directorship was a study of the levels of ozone concentrations in and around Königsberg. Organized and carried out by his good friend, the physician Wilhelm Friedrich Schiefferdecker, this project sought to uncover the relation between ozone level and local outbreaks of disease and illness. (No correlation was found.) The second project was one more directly in the area of public health: measuring the density of milk with a *Milchdichtigkeitsmesser* for the purpose of identifying milk thinned with water, a fraudulent practice in the sale of milk. This project was carried out in the interest of standards control in public health, not unlike the type of control which was found in the administration of weights and measures reform (Anon. 1859; Hilbert 1901: 3-11, 16-17).

More than in other cultural societies, in this medical society Helmholtz reached outward into the community and into public health issues. His active participation in this aspect of Königsberg civic culture compels us to reconsider interpretations of his medical training as a way station en route to a prechosen scientific career. Helmholtz could have easily chosen not to be a part of the medical community when he assumed his first academic position. Yet although he did not practice medicine in Königsberg, he nonetheless welcomed contact with physicians and became a willing participant in their discussions. Was he purely careerist oriented in announcing his opthalmoscope to local physicians? Undoubtedly some part of the effect of his presentation was an enhancement of his own profile. Yet his examination of patients with eye diseases as well as his subsequent support of projects of obvious practical value seem to suggest that the social concerns of the physician guided his actions and that he was motivated at least in part by a medical ethic of responsibility. Helmholtz's continuing interaction with the medical community during the Königsberg period therefore serves as a powerful reminder at this early stage of his career, in his first academic position, his professional identity was a hybrid, one which combined the physician's responsibility with the scientist's goals and methods.

V. Helmholtz's Calling

In the historiography of German science there is presently a tendency to link laboratory sciences, including a science-based medicine, to the emergence of a politically democratic, economically capitalistic and industrial, and socially mobile state in Germany - in short, to link modern science with modernization (e.g., Lenoir 1992). There is also a tendency when viewing pre-1850 science to look for precursors of post-1860 political, economic, and social realities rather than to look for continuities with pre-1800 conditions. Yet life in the period from 1800 to 1850 had more in common with the late eighteenth century than it did with the late nineteenth. A healthy skepticism regarding modernization as an explanatory framework is quite widespread among social historians, who now regard the framework as passé (Sperber 1985 and 1991). Historians of science have much to learn from them.

Helmholtz's early career in Königsberg is in this regard a warning to avoid hasty generalizations linking science around 1850 to post-1860 political, economic, and social contexts. Helmholtz's audience was mixed, and most certainly constituted mostly of traditional social elements. The images he projected were ambiguous; he did not exclusively promote those of a modern machine culture. Helmholtz himself might be said to be betwixt and between: between medicine and science; between a social world based on station and another where mobility was possible; between mercantile and modern economies; and between the pursuit of self-interest and duty to a collective endeavor. He did not argue for transition, but for integration and harmony. His determinate relations in society, although stable and leaning toward the traditional, were nonetheless clearly being tested. But the traditional order was not yet overthrown, or even seriously threatened. That order survived in part by coopting novelties: scientific, technical, and economic. The strongest contextual ties during the Königsberg period go backward, not forward.

By discarding the modernization framework in explaining early nineteenth century German science, context is not eliminated. As Steven Shapin has recently argued, considerations of context cut two ways in historical explanation (Shapin 1993). Context defines a point of conjuncture, a point at which several trends meet. Context could, in this case, suggest precipitating factors that cause change. But as Shapin has shown in his study of Robert Boyle, context is also the domain of historical continuities and of cultural and other traditions. Context in this second sense is a reservoir of possibilities; it defines the space or the domain of what is possible. It is this second sense of context that is most meaningful when trying to understand Helmholtz's participation in the civic culture of Königsberg.

Among the historical issues in early nineteenth century German science begging for closer scrutiny is that of how the scientific calling, or vocation, took shape. The counterpoint to the *ständische* organization of German society in the nineteenth century were the careers and professions, such as those in science, based on rigorous training, qualifying examinations, and expertise. A continuing historical problem is to understand how individuals viewed: their identities and their self-perceptions; their work habits and the ethic that informed them; their sense of obligation, purpose, and duty; the meaning of their labor; the foundation of their claims to truth; and their ethical commitments. All these are involved in the notion of calling. By the late eighteenth century in Germany, the social notion of calling embraced a toleration of individual self-determination and expression, but retained a commitment to a collective imperative. Still under the sway of an ethic of self-denial, this collective imperative expressed itself in terms of active service to others and in shunning of arrogance (LaVopa 1988). This commitment to a collective imperative disappeared only gradually in the nineteenth century. It metamorphosed, in part, into an ethic that assigned personal value and self-worth to duties well-fulfilled. The eighteenth century collective imperative became the disciplinary conformity of the nineteenth (Olesko 1991). The success of *Naturwissenschaft* as *Beruf* - to survive as a viable professional activity - depended heavily on the incorporation into practice of ethical guidelines, including checks on unbridled claims to truth. Guidelines both continued the eighteenth century control of self-expression to a level short of individual arrogance and helped to shape character traits, such as humility, among professionals involved in knowledge construction.[7]

In this essay I have suggested ways of viewing the social meaning of the secular cultures to which Helmholtz belonged. The ethic of Helmholtz's calling came from science and medicine. Its characteristics are complex, forming no simple composite. Although clearly in pursuit of new knowledge in science research, Helmholtz also practiced self-denial in his public commitment to service, the collective, and the usefulness of knowledge in medical circles. From the context of Königsberg's civic culture we see the ways in which he checked

7 Hence the importance of error analysis, especially the method of least squares, at Königsberg. Error quite simply defined the limits of knowledge (Olesko 1991).

self-expression and assumed responsibility. During the Königsberg period Helmholtz was not in pursuit of an academic career defined by the unbridled, self-interested acquisition of natural scientific knowledge as he later portrayed himself (Helmholtz 1891). His later story, in which he projected himself as having been interested in physics from the start, is itself a testament to both the continuing changes in the notion of calling and the enhanced social relevance of the sciences that unfolded over the course of the nineteenth century.

Literature

A. Manuscript

Dahlem I: Acta betr.: Die Errichtung, Unterhaltung und Ausstattung des physiologischen Instituts der Universität zu Königsberg. Rep. 76Va, Sekt. 11, Tit. X., Abt. X, Nr. 37, Bd. I (1848-90). Geheimes Staatsarchiv Preussischer Kulturbesitz (Dahlem).

Göttingen I: Cod. ms. Ernst Christian Neumann. Nr. 2: Vorlesungsnachschriften. 2a: Vorklinische Vorlesungen (wahrscheinlich gelesen von Hermann v. Helmholtz: allgemeine Pathologie; Physiologie des Nervensystems und thierische Elektrizität; spezifische Reizbarkeit der Nerven und Muskelbewegung; Gehör; Gesichtssinn, Gehirn, Rückenmark und Sympathicus). Niedersächsische Staats- und Universitätsbibliothek, Altbau: Abteilung für Handschriften und seltene Drucke, Göttingen.

Olsztyn I: Acta des Königl. Kuratorium der Albertus-Universität zu Königsberg. Die Anschaffung zwei Mikroscope bzw. Gebrauch bei der Vorlesung über Histologie betr. XXVIII/2/Nr. 414/Rep. 99 H32. Archiwum Panstwowe, Olsztyn, Poland.

Olsztyn II: Acta des Königl. Kuratorium der Albertus-Universität zu Königsberg. Die Bewilligung von 300 rth. zur Anschaffung von Instrumenten und Apparaten behufs der Vorlesungen über experimentale Physiologie. XXVIII/2/Nr. 412/Rep. 99 H30. Archiwum Panstwowe, Olsztyn, Poland.

Olsztyn III: Acta des Königl. Kuratorium der Albertus-Universität zu Königsberg. Die Förderung eines gründlichen wissenschaftlichen Studiums der Medizin betr. XXVII/2/Nr. 94/Rep. 99 A142. Archiwum Panstwowe, Olsztyn, Poland.

Olsztyn IV: Universität zu Königsberg. Acta der medizinischen Fakultät von Ostern 1854 - Ostern 1855. Vol 126. Dekan: Helmholtz. XXVIII/1/Nr. 416/Dep. II/730. Archiwum Panstwowe, Olsztyn, Poland.

B. Printed

Anon.: Historische Einleitung und Mitglieder-Verzeichniss. *Schriften der Königlichen physikalisch-ökonomischen Gesellschaft zu Königsberg in Preussen*, 1, 1860, i-xvi.

Anon.: Der Verein für wissenschaftliche Heilkunde in Königsberg während der ersten sechs Jahre seines Bestehens. *Königsberger medicinische Jahrbücher*, 1, 1859, 1-14.

Anon.: Vorwort. *Königsberger medicinische Jahrbücher*, 2, 1860, iii-v.

Baer, Karl Ernst von: *Autobiography of Dr. Karl Ernst von Baer*, ed. Jane M. Oppenheimer. Canton, Mass.: Science History Publications, 1986.

Baer, Karl Ernst von (Ed.): *Vorträge aus dem Gebiete der Naturwissenschaften und der Oekonomie gehalten vor einem Kreise gebildeter Zuhörer in der physikalisch-ökonomischen Gesellschaft in Königsberg. Erstes Bändchen mit Vorträgen der Herrn Argelander, v. Baer, Bujack, Dove, Dulk, M.H. Jacobi, Ernst Meyer, L. Moser.* Königsberg: A. W. Unzer, 1834.

Burdach, Karl Friedrich: *Amtliche Nachrichten über die Feier des dritten Secularfestes der Albrechts-Universität zu Königsberg*. Königsberg: Gräfe und Unzer, 1844.

Bußmann, Walter: *Zwischen Preußen und Deutschland: Friedrich Wilhelm IV. Eine Biographie*. Berlin: Goldmann, 1990.

Cahan, David: Helmholtz and the Civilizing Power of Science. In: David Cahan (Ed.): *Hermann von Helmholtz and the Foundations of Nineteenth Century Science*. Berkeley, Ca./Los Angeles/London: University of California Press, 1993, 559-601.

Cahan, David (Ed.): *Letters of Hermann von Helmholtz to his Parents, 1837-1846*, Stuttgart: Franz Steiner Verlag, 1993.

Collins, Randall: *The Credential Society: An Historical Sociology of Education and Stratification*. New York, N.Y.: Academic Press, 1979.

Elditt: Die Polytechnische Gesellschaft zu Königsberg i.Pr. *Altpreußische Monatsschrift*, 1, 1864, 261-265.

Gause, Fritz: *Die Geschichte der Stadt Königsberg in Preussen*. 3 Teile. Köln/Wien: Böhlau, 1965-71.

Gause, Fritz: *Königsberg in Preussen: Die Geschichte einer europäischen Stadt*. München: Gräfe und Unzer, 1968.

Helmholtz, Hermann von: An Autobiographical Sketch [1891]. In: Russell Kahl (Ed.): *Selected Writings of Hermann von Helmholtz*. Middletown, Connecticut: Wesleyan University Press, 1971, 466-478.

Helmholtz, Hermann von: The Scientific Researches of Goethe [1853]. In: Russell Kahl (Ed.): *Selected Writings of Hermann von Helmholtz*. Middletown, Connecticut: Wesleyan University Press, 1971, 56-74.

Helmholtz, Hermann von: *Ueber das Sehen des Menschen* [1855]. Leipzig: L. Voss, 1855.

Helmholtz, Hermann von: Ueber die Methoden, kleinste Zeittheile zu messen, und ihre Anwendung für physiologische Zwecke [1850]. *Königsberger naturwissenschaftliche Unterhaltungen*, 2, 1851, 169-189.

Helmholtz, Hermann von: Ueber die Natur der menschlichen Sinnesempfindungen [1852]. *Königsberger naturwissenschaftliche Unterhaltungen*, 3, 1854, 1-20.

Helmholtz, Hermann von: Ueber die Wechselwirkung der Naturkräfte und die darauf bezüglichen neuesten Ermittelungen der Physik [1854]. In: Helmholtz, Hermann von. *Vorträge und Reden*. 2 Bde. Braunschweig: Vieweg, 1903, 1:51-83.

Hermann, L. and P. Volkmann: Hermann von Helmholtz. *Schriften der Königlichen physikalisch-ökonomischen Gesellschaft zu Königsberg in Preussen*, 35, 1894, 63-84.

Hilbert, Paul: *Der Verein für wissenschaftliche Heilkunde in Königsberg i. Pr., 1851-1901.* Königsberg: Leupold, n.d. [1901].

Holmes, Frederic L.: The Role of Johannes Müller in the Formation of Helmholtz's Physiological Career. In this volume, 1994.

Horn, A.: Kleines und grosses Königsberg. *Altpreußische Monatsschrift*, 1, 1864, 341-356.

Kaelble, Hartmut: Der Mythos von der rapiden Industrialisierung in Deutschland. *Geschichte und Gesellschaft*, 9, 1983, 106-118.

Kirsten, Christa (Ed.): *Dokumente einer Freundschaft: Briefwechsel zwischen Hermann von Helmholtz und Emil du Bois-Reymond, 1846-1894.* Berlin: Akademie Verlag, 1986.

Koenigsberger, Leo: *Hermann von Helmholtz.* Transl. Frances Welby. New York, N.Y.: Dover, 1965.

Koselleck, Reinhart: *Preußen zwischen Reform und Revolution: Allgemeines Landrecht, Verwaltung und soziale Bewegung von 1794 bis 1848.* Stuttgart: Ernst Klett, 1967.

Kremer, Richard L.: Building institutes for physiology in Prussia, 1836-46: Contexts, interests, rhetoric. In: Andrew Cunningham and Perry Williams, (Eds.): *The Laboratory Revolution in Medicine.* Cambridge: Cambridge University Press, 1992, 72-109.

Kremer, Richard L. (Ed.): *Letters of Hermann von Helmholtz to his Wife 1847-1859.* Stuttgart: Franz Steiner Verlag, 1990.

Kriedte, Peter, Hans Medick, and Jürgen Schlumbohm: Die Proto-Industrialisierung auf dem Prüfstand der historischen Zunft: Antwort auf einige Kritiker. *Geschichte und Gesellschaft*, 9, 1983, 87-105.

LaVopa, Anthony: *Grace, Talent, and Merit: Poor students, clerical careers, and professional ideology in eighteenth century Germany.* Cambridge: Cambridge University Press, 1988.

Lenoir, Timothy: Laboratories, medicine and public life in Germany 1830-1849: Ideological roots of the institutional revolution. In: Andrew Cunningham and Perry Williams (Eds.): *The Laboratory Revolution in Medicine.* Cambridge: Cambridge University Press, 1992, 14-71.

Nipperdey, Thomas: *Deutsche Geschichte 1800-1866: Bürgerwelt und starker Staat.* München: C.H. Beck, 1983.

Olesko, Kathryn M. and Frederic L. Holmes: Experiment, Quantification, and Discovery: Helmholtz's Early Physiological Researches. In: David Cahan (Ed.): *Hermann von Helmholtz and the Foundations of Nineteenth Century Science.* Berkeley/Los Angeles/London: University of California Press, 1993: 50-108.

Olesko, Kathryn M.: *The Meaning of Precision.* Forthcoming.

Olesko, Kathryn M.: The Meaning of Precision: The Exact Sensibility in Early Nineteenth Century Germany. In: M. Norton Wise (Ed.): *Values of Precision.* Princeton: Princeton University Press, 1994.

Olesko, Kathryn M.: *Physics as a Calling: Discipline and Practice in the Königsberg Seminar for Physics.* Ithaca, N.Y./London: Cornell University Press, 1991.

Olesko, Kathryn M.: Resistance, Tolerance, and Consensus: British and German Measures of Electrical Resistance. Unpubl. essay presented at the Dibner Institute, Massachusetts Institute of Technology, April 1993.

Olesko, Kathryn M.: Tacit Knowledge and School Formation. *Osiris* 8, 1993, 16-29.

Pernet, Johannes: Hermann von Helmholtz. 31. August 1821 bis 8. September 1894. Ein Nachruf. *Neujahrsblatt der Naturforschenden Gesellschaft in Zürich*, 97, 1895, 1-36.

Prutz, Hans: *Die Königl. Albertus Universität zu Königsberg im Preussen im 19. Jahrhundert: Zur Feier ihres 350. jährigen Bestehens*. Königsberg: Hartung, 1894.

Rosenkranz, Karl: *Königsberger Skizzen*. 2 Bde. Danzig: Gerhard, 1857.

Schiefferdecker, Wilhelm Friedrich: Bericht über die Thätigkeit der Königlichen ostpreußischen physikalisch-ökonomischen Gesellschaft zu Königsberg. *Altpreußische Monatsschrift*, 1, 1864, 167-177.

Selle, Götz von: *Geschichte der Albertus-Universität zu Königsberg in Preussen*. Königsberg: Kanter Verlag, 1944.

Shapin, Steven: Personal development and intellectual biography: the case of Robert Boyle. *British Journal for the History of Science*, 26, 1993, 335-345.

Sheehan, James J.: *German History 1770-1866*. Oxford: Clarendon Press, 1989.

Sperber, Jonathan: State and Civil Society in Prussia: Thoughts on a New Edition of Reinhart Koselleck's *Preussen zwischen Reform und Revolution*. *Journal of Modern History*, 57, 1985, 278-96.

Sperber, Jonathan: *Rhineland Radicals: The Democratic Movement and the Revolution of 1848-1849*. Princeton, N.J.: Princeton University Press, 1991.

Stieda, L.: Zur Geschichte der physikalisch-ökonomischen Gesellschaft. *Schriften der Königlichen physikalisch-ökonomischen Gesellschaft zu Königsberg in Preussen*, 31, 1890, 38-84.

Tilly, Richard: The Political Economy of Public Finance and the Industrialization of Prussia. *Journal of Economic History*, 26, 1966, 484-97.

Tuchman, Arleen: Helmholtz and the German Medical Community. In: David Cahan (Ed.). *Hermann von Helmholtz and the Foundations of Nineteenth Century Science*. Berkeley/Los Angeles/London: University of California Press, 1993, 17-49.

Wichert, E.: Die Bewegung des altpreussischen Handels im letzten Decennium. *Altpreußische Monatsschrift*, 1, 1864, 426-445, 513-531, 601-617.

How Hertz Fabricated Helmholtzian Forces in His Karlsruhe Laboratory or Why He Did Not Discover Electric Waves in 1887[1]

Jed Z. Buchwald

Reversing the spirit of Charles Dickens's *Christmas Carol* I will begin with a story about something that might have been but never was. On Christmas Eve, 1987 an older chemist whom I shall call G climbed slowly to his attic. His father, himself a well-known physicist in the early years of the century, had long ago told him about a box of papers that was not to be opened until that very day. After many hours of digging through the dust of decades, G found a small, leather-covered box with the initials "*HH*" prominently inscribed on it in gold in the old German script. He carefully opened the case, which contained thirty or so separate pages, each covered with diagrams, numbers and the occasional remark in the same script as the box's cover. G sat in a broken chair by the pale winter light that filtered through an attic window and began to read.

It did not take long for G to realize that he had in his hands the laboratory notes for a completely unknown experiment undertaken by one of his father's closest and long-mourned friends, the great Heinrich Hertz, discoverer of electric waves. G recalled his father's tales of Hertz's glory days, so soon ended

1 This article is based on themes developed at length in Buchwald, 1994. Full references can be found there.

by the terrible blood poisoning that stole from him his rightful place at the helm of German physics as successor to his mentor, Hermann von Helmholtz. In the spring of 1888, G's father had often said, Hertz suddenly and astoundingly proved Maxwell's electric waves to exist by reflecting and refracting them. But the papers in G's hands gave him an uneasy feeling. They seemed to have something to do with waves. There, clearly diagrammed, were Hertz's devices - his oscillator and clever, detecting resonator. Numbers that seemed to be wavelengths appeared in appropriate places. And yet, something did not look quite right, for nowhere could G find the slightest trace of Maxwell's equations or anything even vaguely like them.

The weak light was rapidly fading now, so, puzzled and perturbed, G took the papers downstairs with him. The family had gathered for the evening's celebration, but G could not keep his mind on the festivities. When everyone had gone home, he quickly grabbed the old papers and started reading them again from the beginning, this time with pencil in hand. Every afternoon and evening for the next three weeks G poured over Hertz's lost manuscript again and again. In mid-January he felt that he had grasped its inner meaning. And he also knew that he would never breath a word about it.

The lost manuscript, G now realized, contained an astonishing record of experiments that ran completely counter to the demands of the very theory for radiating dipoles that Hertz had himself developed on the basis of Maxwell's equations in the summer and fall of 1888. These experiments had been done in December of 1887, exactly a hundred years before G was permitted by his father's will to open the sealed box. According to them, the field near the dipole behaves quite differently from the requirements of Hertz's equations. Equally unfortunate, Hertz had apparently measured a substantial difference between the wave's speed in air and its speed when guided by wires, which runs completely counter to Maxwell's theory.

Far from having confirmed Maxwell's theory, G now saw, Hertz's earliest laboratory work confirmed something very different from it indeed, something that had nothing at all to do with fields. G could not quite see what that other thing was, except that Helmholtz had produced it, for he was of course no historian. Hertz, G concluded, must have turned quickly to the experiments on reflection and refraction that had made him famous and then carefully hidden away these early ones, trusting them in the end to the care of G's father, who could not bring himself to burn these last few relics of his closest friend. G felt the warmth of the fireplace behind his back. With only a slight twinge of regret he turned and tossed the manuscript into it. Hertz's reputation was forever secured.

These events never happened. Nor did anything *quite* like them. There is however a chemist named Gerhard Hertz, grand-nephew to Heinrich and recently retired at Karlsruhe, who not long ago uncovered his grand-uncle's laboratory notes. Unlike the G of my story, he made the notes immediately available. By means of them it has been possible to reconstruct in precise detail the course of Hertz's work during a critical three month period from October through December, 1887. Though the events of my story may never have taken place, nevertheless the contents of my fictitious manuscript and the actual discovery document lead to the same conclusion: namely, that Hertz did not at first discover new kinds of waves; he discovered new kinds of forces. Unlike the fictitious Hertz, the real one did not hide his discovery; he trumpeted it loudly in the pages of the Berlin *Sitzungsberichte* and soon thereafter in the *Annalen der Physik* itself. My purpose here is to make clear what Hertz felt he had found, and to explain what Hertz did when he later decided that he had been mistaken.

On Christmas Eve, 1887 the young and intensively competitive Heinrich Hertz became convinced that he had produced and detected propagating electrodynamic force in his Karlsruhe laboratory. To secure his priority and to make his reputation Hertz wrote his mentor, the great and supremely influential von Helmholtz, about his success. Helmholtz communicated Hertz's report to the *Sitzungsberichte* of the Berlin Academy, which received it on February 2. On January 24 a letter from Wiedemann, editor of the *Annalen*, reached Hertz asking for a paper on experiments concerning dielectrics that Hertz had completed earlier in the fall. Hertz wrote back with an alternative proposal. He wanted to give Wiedemann a trilogy, to present his new work in a logical progression. He wanted to explain how his new apparatus worked, then show how he had used it to prove that changing dielectric polarization exerts electromotive force. In the third paper Hertz would report on how the device was used to detect and to produce propagation in air and in wires. With the partial exception of the paper on dielectrics, none of these three reports from the laboratory can be reconciled with Maxwell's equations - with, that is, the mathematical tools (the structure, in fact, of antenna theory) that Hertz himself developed in the summer and fall of 1888 to investigate in a novel way the behavior of his radiating electric dipole. The very experiments which first convinced Hertz that propagation in air occurs were, scarcely a quarter-year later, and despite their prominent publication, thought by Hertz himself to be problematic. To understand what occurred requires grasping why Hertz built his device, how he thought it worked, and what he felt it was good for. Begin with the character of Hertz's experimenting.

In all of his work before 1887, and there was a great deal of it (much of which had nothing to do with electrodynamics) Hertz scarcely ever designed apparatus on the basis of theoretical specifications. He took existing devices, bits and pieces of equipment, or strangely-behaving objects and played around with them in ways aimed at eliciting peculiar, indeed unknown forms of behavior. His effect-probing work in the laboratory, as well as his paper analyses, betray the overpowering influence of Helmholtz himself. I will not here pursue the extraordinary ways in which Helmholtz, or perhaps it would be better to say Helmholtz's *imago*, his living image in Hertz's mind, influenced how Hertz thought and what career moves he made. Suffice it to say that when Hertz experimented, he experimented with a virtual Helmholtz constantly in the audience watching him; when he produced mathematics, Helmholtz and Kirchhoff both sat as silent critics.

By the end of the '70s Helmholtz was especially interested in two effects that had never been directly observed. Experiments that had been performed in his laboratory during that decade seemed to him to imply that his electrodynamics could be sustained only if these two effects existed. The effects at issue were, first, whether dielectrics can be polarized by changing currents in conductors, and, second, whether changing dielectric polarization can exert electromotive force. I shall refer to this pair as the "Berlin effects". Together they linked through to Maxwellian fields (better, to Helmholtz's version of Maxwell's theory), and hence to propagation, in a rather complicated way but they had their own meaning and resonance in the context of Helmholtzian electrodynamics.

Specifically, in Helmholtz's system the ether constitutes a universally-present body that can be electrically polarized by charge and that can be magnetically-polarized by magnets or by currents in conductors, just like ordinary dielectrics and magnetically-susceptible bodies can be. Helmholtz's scheme did not however require *a priori* either that changing currents in conductors must behave like charge, or that changing dielectric polarization must behave like conduction current. It did not, that is, require the Berlin effects to exist. Whether or not they did accordingly remained important but open questions for Helmholtz and his followers. They were by contrast completely closed questions for Maxwellians, since proponents of the field did not distinguish between different kinds of electric forces or between different kinds of magnetic forces - all electric forces contributed to a unitary electric field, and all magnetic forces contributed to a unitary magnetic field.[2] Nevertheless, if both types of effects do in fact occur, then Helmholtz's equations for electrically and magnetically polarizable bodies, such as the ether, imply that waves of polarization will indeed occur in them (albeit ones that, except under certain limiting conditions, are only weak facsimiles of the waves required by Maxwell's scheme).

Helmholtz tried to pressure the neophyte Hertz into performing experiments on the Berlin effects. Hertz resisted because he felt, after a long paper analysis, that they could not fruitfully be addressed with existing apparatus. He ever-after kept the two effects in mind, but they formed for Hertz something quite distinct from propagation. They constituted, as it were, discrete *icons* that represented core issues for the physics promulgated by his mentor, and, as such, they remained *in and of themselves* critically important to him.

This provides one basis for understanding why Hertz would not likely have set out to look for propagation *per se*: in and of itself - i.e. considered apart from the conjunction of the Berlin effects - propagation was not at the center of Helmholtz's electrodynamics. Many Maxwellians, by contrast, were strongly captured by magneto- and electro-optical processes during the 1880s, for they saw the fullest realization of field theory precisely in effects that involved propagation. There is another reason, which is undoubtedly the more significant one. Hertz would not have set out to investigate propagation with his newly-fabricated apparatus because he did not initially think of the device as a radiator-detector pair. He had constructed it piecemeal in response to particular, local problems that had little to do with anything beyond the intense pursuit of an instrumental novelty. This brings me to how Hertz came to build the apparatus in the first place.

2 This meant that, e.g., an electric field capable of producing a current had *de facto* to be capable of polarizing a dielectric since only the state of the field, but not its character, is influenced by the nature of the source. The role of the source remained central to Helmholtz's electrodynamics, whereas it had been relegated to secondary status in field theory.

Hertz tells us in his *Introduction*, and there is no reason to doubt his claim,[3] that he was first caught in 1886 by what he felt to be the unusual inducing power of a pair of Riess or Knochenhauer spirals that he had for demonstration purposes in his Karlsruhe laboratory. This device consisted of a pair of spirally-wound conductors that were placed face-to-face. When a spark is drawn across one of the spirals (by connecting it to, say, a battery or to an induction coil), then sparking occurs in the other one as well, visually demonstrating electromagnetic induction.

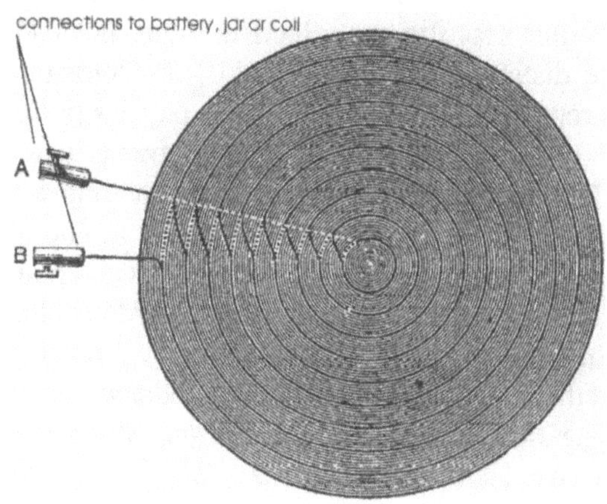

Figure 1. Riess spirals

We should immediately recognize something odd about Hertz's particular interest, because Riess spirals had been used by many people before him without anyone else apparently wondering about their power. But Hertz differed from everyone else in at least two respects. First, he was constantly on the lookout for novel effects. Berlin, which is to say Helmholtz, had taught him that the world was filled with such things and that his job was to find them. Second, Hertz had by this time a very great deal indeed of experience with coupled circuits, both in designing and operating them. The Riess spirals sparked in ways that simply lay outside his experience, and he at once began to chase down their power.

3 Other aspects of the *Introduction* require careful interpretation, because it was designed strategically to make the discovery process seem much more linear and logical than it was.

Hertz took apart the spirals, manipulating, distorting and altering them until he had evolved what we now call the dipole oscillator. At the same time, and in lock-step with the dipole's evolution, Hertz developed the resonant detector. The words 'oscillator' and 'resonant' of course suggest vibrations. He did not at first associate either device with such things, for he thought that whatever electric oscillations do occur on the dipole would hardly be regular enough to be worth thinking about. In the event, Hertz was eventually able to produce stable coupling effects between the dipole and the detector that convinced him he was dealing with synchronous oscillations in the range of tens of millions of cycles per second. This - the production of a new regime of electric oscillations - in itself constituted a tremendous accomplishment which he proudly announced to Helmholtz.

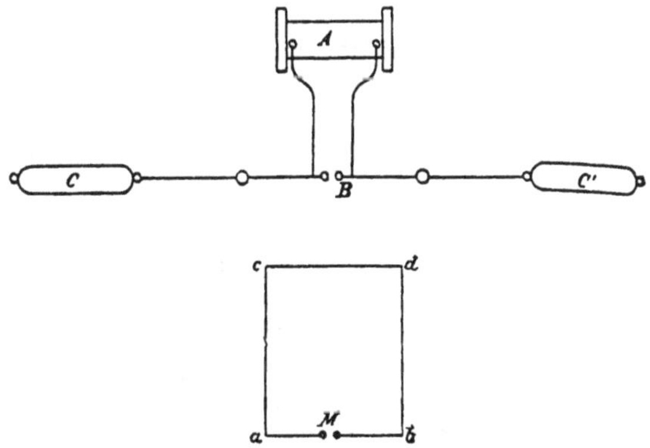

Figure 2. Dipole oscillator and resonator

The coupling Hertz had produced suggested nothing at all about propagation in air, because it operated in his view according to laws of induction that held in every contemporary theory of electrodynamics, from one based on Wilhelm Weber's electric particles, through Helmholtz's uninterpreted electrodynamic potential, to Maxwell's fields. The enormous rapidity of the oscillations did however convince Hertz that he had at hand a device that could be used to produce equally rapid polarization currents in dielectrics, and so to detect their electromotive actions. In pursuing this specific problem Hertz learned how to do something entirely new: he learned how to trace the behavior of inducing forces in space. To grasp what he was soon to do in tracking propagation requires an appreciation of how he did so.

The flat plates labeled *A* and *A'* in the center of Fig. 3 are the termini of Hertz's driven oscillator; *D* is a huge block of dielectric material; *C* is a metal bridge. The large circle, with center *B*, is the resonant detector, which is provided with an adjustable spark gap. The resonator's size in comparison to the oscillator is both noteworthy and critical for Hertz's apparatus to work. By observing the behavior of the sparking as the resonator was rotated about its center Hertz could draw conclusions about the forces that were acting to drive it. He had early learned, purely by manipulation, that the sparking is governed almost entirely by whatever forces act on the resonator parts that are diametrically opposite the spark gap proper - and so that, for purposes of understanding, the resonator could be reduced insofar as the activating forces are concerned to a small, linear piece of metal at point *a'* in Figure 3.

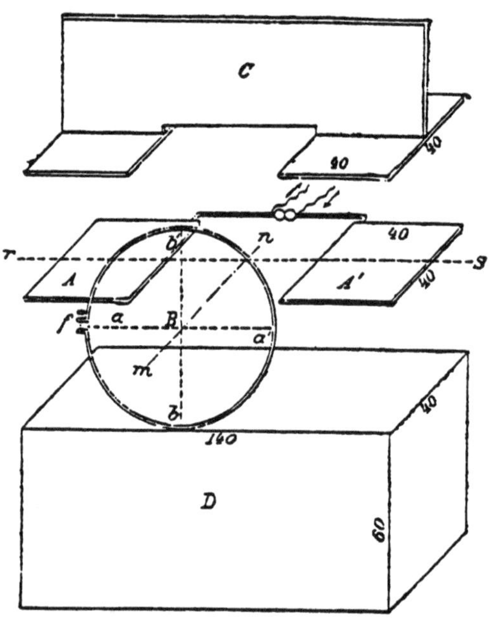

Figure 3. Using the apparatus to probe dielectric behavior

In the fall of 1887 Hertz taught himself through manipulation how the sparking behaves when another conducting plate (*C* in Figure 3) is brought near the oscillator. The huge, rapidly-changing charges that surge back and forth on the oscillator image themselves electrostatically in this conductor, which, like the oscillator itself, is therefore the site of powerful conduction currents. This gave Hertz a way to calibrate the behavior of a dielectric against the effect of something entirely unproblematic - namely, the electrodynamic induction of a changing conduction current. To examine dielectric currents was then altogether simple: remove the conductor *C* and use instead a block *D* of pitch. The oscillator again images itself electrically, this time in the dielectric's polarization, and accordingly engenders polarization currents. All Hertz had to do to find the Berlin effect that involved the electromotive action of changing polarization was to see whether the resonator's sparking behaved in essentially the same way with the dielectric in place as with the conducting plate C in place. It did. The experiment worked quite stably, and Hertz went rapidly into print with it.

Two important points emerge from this. First, the experiment required essentially no theory (and certainly no computations) beyond the widely-assimilated notion of induction by changing currents. Theory might have entered in considering how the resonator worked, but it did not in any important way because Hertz learned to work the resonator by play and not by calculation. Second, Hertz carefully sought to calibrate a novel effect against one that nobody would question - even though, strictly speaking, no one before Hertz had ever detected the action of an imaged conduction current (or, probably, had been interested in doing so). Hertz calibrated against an unproblematic effect, not an already-produced and used one. Both characteristics carry directly over into his propagation experiments.

We are now prepared to understand why Hertz tackled propagation, how he did it, and why he can reasonably be said to have discovered something about forces and not about fields on that Christmas Eve over a hundred years ago. It did not at once occur to Hertz to use his device for propagation experiments. Why should it have? Waves themselves were very far indeed from his mind, which was now intensely concentrated on the highly specific questions concerning dielectric effects raised long ago by Helmholtz. He had however unsuccessfully tried the year before (1886) to use the device to test for the electric polarizing action of changing conduction currents (the first Berlin effect), and he still saw no way to do so. At a certain point, he tells us in his *Introduction* he suddenly realized what in retrospect might seem obvious - namely, that he could as it were find both effects simultaneously by demonstrating propagation in air since, as mentioned above, the latter is a joint implication of the two together, supposing air to act like a dielectric in all respects.

This realization would not have been foremost in Hertz's mind at the time, despite its clear presence in Helmholtz's own work because Hertz had convinced himself in 1884 that the basic principles of Helmholtz's electrodynamics (ones that he claimed were in fact universal) might actually entail propagation *independently* of dielectric effects. Dielectrics might very well not be anything at all like what Maxwellians thought they were, and yet forces ought perhaps still to propagate.

This view, which markedly distinguished Hertz from other Helmholtzians, nevertheless resulted from his having probed more deeply even than Helmholtz himself had the roots of his mentor's electrodynamics. Hertz was certainly aware (probably as early as 1879) that propagation can be obtained by introducing the ether as dielectric, since this is entirely explicit in Helmholtz's work, but he had in 1884 sought to purify Helmholtz's scheme of any reliance on a universal dielectric because the latter could at best be fit phenomenologically into Helmholtz's system, by which I mean that there was no expression for the energy of dielectric-conductor interactions except when both are statically charged - and expressions for interaction energies constituted the core of Helmholtz's electrodynamics. From 1884 on, therefore, dielectrics *per se* were not thought of by Hertz in terms of their putative link to propagation. It required an effort on his part to put back together what he had sundered in 1884.[4] His search in the fall of 1887 for propagation in air was accordingly not motivated by a desire to find propagation *per se*. It was instead governed by his wish to satisfy a nearly decade-old demand by his mentor. Propagation was a tool; it was not an end in itself. This goes very far in making familiar Hertz's way of understanding his discovery experiments, for, as we shall now see, he initially paid no attention whatsoever to the possibility that the propagating force in air might have wavelike attributes beyond mere motion.

Turn now to the apparatus with which Hertz produced propagation and to how he understood its working. This will put us in a position to see how propagation first became real to him. Following the same pattern as in his dielectric experiments, Hertz decided to set off a novel effect (air propagation) against an effect that no one would question but that no one had ever produced in the laboratory: extremely high-frequency waves in wires. To do so first required persuasive evidence that the metal waves existed, which meant showing that his resonator could reliably detect them. Hertz spent a great deal of time doing just that. Once these waves became real - they were already un-

---

4 Full details can be found in Buchwald, 1994.

problematic - Hertz had somehow to play them off against the problematic action (air propagation). To do so he decided to make the force from the metal wire and the force from the oscillator act together to drive the resonator. Figure 4, which Hertz published, diagrams his apparatus for doing so.

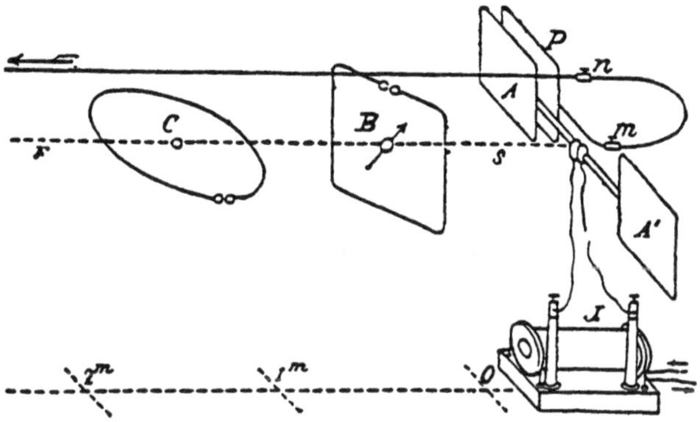

Figure 4. Hertz's published diagram for his interference experiments

The essence of Hertz's experiment lay precisely in its direct reliance on interference between *driving forces* - and not, say, on the composition of fields. His forces were in no respect whatsoever different in kind from the forces that act between conductors with changing currents, by which I mean that for Hertz they had every one of the practical attributes of such things that he had learned in a decade of play, and no others. Or, rather, with precisely one other attribute - namely, that the force exerted by the oscillator might take time to reach the resonator, just as the wave sent down the metal wire takes time to travel. But for Hertz at this point a time delay merely complicated an otherwise well-understood, indeed utterly simple, situation, one in which two forces act at the same time on the same object to produce a net effect. The experiment was designed to map this altogether well-understood addition of forces from point to point along the wire. As far as Hertz was concerned he was looking for a new kind of force, one that propagates, and not for an altogether new electromagnetic structure. Every prior characteristic of electromotive force as he understood it accordingly carried over directly into his initial work.

The device detected the composition of forces in the following way. Figure 5 represents schematically the wire above, the oscillator on the right, and the active part of the resonator (*gh*) in the center. Both the wire and the oscillator are to be treated as nothing more than bits of wire that bear rapidly changing currents. These currents exert inductive forces that are essentially parallel to their directions. The wire and oscillator consequently act on the resonator *gh*, Hertz reasoned, with the respective electromotive forces f_w , f_o. From this it seemed to Hertz quite simple to see how the apparatus would work to detect interference.

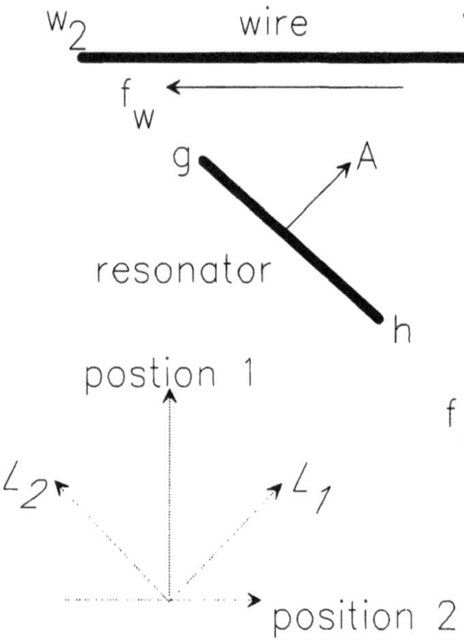

Figure 5. Hertz's interfering forces

Suppose that at some point along the wire the two forces act in the directions specified by the arrows in Figure 5, and that the resonator is oriented so that its normal points in the direction L_1. In that case the force from the wire will generate a current from *h* to *g* in the resonator. The force from the oscillator will, on the other hand, drive a current in the opposite direction, namely from *g* to *h*. Turn the resonator so that its normal now points in the direction L_2. The wire will still drive a current in it from *h* to *g*. The oscillator, however, will now drive a current in the same direction as the wire does. Consequently in turning the resonator from L_1 to L_2 we have changed the interference - the composition of forces, as Hertz saw it - from mutually opposing to mutually supporting. To work the device Hertz would set it first to L_1, observe the sparking in that position, turn it to L_2, and finally observe whether the sparking intensity stayed the same, increased or decreased. He would label the particular result 0, +, or – respectively.

By moving the resonator down the wire and noting the character of the change in sparking between his two canonical positions, Hertz felt that he could readily draw conclusions about propagation. He already knew that the force exerted by the wire moves along it, as it were, because he had mapped the wire waves themselves.[5] Suppose for example that the oscillator's force does not propagate - that only its magnitude changes with distance. In that case the character of the sparking will simply keep in step with the wire wave itself because only the latter travels. If, *per contra*, the force in air travels at precisely the same rate as the wire wave, then the sparking will have to be of exactly the same type at every point along the wire (what type that is will depend upon the initial phase difference between the two propagations). Anything else will produce sparking variations that can be calculated given the ratio of the two speeds. Hertz unfortunately did not provide any mathematics to accompany the conclusions he drew from his observations, but appropriate formulae can be easily reconstructed. He always rotated the resonator through 90°, and under these circumstances the change in sparking strength between the two orientations can be represented by means of the following formula:

$$\text{change in sparking strength} = \cos\left[2\pi z\left(\frac{1}{\lambda_M}-\frac{1}{\lambda_A}\right)+2\pi\left(\frac{z_M}{\lambda_M}-\frac{z_A}{\lambda_A}\right)\right]$$

Here λ_M, λ_A are respectively the wavelengths of the wire and oscillator forces; z_M, z_A are the distances of their origins from the zero-point of measurement. In his critical experiments, Hertz actually held the resonator fixed in a given position and delayed the phase of the wire wave by adding wire between points *m* and *n* in Figure 4, which increases z_M.

5 This is not, of course, a true propagation of force in the sense that propagation in air would be: the force varies in a wavelike way along the wire because the current that exerts it moves as a wave down the wire, and not because the force *per se* propagates through the wire.

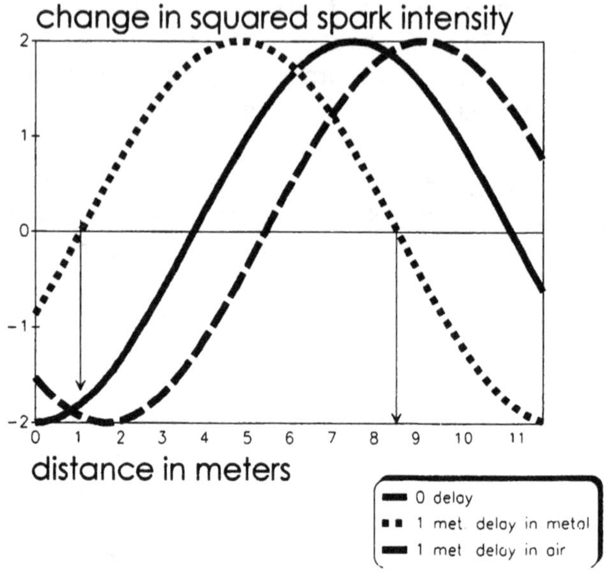

Figure 6. Possible spark-change curves

What, now, can we expect from this scheme? Figure 6 gives several possible situations and nicely represents what Hertz thought would happen if, say, the force in air propagated 1.6 times as fast as the force in the metal wire. As the figure shows, a given spark-intensity locus will under these circumstances be shifted *towards* the origin by retarding the slower, metal action, and it will be shifted *away from* the origin by retarding the faster, air action. From observations done on successively larger metal retardations Hertz could also deduce the ratio of the speeds of force propagation through the following formula (which he did not however provide):

$$\frac{v_A}{v_M} = \frac{\Delta z}{\Delta z - \frac{\lambda_M}{2}}$$

Here Δz represents the distance by which the metal wave must be retarded in order for the spark-change curve to alter sign.

Turn now to what Hertz found. I will not here reproduce the tables he drew up in his laboratory notebook, and that he subsequently printed with some modifications, instructive though they are. Every one of these tables led to the same result: before Christmas Eve, Hertz could not find any definite, or even suggestive, indication that the spark-change pattern did *not* track the wire wave, which meant that the air propagation did *not* have a detectably finite speed. Even granted an extremely wide margin for error - though Hertz did not explicitly consider any such thing - his experiments could not possibly be used to persuade skeptical contemporaries that he had found propagation in air, and Hertz knew it.

During these weeks Hertz talked himself into accepting what seemed to be this unavoidable negative result. He had however never been happy with negative consequences. His training placed a tremendous premium on producing positive results, which is to say new effects, but there seemed to him to be no way out. He accordingly decided to write a paper detailing these results, one that would count as a negative answer to the second Berlin effect, and so also against Maxwell's theory. But he had to be certain. He had to check his results, and in particular to make his apparatus utterly convincing as an interference-detector, since everything depended upon its not being subject to doubt.

Hertz knew that he had to convince an audience he had himself only recently introduced to very high frequency oscillations that a detector built to respond to such things was a reliable indicator of the composite force exerted by these oscillations and by equally high-frequency waves in wires. He had to be able to argue persuasively that this entirely new device, which he had just invented, and with which he had already produced novel and difficult effects, could now be deployed as an instrument of such eminent trustworthiness that it was capable of detecting something that lay at the extreme boundaries of contemporary theoretical speculation, and certainly beyond the boundaries of accepted instrumentation. Hertz accordingly worked to turn his device into an appliance for showing interference. On December 17 he set out on a course of 35 numbered experiments designed to stabilize the apparatus.

He completed this work on December 21. On the 22nd and the 23rd he again examined the interference between the wire and oscillator forces in order to corroborate his earlier finding that the force in air does not propagate. In these experiments, as in his earlier ones, the plane of the resonator was perpendicular to the plane formed by the wire and the oscillator, and its gap was vertical (as in *B* in Figure 4). The resonator was therefore driven entirely by the forces acting on its lower portion.

This position was the most sensitive to interference, but it was not the only one possible, and indeed it was not the one used by Hertz to map the wire wave. To do that he had used the position *C* in Figure 4. Here the plane of the resonator includes both the wire and the oscillator, and its spark gap is parallel to the wire. Hertz had not used this position to detect interference before for one very important, and essentially pragmatic, reason - there was no easy way to reverse the interference of the forces from wire and oscillator: there were no simple analogues of the positions L_1 and L_2.

On December 24th Hertz nevertheless decided to try interferences in this inconvenient position. His reason for doing so is particularly instructive, because it shows just how thoroughly bound his laboratory work at this point was to the pure image of interfering forces. In position *B*, Hertz knew from experience, the resonator responds to more than the electrodynamic, inducing forces exerted by the wire and the oscillator. It responds as well to electrostatic forces exerted by the huge charges on the oscillator's termini. Hertz wanted an experiment in which *only* the electrodynamic force acted to be certain that nothing troubled the observation. In position *C*, Hertz knew from experiment and from analysis, the resonator responds only to the oscillator's electrodynamic action, as well as to the wire's force. He therefore went to the considerable effort of replicating the interference effect for this inconvenient position, one he had used for his device-testing experiments between the 17th and the 21st. For any given position of the resonator he had to reverse one of the two forces, and the only way to do that here was to physically pick up the wire and move it from one side of the resonator to the other. In three experiments done on the 24th, numbered 50 through 52, Hertz set the resonator in a given position, observed the spark character, flipped the wire to the other side, and observed any change in sparking.

To understand what was involved here suppose that the oscillator's action propagates (as Hertz now thought) vastly more rapidly than that of the wire's. Hertz's wire wave had a half-length somewhere between 2.7 and 3.1 meters. At his zero-point of measurement the oscillator and the wire interfered constructively with one another. About 1.5 meters past this point the interference should have begun to change sign, until sparking occurred very strongly in an opposite fashion at about 3 meters distance. In terms of Hertz's device, under this assumption he should have seen something like the following. At the zero-point a wire flip kills sparking. Move the resonator down the wire several hundred centimeters. Sparking should weaken, but a wire-flip should still obliterate the effect. Near 1.5 meters the sparking should be very weak, and a wire flip should have nearly no effect. Several hundred centimeters further on weak sparking should occur with the wire in its previously-flipped position, with obliteration now taking place when the wire is on the previously-reinforcing side of the resonator. At about 3 meters distance the effect must be very marked and indeed just as unmistakable as the strong sparking and obliteration or weakening that takes place at the zero position, but now reversed. This would have constituted a replication of the effect Hertz had observed with the resonator in position *B*.

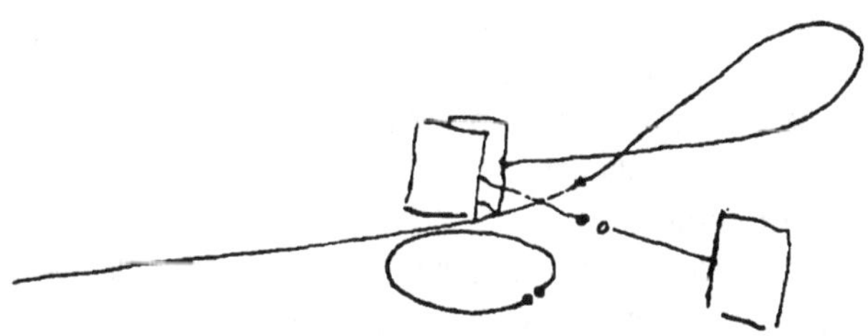

Figure 7. The critical arrangement, from Hertz's laboratory notebook

On, in Hertz's words, "the night before Christmas" he performed experiment 51 (Figure 7). He placed the resonator 3 meters away from the spark-strengthening reference point established in experiment 50. The wire-flip should now, and quite markedly, weaken instead of strengthen the spark. *It did not do so.* In the words of Hertz's laboratory notes, strong sparking took place with the wire "on the same side as when close [to the oscillator at the null point]" - the original reads, with emphatic punctuation, "Also von derselben Seite wie in der Nähe!". He immediately pursued the discovery in experiment 52: at 1, 2, 3 and even 4 meters from the null point the effect remained the same in kind, though it was sufficiently weak at 4 (and 5) meters that things became "doubtful".

This completely unambiguous, and before the 24th entirely unexpected, result could in Hertz's conception of interfering forces have only one interpretation: the oscillator force must itself propagate at a speed not undetectably different from that of the wire wave. This of course raised the question of what had gone wrong in the experiments with the resonator in position *B*. Hertz had an answer. Obviously, he reasoned, the electrostatic force must propagate vastly more rapidly than the electrodynamic force. Closer to the oscillator, where the earlier experiments had given infinite speed for electrodynamic action, the electrostatic force is quite powerful. There it overwhelmed the electrodynamic action and falsely gave the latter infinite speed as well. This meant that the *B* experiments now constituted presumptive evidence for the high, perhaps infinite speed of electrostatic force, whereas the *C* experiments gave evidence for the finite speed of electrodynamic action. The two forces manifested themselves quite directly and in markedly different ways.

There was more. On Maxwell's theory, Hertz was at this point vaguely aware, the speeds of air and wire propagation had to be the same. On Helmholtz's principles the speeds can be different, although Helmholtz did not explicitly carry through a full analysis on this point. Hertz's experiments could be combined to deduce the ratio of speeds according to the formula I gave above. Hertz observed that at about 7.5 meters the interference changes sign; using 2.8 meters for the half-length of the wire wave, Hertz asserted in print that the ratio of the speed of the oscillator's electrodynamic action to that of the wire wave is as 75 to 47, or about 1.6 to 1.

It will come as no surprise that Hertz had later to backtrack over these first results in a manner I will shortly discuss. But he rushed into print with them, writing his artfully-constructed *trilogy* to persuade his German colleagues of their instrumental cogency. Hertz's rhetoric nevertheless failed to work as he wished; few among his colleagues greeted this early demonstration of propagation with eager approval. There are at least two reasons for this less than overwhelming reaction. First, Hertz's own credibility was apparently not extremely high, at least among some people; he had yet to establish a secure reputation. Second, the experiments required a tremendous amount of interpretation to be understood, and indeed were best grasped when Hertz himself physically performed them with accompanying explanation.

Hertz recognized at least the second of these reasons for the disappointing general reaction that his discovery received. In the spring he accordingly set out to find a more convincing way to show propagation, and, not without difficulty, he hit upon the notion that the force in air might be like a wave in the fullest sense of the term. It occurred to him that he might be able to produce standing-waves in air by reflection, just as he had produced them in wires. This led to experiments that were utterly transparent in comparison with the ones that had led him to believe in propagation in the first place. These new experiments, unlike the original ones, were rapidly influential because the overall features of standing waves were widely-understood from optics and from mechanics, and because Hertz had already produced electric ones in wires.

Having decided, and then shown, that the force in air behaves in the fullest sense like a wave, Hertz worked in the summer of 1888 to provide an appropriate mathematics that went beyond the composition of forces with which he had begun. That composition, he now felt, should nevertheless continue to work well as an approximation to the true state of affairs. Hertz accordingly set out explicitly to solve his version of Maxwell's equations for the dipole oscillator. His goal would now be to see just what these equations required in his discovery experiments, anticipating at best minor variations from the implications that he had previously drawn by compounding forces.

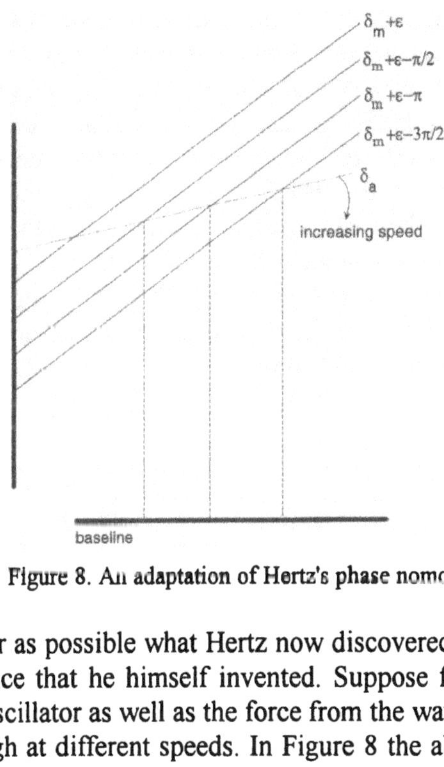

Figure 8. An adaptation of Hertz's phase nomograph

To make as clear as possible what Hertz now discovered on paper I will adapt a graphical device that he himself invented. Suppose for a moment that the force from the oscillator as well as the force from the wave on the wire do both propagate, though at different speeds. In Figure 8 the abscissa represents distance along the wire, and the ordinate represents time. If we assume, as Hertz did in his discovery experiments, that a propagating force simply moves away from its origin at a constant speed, then we may represent both the wire wave and the action from the oscillator by means of straight lines in this diagram.

Here, for example, the line δ_a represents the motion of a given phase of the force in air as it moves point by point down the wire. Suppose that at some point along the wire the force in air differs in phase from that of the metal wave by an amount ε, and let a line $\delta_m + \varepsilon$ through that point represent the metal wave. Draw a sequence of lines parallel to this one but differing from it by integral decrements of 90°. The points at which these lines intersect the line δ_a determine the loci on the baseline where the phase difference between the metal and air waves increases at each step by a constant 90°. These loci can then be directly correlated with Hertz's three types of observed interference (designated +, 0, –).

This diagram constitutes in effect a graphical realization of the original interpretation that governed Hertz's discovery experiments. It operates in the following way. One can vary either, or both, the speeds of the air and metal forces in order to see how the interference will change, but for simplicity let us hold the speed of the metal wave constant and increase that of the force in air by tilting its line δ_a towards the horizon. Then the distance between successive points of intersection, projected down to the baseline, decreases until the "interference" simply tracks the metal wave. Conversely, moving the air line up increases the distance between equispaced points, until the air line parallels the metal line. The phase difference will then have the same value everywhere, and this would correspond to the two forces having the same speed. Inclinations in between correspond to Hertz's discovery experiments, with the air speed being greater than the metal speed.

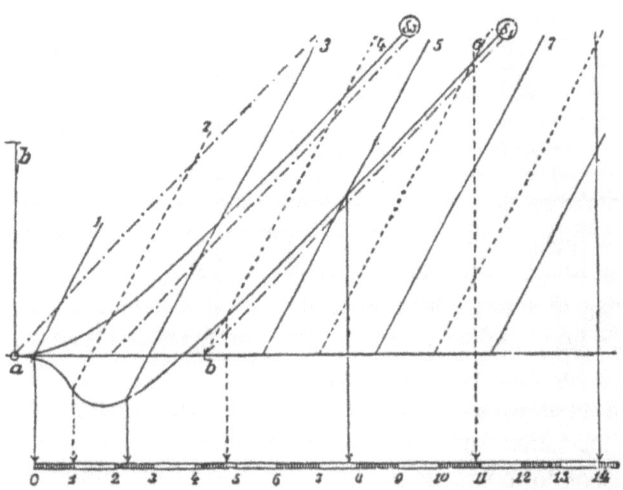

Figure 9. Hertz's phase diagram for the field

We can, with Hertz, use the same graphical device to see what ought to happen according to Hertz's solutions to his version of Maxwell's equations. It was precisely here, within a specific, technical problem, that the profound difference between the concept of a *propagating force* and the demands of the *field* first became strikingly clear to him. Figure 9 is Hertz's own phase diagram for the dipole field, drawn from his solution. This is the first depiction by anyone of phase behavior in electromagnetic radiation.

Here the baseline is drawn thickly on the bottom, and the curved line that dips below the upper horizontal represents the phase that would be detected in Hertz's first experiments, wherein the plane of the resonator was vertical. The straight lines *1, 2, 3, 4* represent the succession of metal-wire phases. To find the loci of interferences that differ in phase by 90° now requires finding the intersection of these lines with the curved line for the air phase. Below about a wavelength the points are by no means equispaced, which, Hertz argued, nicely fits the results he had obtained in December. The indications of near-infinite wave speed that occurred below about a half-wavelength are due to the dip in the phase-curve: at the curve's minimum point its tangent is horizontal, which corresponds to an infinitely-high speed in the image of propagating forces that Hertz held at the time of the discovery experiments.

The curved line that does not dip below the horizon represents the phase that (on Maxwell's equations) Hertz should have detected in the very experiments that had first shown him propagation in air. This line, unlike the first curve, is not inflected, and so it will not entail apparently-infinite wave speeds at any point. It is nevertheless quite markedly bent within 3 meters or so of the dipole. Projecting its intersection with the straight lines for the sequence of wire-wave phases indicates that the interference should have changed character at about 2 meters, and then have stayed the same in type until about 4.5 meters. Instead, *Hertz had found no change at all until about 3 meters.* Had Hertz observed what his calculations now indicated to him that he should have (but did not), then, thinking in terms of forces, he would have concluded once again that propagation does not occur. He would certainly have gone into print with the claim, since that was his intention!

Hertz decided that his observations were nevertheless to be trusted, and that his solution to the dipole equations must itself be inaccurate as a result of approximations involved in supposing the wire to be infinitely thin and perhaps also to incorrect values for the wavelength in the wire. Between the winter of '87-'88 and the fall of '88, then, an observation had changed from a triumphant confirmation into a vexing problem. That change coincides with

Hertz's first attempts to replace Helmholtzian propagating forces by wave transmission through fields. Let me quote his own words as an exemplification of the difficult mutations in understanding (and in practice) that this replacement made necessary: "In the sense of our theory we more correctly represent the phenomenon by saying that fundamentally the waves which are being developed do not owe their formation solely to processes at the origin, but arise out of the conditions of the whole surrounding space, which latter, according to our theory, is the true seat of the energy". These were waves utterly unlike any that physicists were used to treating.

Hertz's state of mind between 1888 and 1890 bears every mark of the instability that usually characterizes a period of profound change. He sought at first to understand fields as something like propagating forces, but he found that he could not carry through into Maxwell's equations the connections that such things demanded. Hertz struggled to mix the immiscible - propagating forces with fields - until he ultimately fashioned a purified structure by 1890 that had been purged of forces, and indeed of the practical and conceptual emphasis on sources that this previous way of working and thinking required (Hertz, 1962: 195-240). During this transition period, while forces struggled with fields for supremacy, Hertz did not think Maxwell's equations to be entirely correct. He remained convinced that air waves propagate faster than metal waves. Indeed, it seems that, shortly before he had solved Maxwell's equations (and so before he had fully perceived the differences between forces and fields), but after he had already done the reflection experiments, this very discrepancy seemed to Hertz to be his most significant discovery. He wrote Helmholtz about it on March 19, 1888 in the following words:

> "It seems to me that the most important point is now the certainty that propagation through conducting wires is perceptibly slower than in air, whereas all previous theories, including Maxwell's, led to the conclusion that propagation in wires took place with the velocity of light" (J. Hertz, 1977: 255).

Hertz cannot be said thoroughly to have replaced compounding forces with unitary fields until he became convinced that wire waves do not exist at all except as special cases of air waves, and that contrary experiments had to be ruled out of court or else reinterpreted. One might put it this way. In 1887 and 1888 Hertz thought that propagation in air involved a special kind of electromotive force, and that propagation in metals was a distinct process. By 1890 he thought instead that propagation in air involved the only kind of electric field that exists, and that propagation involving a wire actually took place in the surrounding air, with the wire acting as a wave guide.

I will close with a different ending to my story about G and the box of old papers. Being something of a philologist G did not after all toss them into the fire. He published them with a few remarks about the ability of great genius to see beyond momentary disappointment to preordained success. Soon Hertz's forgotten laboratory notes would have become records of hidden failure, pored over for indications of pathological behavior by inquisitive historians. They are instead the remnant traces of an original success that became an acknowledged embarrassment.

Bibliography

Buchwald, Jed Z.: *The Creation of Scientific Effects. Heinrich Hertz and Electric Waves.* Chicago: The University of Chicago Press, 1994.

Hertz, H.: *Versuchsprotokolle*, 1887, courtesy of Prof. Dr. H. G. Hertz, Karlsruhe and Prof. M. Doncel, Barcelona.

Hertz, H.: *Electric Waves.* Translated by D. E. Jones. Reprint of the 1893 edition from MacMillan and Co. New York: Dover Publications, 1962.

Hertz, J.: *Heinrich Hertz. Memoirs. Letters. Diaries.* Second enlarged edition, prepared by M. Hertz and C. Susskind. English translation by L. Brinner, M. Hertz, and C. Susskind. San Francisco: San Francisco Press, 1977.

Hermann von Helmholtz' Beziehungen zu russischen Gelehrten

Annette Vogt

Hermann von Helmholtz (1821-1894) unterhielt sehr enge Wissenschaftsbeziehungen zu englischen Fachkollegen,[1] zu französischen Wissenschaftlern, insbesondere zu Ophthalmologen,[2] zu österreichischen Physikern, vor allem zu Ludwig Boltzmann (1844-1906),[3] zu amerikanischen Wissenschaftlern und nicht zuletzt zu russischen.

Untersucht man die Beziehungen von Helmholtz zu russischen Gelehrten, fällt auf, daß es sich vorrangig um Lehrer-Schüler-Beziehungen handelte. Zu dem Professor in Deutschland kamen, zunächst nach Heidelberg, später nach Berlin, junge russische Wissenschaftler verschiedenster Fachgebiete. Entweder sie wollten ihre Ausbildung beginnen oder sie waren sich der Lücken ihrer erworbenen Kenntnisse bewußt und kamen zu einem der Großen ihres Faches, um ihr Wissen und Können zu vervollständigen. Je größer der Ruhm Helmholtz' wurde, desto größer wurde die Anziehungskraft seiner Person. Für seine Studenten bedeutete es eine nicht unerhebliche Hilfe für ihre weitere wissenschaftliche Karriere, Gutachten vom berühmten Helmholtz den heimatlichen Behörden vorlegen zu können.

---

1 Vgl. Koenigsberger 1903, Bd. 2.
2 Vgl. *Festschrift* 1891.
3 Vgl. zu den Beziehungen Helmholtz-Boltzmann Hörz/Laaß 1989; Hörz 1993, 77-95.

1. Zum Stand der Wissenschaften im zaristischen Rußland (1800-1900)

Nach vielen Jahrzehnten einer Stagnation in Rußland entstanden im 19. Jahrhundert vielfältige Bemühungen, den Anschluß an den internationalen Wissenschaftsstandard, wie er zunächst in Paris und ab der zweiten Hälfte des 19. Jahrhunderts an mehreren deutschen Universitäten geprägt wurde, herzustellen.

Bis zum Beginn des 19. Jahrhunderts existierten nur zwei - auch auswärts - bekannte wissenschaftliche Institutionen: die 1724/1725 von Zar Peter I. (1672-1725) ins Leben gerufene Akademie der Wissenschaften (AdW) zu Sankt Peter(s)burg und die 1755 von Michail Vasil'evič Lomonosov (1711-1765) gegründete erste russische Universität in Moskau (s. Tabelle 1). Die Petersburger Akademie spielte im 18. Jahrhundert eine bedeutende Rolle für die Entwicklung einer eigenen russischen Naturwissenschaft und zählte zeitweise Gelehrte von Weltrang zu ihren Mitgliedern; erinnert sei nur an Leonard Euler (1707-1783).[4] Moskau und St. Petersburg blieben für viele Jahrzehnte die einzigen Wissenschaftszentren in Rußland.

Mit dem Sieg über die Heere Napoleons 1812/13 stieg Rußland zu einer der europäischen Großmächte auf. Die Zaren herrschten in Polen, in Finnland, bis an die Grenzen im fernen Osten; bereits 1801 war der Kaukasus, 1804 waren Teile Mittelasiens hinzugekommen. Im Innern gab es große wirtschaftliche und politische Probleme. 1825 fand der Dekabristenaufstand statt, polnische Aufstände 1830/31 und 1862/64 wurden blutig niedergeschlagen; im Krimkrieg (1853-1855/56) zwischen Rußland und der Türkei um Einflußsphären im Nahen Osten und in Südosteuropa erlitt Rußland eine schwere Niederlage, nicht zuletzt, weil Frankreich und England das Osmanische Reich unterstützten. Der Krimkrieg verdeutlichte vor aller Welt die enorme ökonomische Rückständigkeit des Riesenreiches und initiierte eine Reihe von Reformen, die die industrielle, aber auch die wissenschaftlich-kulturelle Entwicklung heben sollten. Es erfolgte 1861 die Aufhebung der Leibeigenschaft.

Für die Wissenschafts- und Technikentwicklung zeitigten die Reformen positive Folgen. Es entstanden neue Universitäten und weitere wissenschaftli-

4 Vgl. Grau 1988, 120-133 sowie 180-182.

che Einrichtungen wie Institute, Observatorien und medizinische Einrichtungen.

Man übernahm ein Prinzip der Förderung begabter russischer Studenten, das schon einmal im 18. Jahrhundert erfolgreich gewesen war. Talentierte junge Leute delegierte man zu weiterbildenden Studien ins Ausland, insbesondere nach Frankreich (hier im wesentlichen nach Paris) und Deutschland, und vertraute ihnen bei ihrer Rückkehr Lehrstühle oder gar Institute an. Sie sollten - ausgerüstet mit modernstem Wissen - ihr Können an die "Daheimgebliebenen" weitergeben und so für eine Reproduktion des wissenschaftlichen Nachwuchses sorgen. Dieses Prinzip erwies sich als erfolgreich, obwohl einschränkend angemerkt werden muß, daß es nicht nur gezielt angewandt wurde, sondern oft den Zufällen oder Zwängen des Lebens unterlag.

Die russischen Wissenschaftler versuchten so, den Anschluß an die internationale Entwicklung zu gewinnen. Die Wissenschaftsbeziehungen zwischen Deutschland und Rußland in jenen Jahrzehnten bestanden somit meist aus einem Transfer wissenschaftlicher Kenntnisse *von* Deutschland *nach* Rußland; umgekehrte Transferbeziehungen entwickelten sich partiell besonders Ende des 19. und Anfang des 20. Jahrhunderts.

Die herausragende wissenschaftliche Institution in Rußland blieb die Petersburger Akademie, die als höchste wissenschaftliche Einrichtung wirkte. Sie war auch international als Forschungszentrum anerkannt, besonders durch die durchgeführten Expeditionen und die astronomischen Arbeiten an der nach neuesten Gesichtspunkten errichteten Sternwarte Pulkovo, die am 1.1.1839 unter der Leitung von Friedrich Georg Wilhelm Struve (1793-1864) eröffnet wurde.

Zwei Aspekte dürfen nicht unberücksichtigt bleiben: Rußland blieb im 19. Jahrhundert trotz aller Anstrengungen und herausragender Einzelleistungen in Wissenschaft und Kultur ein rückständiges Land, in dem über 80 % Analphabeten lebten.

Die zaristische Regierung betrieb das ganze 19. Jahrhundert über eine sehr restriktive antisemitische Politik. Es war Juden meist verwehrt, russische Universitäten zu besuchen, so daß besonders im letzten Drittel des vorigen Jahrhunderts jüdisch-russische Studenten an deutschen Universitäten studierten, um dem diskriminierenden "numerus clausus" zu entgehen. Unter diesen Studenten standen viele den Volkstümlern, den Sozialrevolutionären oder später den Sozialdemokraten und Bolševiki nahe. Insbesondere in Preußen widmete sich folglich die Polizei auch der Überwachung der "gefährlichen" Studenten.[5]

So gab es im wesentlichen Vertreter dreier Gruppen, die bei Helmholtz studierten:

1. die von Rußland delegierten ("abkommandierten" (russ. "komandirovanye")) jungen Leute, die bereits an russischen Universitäten studiert, das Magisterexamen bestanden hatten und nun eine höhere Ausbildung erlangen wollten;
2. junge Studenten, die frühzeitig erkannten, daß die Koryphäen auf ihrem Spezialgebiet nur im Ausland tätig waren und die auch ihre Doktorarbeiten in Deutschland vorbereiteten;
3. schließlich jene Gruppe jüdischer Studenten, die von vornherein im Ausland studieren mußten. Ihre weiteren Karrieremöglichkeiten in Rußland blieben ebenfalls begrenzt; oft errichteten sie ein privates Labor. Die Mediziner unter ihnen hatten etwas bessere Chancen und gründeten häufig eigene Krankenhäuser. Es existierte außerdem auch in Rußland noch die Möglichkeit der Taufe "als Eintrittsbillet in die bürgerliche Gesellschaft" (H. Heine).

2. Die russischen Studenten bei Hermann von Helmholtz

Man muß deutlich zwischen der Zeit, die Helmholtz in Heidelberg lehrte, und der, in der er in Berlin der "Reichskanzler der deutschen Physik" genannt wurde, unterscheiden. Hervorzuheben ist, daß Helmholtz russische Schüler verschiedenster Fachgebiete hatte, die auf jedem dieser Gebiete zu Begründern russischer wissenschaftlicher Schulen wurden, und dies nicht nur in Moskau und St. Petersburg, sondern auch in Kazan' und Char'kov, in Kiev und Odessa. Helmholtz' Schüler studierten die Fachgebiete Chemie, Physiologie, Medizin (besonders Ophthalmologie) und schließlich Physik (s. Tabelle 2).

In Heidelberg gab es sogar einen besonderen "russischen Zirkel", und der Chirurg und Anatom Nikolaj Ivanovič Pirogov (1810-1881), der selbst bereits

5 Vgl. den "polizeilichen Überwachungsbericht der Politischen Polizei an den Berliner Polizeipräsidenten" vom 8.3.1902 über Anna Elisarova geb. Uljanova (1864-1935), in: HUB, Dokumente 1985, 41.
Vgl. auch S.V. Kovalevskaja und ihre Beziehungen zur deutschen Sozialdemokratie, die sie selbst vor K. Weierstraß verbarg.

1833-1835 in Deutschland studiert hatte, war eigens aus der Hauptstadt St. Petersburg "abkommandiert" worden, um die russischen Studenten in Deutschland zu betreuen. Auch die später berühmte Mathematikerin Sofja Vasil'evna Kovalevskaja (1850-1891) gehörte in Heidelberg zu Helmholtz' Studenten. Anders als in Berlin, durfte sie als Frau die Heidelberger Universität besuchen.[6]

Unter Helmholtz' Studenten aus Rußland waren einige, die später weltweit Anerkennung fanden, und solche, die für die Entwicklung der Medizin und der Naturwissenschaften in Rußland besondere Bedeutung erlangten. Oft gründeten sie Institute, aus denen im 20. Jahrhundert international anerkannte Fachleute kamen. Von allen bekannten Helmholtz-Studenten (Tabelle 2) sollen hier besonders die folgenden erwähnt werden:
- die Chemiker D.I. Mendeleev (1834-1907) und A.P. Borodin (1833-1887), letzterer heute oft nur als Komponist bekannt;
- der Physiologe I.M. Sečenov (1829-1905) und der Landwirtschafts-Physiologe K.A. Timirjazev (1843-1920);
- die Ophthalmologen E.A. Junge (1833-1898) und L.L. Hiršman (1839-1921), die in St. Petersburg und Char'kov spezielle Kliniken für Augenkrankheiten gründeten und leiteten;
- der Physiker A.G. Stoletov (1839-1896) sowie seine Schüler R.A. Kolli (1845-1891), P.A. Zilov (1850-1921), A.P. Sokolov (1854-1928) und P.N. Lebedev (1866-1912).

Anläßlich des 70. Geburtstages von Helmholtz sandten mehrere russische Universitäten Grußschreiben, in denen sie Helmholtz als Lehrer russischer Wissenschaftler rühmten und ihm dankten.[7] Aus demselben Anlaß beteiligten sich Moskauer Wissenschaftler an einer speziellen Vorlesungsreihe, deren Erlös - immerhin 1000 Rubel - dem neu gestifteten Helmholtz-Fond zugeführt wurde.[8]

---

6 Vgl. Vogt 1988, 271.

7 Vgl. *Ansprachen und Reden* 1892, 61-63. Im einzelnen schrieben: Kaiserl. Universität Kiev, Kaiserl. Universität Moskau, Kaiserl. AdW St. Petersburg, Russische physikalisch-chemische Gesellschaft zu St. Petersburg, Kaukasische medizinische Gesellschaft in Tiflis, medizinische Gesellschaft in Warschau und die Russische medizinische Gesellschaft in Warschau. Außerdem erhielt Helmholtz Urkunden über die Ernennung als Ehrenmitglied der Universitäten Char'kov, Dorpat, Kazan', Moskau, Odessa, St. Petersburg und Warschau.

8 Vgl. Lebedinskij u.a., 1966, 170.

Mehrere Helmholtz-Schüler erinnerten sich ihrer Erlebnisse, schilderten seine Vorlesungen, seine Art des persönlichen Umgangs mit Studenten und Assistenten, berichteten über die Ausstattung seiner Laboratorien in Heidelberg bzw. in Berlin.

So hielt Helmholtz in Heidelberg für Studenten aller Fakultäten Vorlesungen über "Allgemeine Ergebnisse der Naturwissenschaften", an die Erinnerungen von F.P. Šeremetevskij aus dem Jahre 1891 existieren. Sonja V. Kovalevskaja schilderte in ihren *Erinnerungen an meine Kindheit*, wie Helmholtz' Ruhm bis ins ferne dörfliche Rußland gelangte.[9] Onkel Petr Vassil'evič "verführte" die Schwestern Anjuta und Sonja zum Lesen und liebte es, bei Tisch über neue wissenschaftliche Entdeckungen mit großer Begeisterung vorzutragen.

> "Ich erinnere mich noch, welchen Sturm zwei Aufsätze in der "Revue des deux Mondes" bei uns entfesselten; der eine behandelte die "Einheit der physikalischen Kräfte" und berief sich dabei ausführlich auf eine Broschüre von Helmholtz; der andere berichtete von "Claude Bernards Versuchen mit einem herausgeschnittenen Taubengehirn". Sowohl Helmholtz wie Claude Bernard wären wohl recht erstaunt gewesen, hätten sie erfahren, daß sie mit ihren Schriften sozusagen zum Zankapfel für eine im morastischen Gouvernement Witebsk lebende russische Familie geworden waren" (Kovalevskaja 1961, 70-71).

An die Heidelberger Zeit erinnerte sich der Physiologe K.A. Timirjazev:

> "Ich sehe sie noch vor mir, die berühmte Heidelberger Universität in jener glänzenden Epoche, als Bunsen, Kirchhoff, Helmholtz und der einer breiten Öffentlichkeit weniger bekannte Hofmeister[10] an ihr lehrten, deren Ruhm fortgesetzt wuchs. Die Vertreter der Badener Regierung schockierte nicht die Kosten, die für jene Zeit noch nicht dagewesen waren, die aber jetzt mit einem nachsichtigen Lächeln aufgenommen werden, die für die Ausrüstung der notwendigen Laboratorien ausgegeben werden mußten. Mit dem heutigen Blick gesehen [d.i. 1909 - A.V.], zählte Bunsens Laboratorium in jenen Jahren zu den besten in Deutschland, aber es war nicht vergleichbar mit den Laboratorien von Kirchhoff und Helmholtz (damals noch für Physiologie), letzteres in einem zweistöckigen Ge-

9 Vgl. Šeremetevskij 1891: Šeremetevskij, F.P.: Gel'mgol'c kak fiziolog i značenie ego dlja psichologii. In: "German fon Gel'mgol'c", Moskva 1892, str.107-108, zitiert in: Lebedinskij u.a., 1966, 104.

10 Es handelt sich um den Botaniker Wilhelm Friedrich Benedikt Hofmeister (1824-1877), der 1863 an die Heidelberger Universität berufen wurde und 1872 an die Universität Tübingen ging, A.V.

bäude mit großen Fensterfassaden, das unter dem Namen "Natur-Palast" bekannt war".[11]

Über die Heidelberger Zeit schrieb der berühmte russische Physiologe Ivan Michailovič Sečenov in seinen "autobiographischen Bemerkungen":

> "Er [Helmholtz, A.V.] las die öffentlichen Vorlesungen vor den Medizin-Studenten, die ich hörte, nicht schön, er las elementar, ohne jegliche Mathematik. Wahrscheinlich langweilten sie ihn, denn als ich einmal an einer Abendveranstaltung der Heidelberger Wissenschaftlichen Gesellschaft teilnehmen konnte, auf der er über den Schall referierte, las er hier direkt fröhlich, wählte das Schicksal des anwesenden tauben Bunsen zur Demonstration seines Gerätes, was dieser mit einem guten Lächeln geschehen ließ.
>
> Im Laboratorium (es war nicht groß, es hatte ein Einzelzimmer für den Professor, aber kein eigenes Zimmer für seinen damaligen Assistenten Wundt) arbeiteten 4 Personen: 2 Ophthalmologen - Junge (mein Kamerad) und Knap(p) - , ein Deutscher mit sehr schrägstehenden Augen, dessen Name ich vergessen habe (und der sich mehrere Monate mit dem Myograph Helmholtz' abplagte), und ich. Wundt saß faktisch das ganze Jahr über in seiner Ecke über irgendwelchen Büchern, ohne irgend jemandem Aufmerksamkeit zu schenken, und sprach mit Niemandem auch nur ein Wort. Ich hörte kein einziges Mal seine Stimme. Helmholtz sahen wir vorbeihuschen. Täglich kam er einmal ins Arbeitszimmer, besuchte alle Arbeitenden, fragte jeden, was es gegeben hatte und gab Hilfestellung, wenn es notwendig war".[12]

Von den russischen Physikern, die bei Helmholtz, insbesondere in seiner Berliner Zeit, studierten, hinterließen einige Erinnerungen bzw. Reiseberichte. So beschrieb R.A. Kolli das Berliner Laboratorium. Ein Vergleich mit Sečenovs Schilderung der Heidelberger Einrichtung zeigt, wie es Helmholtz in Berlin schließlich gelang, seine Arbeitsbedingungen und die seiner Assistenten und Studenten zu verbessern. Kolli schrieb über seinen Aufenthalt im Wintersemester 1881/1882:

> "Die experimentelle Arbeit stand im Zusammenhang mit den Forschungen zum Energieerhaltungssatz. Lebhaft erinnere ich mich an die Besuche des bedeutenden Gelehrten. Es ist wahr, nicht immer war er in Stimmung, wie man zu sagen pflegt. Aber wenn er es war, interessierte ihn dieses oder jenes Stadium im Verlauf der Untersuchung, er gab Beispiele, wie man sich dem Thema nähern könnte, verwies auf Analogien und Beziehungen der gegenwärtigen Frage mit anderen, manchmal aus völlig fernliegenden Gebieten, und du weißt nicht, wohin das führen wird - ob zu einer Erweiterung des Wissens oder zu einer Vertiefung und

11 Timirjazev 1909: Timirjazev, K.A.: God itogov i pominok. Iz naučnoj letopisi 1909 g. Sočinenija, t. IX. Sel'chozgiz, 1939, str.119-120, zitiert in: Lebedinskij u.a. 1966, 105-106.

12 Sečenov 1945: Sečenov, I.M.: Avtobiografičeskie zametki. Moskva 1945, str.94-95, zitiert in: Lebedinskij, 108-109.

> größeren Originalität der Idee. So hinterließ jeder dieser Besuche im Kopf der Beteiligten den Eindruck, daß eine Masse neuer Fragen aufgetaucht waren, neue Themen für interessante experimentelle und theoretische Arbeiten. So ist es nicht rätselhaft, daß in dem allgemeinen physikalischen Institut viele Dutzende Arbeitende versammelt waren, und für jeden gab es eine interessante und geeignete Arbeit. Die Physiker kamen aus allen Ecken der Welt und versammelten sich im Institut an der Neuen Wilhelmstraße".[13]

An die Berliner Zeit erinnerte sich auch Petr Alekseevič Zilov:

> "In den ersten Jahren der Tätigkeit von Helmholtz in Berlin war sein Laboratorium in der Universität untergebracht. An den Türen des Laboratoriums stand deutlich "Physikalisches Institut", aber mit diesem gewaltigen Namen verband man noch nicht jene Vorstellung, die wir heute [d.i. 1911, A.V.] davon haben. Der allgemeine Charakter des Laboratoriums war dunkel, ärmlich, und es fehlten viele notwendige Apparaturen. Das Laboratorium besaß 4 bis 5 Zimmer (jedes mit einem Fenster) im linken Flügel des Hauptgebäudes der Universität, vor dem jetzt die Marmorstatue des berühmten Gelehrten steht. In jedem der Zimmer arbeiteten 3 bis 4 Leute (insgesamt gab es nur 14 Praktikanten), die sich bei dieser Enge ständig gegenseitig störten. Das Universitätsgebäude selbst erzitterte vom Fahren der Equipagen, und täglich um 12.00 Uhr, wenn die Militärkolonne zum Schloß zog und Salut schoß, erzitterten die Geräte, und die Galvanometer und Elektrometer fingen zu springen an, so daß wir unsere Arbeiten unterbrechen und auf das Ende des Schießens warten mußten".[14]

Jahre nach seiner Studienzeit besuchte Zilov wiederum Berlin und konnte den stattlichen Neubau des physikalischen Instituts (1876 eröffnet) bewundern. Dies illustrierte er in seinem "Reisebericht" von 1889:

> "Das physikalische Laboratorium der Berliner Universität befindet sich seit 1877 in einem eigens dafür errichteten Gebäude, das 4 Etagen mit 70 Zimmern und allen notwendigen Apparaturen besitzt. Es gibt hier 2 Hörsäle (einen großen und einen kleinen), einige Säle für Übungen mit Anfänger-Studenten und 24 Zimmer für Studenten, die selbständige Arbeiten ausführen, außerdem einen Raum für Gewichte, einige Räume für optische Untersuchungen, Zimmer für Magnet-Arbeiten, die Bibliothek, Räumlichkeiten, in denen die "Fortschritte der Physik" vorbereitet werden, und schließlich noch die Wohnungen für den Direktor, die Assistenten und Gäste. Das Gebäude ist innen und außen sehr schön hergerichtet und befindet sich natürlich in einer mustergültigen Ordnung. [...]

---

13 Kolli 1892: Kolli, R.A. Sbornik "German fon Gel'mgol'c". Publičnye lekcii, čitannye v Moskovskom universitete. Moskva 1892, str.83, zitiert in: Lebedinskij, 119-120.

14 Zilov 1911: In: Kosonogov, I.I.: Nikolaj Nikolaevič Šiller. In: Žurnal Russkogo fiziko-chimičeskogo obščestva, 1911, c. fiz., t.43, vyp. 9. Otd. ottisk: St. Petersburg 1911, str.2, zitiert in: Lebedinskij, 149.

Die einzelnen Zimmer für die Studenten sind nicht groß, aber mit allem Notwendigen (Gas, Wasser, Abwasser usw.) ausgestattet; in jedem gibt es steinerne Konsolen für die Instrumente. Hier arbeitete Helmholtz bis 1888".[15]

3. Die Verbreitung der Helmholtzschen Arbeiten in Rußland

Die sowjetischen Physiker Andrej Vladimirovič Lebedinskij, User Ionovič Frankfurt und Aleksandr Moiseevič Frenk veröffentlichten 1966 in Moskau eine Helmholtz-Biographie und erarbeiteten eine spezielle Bibliographie, in der alle Arbeiten von Helmholtz und die über ihn, die in russisch erschienen waren, zusammengestellt wurden (Lebedinskij u.a., 1966, 302-306). Daraus geht hervor, daß Arbeiten von Helmholtz seit 1864 in russischer Sprache publiziert wurden. Es handelte sich dabei um physiologische, medizinische aber auch philosophische und populärwissenschaftliche Aufsätze. Als Übersetzer waren kaum Helmholtz-Schüler zu finden. Man sollte noch genauer untersuchen, warum bestimmte Arbeiten von Helmholtz relativ schnell übersetzt wurden.

Unter den Übersetzungen gibt es eine, zu der ein Brief an Helmholtz gefunden werden konnte. Im Nachlaß von Helmholtz, der im Archiv der Berlin-Brandenburgischen Akademie der Wissenschaften aufbewahrt wird, sind nur wenige Briefe russischer Gelehrter erhalten geblieben (s. Tabelle 3). Aber zwei Briefe ragen doch heraus, zumal es sich bei den Verfassern um bekannte russische Wissenschaftler handelte.

Am 1.(13.)12.1892 hatte der Mathematikprofessor Aleksandr Vasil'evič Vasil'ev (1853-1929) aus Kazan' an Helmholtz geschrieben (s. Anhang 1) und ihm von einem Vortrag für Mathematik- und Physiklehrer bei der Eröffnung der Kazan'er physikalisch-mathematischen Gesellschaft berichtet, in dem er auch über die Arbeit von Helmholtz "Zählen und Messen" (1887) sprach. Die Lehrer baten ihn daraufhin, so Vasil'ev an Helmholtz, diese Arbeit ins Russische zu übersetzen, was er gern tun würde (zusammen mit L. Kroneckers Aufsatz "Über den Zahlbegriff", ebenfalls 1887 in dem Band zu Ehren E. Zellers

15 Zilov 1889: Zilov, P.A.: Berlinskie i parižskie fizičeskie instituty i električeskij kongress. Varšava, 1889, str.3, zitiert in: Lebedinskij, 152-153.

veröffentlicht). So bat er Helmholtz um dessen Erlaubnis.[16] Es ist unklar, ob er sie erhielt, aber Vasil'evs Übersetzung erschien jedenfalls bereits 1893 in Kazan'. Er muß übrigens im fernen Kazan' von Helmholtz' Augenproblemen gewußt haben, denn der Brief ist mit besonders großen (deutschen) Buchstaben verfaßt worden. Ebenfalls 1893 erschien in Kazan' auch Vasil'evs Übersetzung der Helmholtzschen Arbeit "Über die Tatsachen, welche der Geometrie zu Grunde liegen" (1868), die er in seinem Buch *Über die Grundlagen der Geometrie* erstmals in Russisch publizierte.

Einige Jahre zuvor, am 20.11.(2.12.)1886, hatte der "Vater der russischen Luftfahrt", der berühmte Nikolaj Egorovič Žukovskij (1847-1921) an Helmholtz geschrieben und ihm seine Überlegungen zu der Helmholtzschen Veröffentlichung "Über die physikalische Bedeutung des Princips der kleinsten Wirkung", die 1886 im 100. Band des Crelle-Journals (*Journal für die reine und angewandte Mathematik*) erschienen war, dargelegt. Žukovskij hoffte, seine "Bemerkung zur zweiten Variation" werde von Helmholtz "der Oeffentlichkeit übergeben", wie er es formulierte (s. Anhang 2). Leider fand sich nur dieser singuläre Brief.[17]

Hermann von Helmholtz hat mit seinen Arbeiten, seiner Methodik und seinen Ideen, aber insbesondere durch seine Vorlesungstätigkeit praktisch eine ganze Generation russischer Naturwissenschaftler ausgebildet oder mindestens ihr Denken und Arbeiten geprägt. Russische (und später sowjetische) Wissenschaftler erinnerten sich immer dankbar seines Wirkens. In Moskau existiert bis in die Gegenwart das Helmholtz-Institut für Ophthalmologie.

Er gestaltete damit ein Kapitel russisch-deutscher Wissenschaftsbeziehungen, vergleichbar seiner österreichisch-deutschen oder französisch-deutschen Beziehungen, und festigte die Tradition dieser Kontakte, die im 20. Jahrhundert eine besondere Bedeutung gewannen.

16 Vasil'ev (Wassilieff) am 1.(13.)12.1892 an Helmholtz, in: Archiv der BBAdW, Nachlaß Helmholtz, Nr. 494. Brief als Anhang 1 hier erstmals publiziert.

17 Žukovskij (Joukovsky) am 20.11.(2.12.) 1886 an Helmholtz, in: Archiv der BBAdW, Nachlaß Helmholtz, Nr. 223. Brief als Anhang 2 hier erstmals publiziert.

Tabelle 1

Gründungen von Universitäten und anderen wissenschaftlichen Institutionen im russischen Reich im 18. und 19. Jahrhundert

Sankt Petersburg	Artillerie-Hochschule	1721
	Kaiserliche **Akademie der Wissenschaften**	**1724/1725**
	Medizinisch- chirurgische Akademie	1733 (Schule) 1786 (Hochsch.) 1798
	Pädagogisches Institut	1804
	Universität	**1819**
	Land- und Forstakademie	1811/1864
	Praktisch-technisches (technologisches) Institut	1828
	Ingenieur-Hochschule	1832
	Arzt-Kurse für Frauen	1872
Moskau	**Universität** (erste russische Universität)	**1755**
	Landwirtschaftliches Institut (K.A. Timirjazev-Akademie)	1892
Dorpat (Tartu)	Universität	1802
Kazan'	Universität	1804
Char'kov	Universität	1805
Warszawa	Universität Russische Universität	1816 1869 - 1917
Kiev	Universität	1834
Odessa	Universität (Novorossijskij Universität)	1865
Tomsk	Universität	1880

Tabelle 2

Russische Studenten bei H. v. Helmholtz in Heidelberg und in Berlin

1859-67	Marija Aleks. Vilinskaja-Markovič (Marko Vovčok) (1834-1907)	Schriftstellerin
1869	Sof'ja Vasil'evna Kovalevskaja (1850-1891)	Mathematikerin; 1884 Prof. Univ. Stockholm.
<u>Chemiker</u>		
1859 - 1861	Dmitrij Ivanovič Mendeleev (1834-1907)	Ab 1864 Prof. St. Petersburg; 1876 KM AdW Petersburg
1859 - 1862	Aleksandr Porfir'evič Borodin (1833-1887)	Auch Komponist; ab 1864 Prof. St. Petersburg; einer der Organisatoren der med. Frauenausbildung; hielt für die Frauen Chemie-Vorlesungen.
Mitte 1860 1872/73	Vladimir Fedorovič Luginin (1834-1911)	1874-1881 eigenes Labor in St. Petersburg; dann an med.-chirurg. Akademie; 1889-1905 Univ. Moskau.
<u>Physiologen</u>		
1856 - 1860	Ivan Michajlovič Sečenov (1829-1905)	Ab 1860 Adj.-Prof. an med.-chirurg. Akademie; 1869 KM AdW Petersburg; 1870-76 Prof. Univ. Odessa; 1876-88 Prof. Univ. Petersburg; ab 1889 Univ. Moskau; 1904 EM AdW Petersburg.

1862 - 1867	Nikolaj Ignat'evič Bakst (1843-1904)	Ab 1867 Privatdozent Univ. Petersburg; Vorlesungen f. Physiologie an med. Frauenkursen. Seit 1886 Mitgl. des Wissenschaftler-Komitees beim Min. f. Volksbidung. Grundlegende Arbeiten zur Nervenphysiologie (s. Monatsber. Bln. AdW 1871).
um 1863	F. P. Šeremetevskij	1891 Prof. Univ. Moskau
1868 - 1870	Kliment Arkad'evič Timirjazev (1843-1920)	1871-75 a.o. Prof. an Land- u. Forstakademie Petersburg; 1875-77 dort o. Prof.; 1877-1911 Prof. Univ. Moskau; 1919 Mitorganisator d. Komm. Akademie.
Mediziner		
1856 - 1860	Eduard Andreevič Junge (1833-1898)	Ophthalmologe; 1860 Diss. u. Lehrstuhl für Ophtalmologie an der med.-chirurg. Akademie Petersburg; 1861-1868 Auslandsaufenhalt; 1869 Prof. Ophthal. u. Leiter d. Augenabt. d. Krankenhauses. Bildete Schule russ. Ophthalmologen.
1862 - 1866	Nikolaj Ivanovič Pirogov (1810-1881)	Chirurg und Anatom; studierte 1833-35 in Deutschland; 1862-66 abkommandiert zur Betreuung russ. Studenten in Deutschland.
1863 - 1869	Ivan Michajlovič Dogel' (1830-1916)	Pharmakologe; ab 1869 Prof. Univ. Kazan'.
1863 - 1868	Emilijan Valentionovič Adamjuk (1839-1906)	Ophthalmologe; 1868-1870 Privatdozent; 1871-1901 Prof. Univ. Kazan'
1864 - 1868	Leonard Leopol'dovič Hiršman (1839-1921)	Ophthalmologe; 1875 a.o., ab 1884 o. Prof. an Univ. Char'kov (bis 1905). Gründete in Char'kov Klinik f. Augenkrankheiten; ab 1908 Arbeit im Krankenhaus; landesweit bekannter Arzt.
1864 - 1872	Sergej Ivanovič Lamanski (geb.1841)	1874 Doz. an d. med.-chir. Akademie in Petersburg; dann Warszawa; 1879 in Paris; 1881 Prof. am ph. Centr.-Observatorium in Petersburg.

Physiker

1862 - 1866	Aleksandr Grigor'evič Stoletov (1839-1896)	Univ. Moskau, Begründer russ. Physiker-Schule in Moskau, Schüler Kolli, Kasterin. 1870 physikalisches Labor an Univ.; ab 1873 Prof.; Mitorganisator des Physikalischen Inst. an d. Univ. Moskau.
Mitte 1860er	Michail Petrovič Avenarius (1835-1895)	1876 KM AdW Petersburg; Prof. an d. Univ. Kiev; 1881 bei Pariser Elektr.-Ausstellung.
1881/1882	Robert Andreevič Kolli (1845-1891)	Stoletov-Schüler; Univ. in Moskau u. Kazan'.
1872 - 1874	Nikolaj Nikolaevič Šiller (1848-1910)	Thermodynamiker; 1876 Prof. Univ. Kiev; ab 1903 Dir. d. Technol. Inst. in Char'kov. Mitgl. im Kollegium für Volksbildung.
1871 - 1875/77	Petr Alekseevič Zilov (1850-1921)	Stoletov-Schüler; 1877-84 Doz. f. Physik an d. Techn. Hochschule Moskau; ab 1884 Univ. Warszawa; ab 1900 Hrsg. d. Zeitschrift "Fizkoe obozrenije".
vor 1882	Aleksej Petrovič Sokolov (1854-1928)	Stoletov-Schüler; ab 1882 Leiter d. physik. Labors an Univ. Moskau; ab 1884 Prof. Univ. Moskau; Mitorg. Physik. Inst. (1904).
1871 - 1872	Nikolaj Aleksandrovič Hezehus (1845-1918 o.19)	1. Rektor (1888-89) der neu eröffneten Univ. Tomsk; davor Privatdoz. Univ. Petersburg; 1889-1918 Prof. am technol. Inst. in Petersburg; 1911-18 Red. d. Zeitschrift "J. Russ.kogo fiz.-chim.ogo obscestva".
1883 - 1884	Vladimir Aleksandrovič Michel'son (1860-1927)	Ab 1884 an Moskauer Inst. f. Land- u. Forstwissenschaft-(Timirjazev-Akademie); später Meteorologe.
1889 - 1890	Petr Nikolaevič Lebedev (1866-1912)	Assist. bei Stoletov; 1887-88 Straßburg; 1890-91 Straßburg; 1891 Dissertation; 1900 Prof. Univ. Moskau. Organisierte als erster Gruppe russ. Physiker; 1911 Rücktritt aus Protest; Einladung nach Stockholm.

Tabelle 3

Briefpartner Helmholtz'

(Nachlaß Helmholtz, in: Archiv der BB AdW)

1858, 1859	Eduard Junge	4 Briefe (Probleme v. Studenten)
1872	A. Kosloff (Kozlov, Petersburg)	Bitte um zwei Gutachten
nach 1874	Graf A. K. Tolstoi	Bitte um Besuch
1879, 1890	E. Jaesche (Dorpat)	2 Briefe (Fragen zur Optik und physiologischen Optik)
1886	Joukovsky (Žukovskij)* (Moskau)	Anmerkungen zu einer Arbeit von Helmholtz aus dem Jahre 1886
1892	A. Wassilieff (Vasil'ev)** (Kazan')	Zu einer Übersetzung einer Arbeit von Helmholtz

* Žukovskij, Nikolaj Egorovič (1847-1921), "Vater der russischen Luftfahrt", seit 1872 Dozent an der Moskauer Technischen Hochschule, hier 1910 aerodyn. Laboratorium, 1894 KM AdW Petersburg, Dez. 1918 Leiter d. neugegründeten ZAGI.

** Vasil'ev, Aleksandr Vasil'evič (1853-1929), Mathematiker; ab 1875 Privatdozent, ab 1887 ordentlicher Prof. an der Universität Kazan'; Mitbegründer der physico-mathematischen Ges. in Kazan'; schrieb Biographie über Lobačevskij (1914).

Anhang 1

Kasan 1 [13] Decemb[er]92.

Hochgeehrter Herr Geheimrath

In einer Rede bei der Eröffnung der physico-mathematischen Gesellschaft in Kasan[18] habe ich die hohe Bedeutung für die Philosophie, Mathematik und Pädagogik Ihrer Abhandlung: "Zählen und Messen" hervorgehoben.[19] Nachher habe ich von vielen Bekannten besonders aus dem Lehrerkreise Anforderungen diese schätzenswerthe Abhandlung in's russische zu übersetzen erhalten. Dies habe ich jetzt erfüllt; aber bevor der Publication fühle ich mich verpflichtet die gnädige Erlaubniss Eurer Excellenz zu erhalten. Falls ich diese Erlaubniss erhalte, beabsichtige ich dieser Uebersetzung auch die Uebersetzung des Aufsatzes meines unvergesslichen grossen Lehrers Prof. L. Kronecker "Ueber den Zahlbegriff" beizufügen.[20] Diese zwei Abhandlungen müssen nach meiner Ueberzeugung sehr zur Aufregung des geistlichen Lebens in unserem Lehrerkreise mitbringen; mich werden sie immer an den Tag meines Lebens, "roth bezeichnet", wie ein Grieche gesagt hätte, erinnern in

18 Die physikalisch-mathematische Gesellschaft in Kazan' wurde 1880 als Teil der Naturforscherversammlung an der Universität Kazan' gegründet. Mitglieder waren u.a. Aleksandr Vasil'evič Vasil'ev (1853-1929), ab 1887 N.E. Žukovskij (1847-1921). Zwischen 1883 und 1890 erschienen 8 Bände der Protokolle der Gesellschaft. Seit 1890 hieß sie physikalisch-mathematische Gesellschaft.

19 Helmholtz' Arbeit "Zählen und Messen, erkenntnistheoretisch betrachtet" erschien erstmals in: *Philosophische Aufsätze, Eduard Zeller zu seinem fünfzigjährigen Doctorjubiläum gewidmet*, Leipzig 1887, 17-52. (Auch abgedruckt in: *Wissenschaftliche Abhandlungen*, III, 1895, 476-504). Eduard Gottlob Zeller (1814-1908) war ab 1872 Professor für Philosophie an der Berliner Universität.

20 Leopold Kronecker (1823-1891) hatte als Ordentliches Mitglied (seit 1861) der Berliner AdW das Recht, Vorlesungen zu halten und war ab 1883 a.o. Professor für Mathematik an der Berliner Universität. Er bildete mit Ernst Eduard Kummer (1810-1893, ord. Professor von 1855 bis 1883) und Karl Theodor Wilhelm Weierstraß (1815-1897, a.o. Professor 1856, ord. Professor 1864) den Anziehungspunkt für Studenten und Kollegen aus vielen Ländern. Seine Arbeit "Ueber den Zahlbegriff" erschien 1887 ebenfalls in dem E. Zeller gewidmeten Band. Zur "Ära Kronecker-Kummer-Weierstraß" vgl. Vogt, Annette. 750 Jahre Berlin. 4. Die glanzvollen Jahre. In: *Math. in der Schule* 25 (1987) 4, 217-227.

welchem ich die Ehre hatte einige Stunden in der Gesellschaft Eurer Excellenz in dem Hause meines Lehrers zu verbringen.

In ausgezeichnetster Hochachtung
Eurer Excellenz
ergebener

A. Wassilieff.
Professor an der Universität zu Kasan

Quelle: Archiv der Berlin-Brandenburgischen Akademie der Wissenschaften, Nachlaß Helmholtz, Nr. 494

Anhang 2

Universität Moskau
20 November
[2 December] 1886.

Hoch zu verehrender
Herr Geheimrath!

In Ihrem höchst interessanten Werke "Ueber die physikalische Bedeutung des Princips der kleinsten Wirkung" stellen Sie folgendes Prinzip auf: "Der für gleiche Zeitelemente berechnete Mittelwerth des kinetischen Potentials ist auf dem wirklichen Wege des Systems ein Minimum im Vergleich mit allen anderen benachbarten Wegen, die in gleicher Zeit aus der Anfangslage in die Endlage führen".[21]

---

21 Helmholtz, H. von. Ueber die physikalische Bedeutung des Princips der kleinsten Wirkung. In: (Crelle-) *Journal für die reine und angewandte Mathematik* Bd. 100, 1886, 137-166 und 213-222. Auch in: *Wissenschaftliche Abhandlungen*, III, 1895, 203-248. Das Zitat ist in: *WA*, III, 1895, 205.

Nach meiner Ansicht ist der Mittelwerth des kinetischen Potentials im vorliegenden Falle nicht ein *Minimum,* sondern ein *Maximum*.

Das kann auf folgende Weise bewiesen werden. Wollen wir uns, der Einfachheit halber, nur mit dem Fall eines materiellen Punktes beschäftigen (mit zwei oder drei Graden von Freiheit) und seine wirkliche Bewegung auf dem Wege *abc* vergleichen mit der anderen kinematisch-möglichen Bewegung, die zu derselben Zeit auf dem Wege *ahc* vor sich geht.

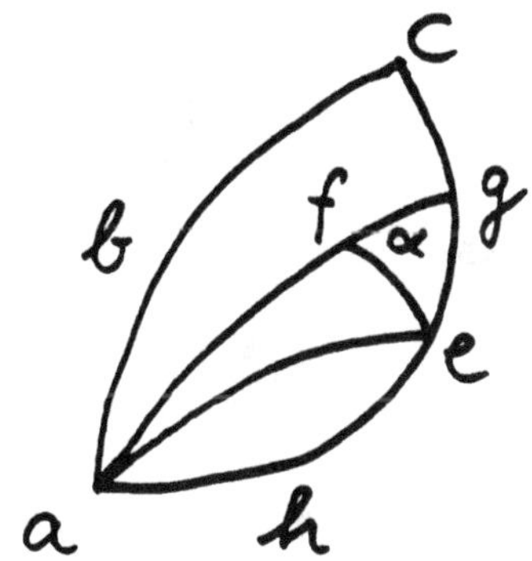

Nehmen wir auf dem Wege *ahc* einen Punkt *e* an und suchen wir eine solche wirkliche Bewegung unseres materiellen Punktes, bei welcher er auf dem Wege *ae* von *a* nach *e* gelangt zu derselben Zeit wie bei der gedachten Bewegung auf dem Wege *ahe*. Wenn wir den Punkt *e* allmählig von *a* nach *c* verrücken, so wird der Weg *ae* von Null an bis zum Werth *abc* wachsen. Nehmen wir an, dass der materielle Punkt in einer unendlich kleinen Zeit *dt* in der gedachten Bewegung von *e* nach *g* und in der wirklichen Bewegung auf dem Wege *ag* von *f* nach *g* gelangt. Da die wirklichen Bewegungen auf den Wegen *af* und *ae* gleichzeitig sind, so haben wir:

$$\int^{ae}(F-L)\,dt - \int^{af}(F-L)\,dt = mv\; csa\, d\delta,$$

wo *F* und *L* die von Ihnen angenommenen Bedeutungen haben, *m* die Masse des materiellen Punktes ist, *v* seine Geschwindigkeit im Punkte *f*, $d\delta$ die unendlich kleine Linie *fe* und α der Winkel *gfe*.

Wenn wir $fg = ds$ und $eg = dl$ setzen, so folgt aus dem unendlich kleinen Dreiecke *efg*, dass

$$dl^2 = ds^2 + d\delta^2 - 2ds\, d\delta\, cs\alpha.$$

Hieraus ergiebt sich die Ungleichheit:

$$\frac{m}{2}\left(\frac{dl}{dt}\right)^2 dt\rangle\ \frac{m}{2}\left(\frac{ds}{dt}\right)^2 dt - m\frac{ds}{dt}d\delta\, csa,$$

welche auf Grund der oben erwaehnten Formel in

$$\frac{m}{2}\left(\frac{dl}{dt}\right)^2 dt \rangle \frac{m}{2}\left(\frac{ds}{dt}\right)^2 dt + \int^{ae}(F-L)\,dt - \int^{af}(F-L)\,dt$$

übergeht.

Fügen wir zu beiden Theilen dieser Ungleichheit $F_g\,dt$, wo F_g die potentiale Energie im Punkte g ist, und geben ihr folgende Form:

$$\int^{af}(F-L)\,dt + \left(F_g - \frac{m}{2}\left(\frac{ds}{dt}\right)^2\right) dt - \int^{ae}(F-L)\,dt \rangle \left(F_g - \frac{m}{2}\left(\frac{dl}{dt}\right)^2\right) dt$$

oder

$$\int^{ag}(F-L)\,dt \; - \; \int^{ae}(F-L)\,dt \rangle \left(F_g - \frac{m}{2}\left(\frac{dl}{dt}\right)^2\right) dt.$$

Ziehen wir die Summe solcher Ungleichheiten, welche für alle Elemente dt der gedachten Bewegungen genommen sind:

$$\int^{abc}(F-L)\,dt \rangle \int^{ahc}(F-L)\,dt.$$

Nachdem wir beide Theile dieser Ungleichheit durch die Zeit t, während welcher die vorliegenden Bewegungen auf den Wegen *abc* und *ahc* vor sich gehen, getheilt haben, erkennen wir, dass der Mittelwerth des kinetischen Potentials der wirklichen Bewegung grösser ist als der Mittelwerth des kinetischen Potentials der gedachten Bewegung.

Dieser Beweis kann leicht auf jedes System der materiellen Punkte ausgedehnt werden. Die Richtigkeit meiner Behauptung kann auch durch Beispiele bestätigt werden.

Stellen wir uns einen materiellen Punkt vor, welcher mit der Geschwindigkeit v_0 in einem Winkel Φ zu der horizontalen Ebene geworfen wird und sich parabolisch bewegt unter der Wirkung seiner Schwerkraft, und vergleichen wir seine Bewegung mit der gleichförmigen Bewegung, die auf der geraden, welche die Schneidungspunkte der Parabel und der horizontalen Fläche verbindet, zu derselben Zeit vor sich geht; so ergiebt sich der Mittelwerth des kinetischen Potentials für die geradlinige Bewegung

$$\tfrac{1}{2}mv_0{}^2 sn^2\Phi - \tfrac{1}{2}mv_0{}^2,$$

welcher offenbar geringer ist als der Mittelwerth des kinetischen Potentials

$$\tfrac{2}{3}mv_0^2 sn^2\Phi - \tfrac{1}{2}mv_0^2$$

für die parabolische Bewegung.

Freilich bezieht sich meine Bemerkung auf die zweite Variation und hat keine Bedeutung für die wichtigen Resultate Ihrer Arbeit, welche sich alle auf die Eigenschaft der ersten Variation des Mittelwerthes des kinetischen Potentials stützen.
Indem ich es Ihrem Ermessen überlasse diese Bemerkung der Oeffentlichkeit zu übergeben,[22] bitte ich, hoch geehrter Herr Geheimrath, den Ausdruck meiner ausgezeichneten Hochachtung zu genehmigen.

Nikolaus Joukovsky
Professor der Mechanik an der Universität
Moskau.

Quelle: Archiv der Berlin-Brandenburgischen Akademie der Wissenschaften
Nachlaß Helmholtz, Nr. 223

22 Hermann von Helmholtz hielt an der Berliner Akademie noch zwei Vorträge zu diesem Thema, am 10.03.1892 und am 14.06.1894. In beiden Vorträgen erwähnte er Žukovskij *nicht*. Vgl. Helmholtz, H.v. In: *Sitzungsberichte*, 1892, I, 459-475 und Helmholtz, H.v. In: *WA*, III, 1895, 597-603.

Literatur:

Ansprachen und Reden, gehalten bei der am 2. November 1891 zu Ehren von Hermann von Helmholtz veranstalteten Feier. Nebst einem Verzeichnis der überreichten Diplome und Ernennungen sowie der Adressen und Glückwunschschreiben. Berlin, Hirschwald, 1892. 63 S.

Большая Советская энциклопедия 2ое изд., Moskau, Bd. 1 - Bd. 50, 1949-1957.

Брокгаузъ и Эфронъ. Энциклопедический словарь. St. Petersburg, Bd. 1 - Bd. 82, 1890-1904. 4 Ergänzungsbände, 1905-1907.

Festschrift zur Feier des 70. Geburtstages von Hermann von Helmholtz. Stuttgart, Union deutsche Verlagsgesellschaft, 1891. 91 S. mit Abb.

Grau, Conrad: *Berühmte Wissenschaftsakademien. Von ihrem Entstehen und ihrem weltweiten Erfolg*. Leipzig: Edition, 1988.

Hörz, Herbert / Laaß, Andreas: *Ludwig Boltzmanns Wege nach Berlin: ein Kapitel österreichisch-deutscher Wissenschaftsbeziehungen*. Berlin: Akademie-Verlag, 1989.

Hörz, Herbert: Hermann von Helmholtz und Österreich. In: Engelbert Broda (1910-1983). *Wissenschaft und Gesellschaft*, herausgegeben von der Zentralbibliothek für Physik in Wien. Wien: Universitätsverlag Wagner, 1993, 77-95.

Humboldt-Universität zu Berlin. Dokumente 1810 - 1985. HUB. Hg. Helmut Klein. Berlin: Deutscher Verlag der Wissenschaften, 1985.

Koenigsberger, Leo: *Hermann von Helmholtz*. Bd. I - III, Braunschweig 1902-1903.

Kowalewsky, Sonja: *Erinnerungen an meine Kindheit*. Weimar: Gustav Kiepenheuer Verlag, 1961.

Lebedinskij, A.V., Frankfurt, U.I., Frenk, A.M.: *Гельмгольц* (1821-1894). Moskau: Izdatel'stvo "Nauka", 1966.

Vogt, Annette: Wissenschaft und Persönlichkeit. Karl Weierstraß und Sonja Kovalevskaja. In: *Wiss. Zeitschrift der FSU Jena, Naturwiss. Reihe*, 37, 1988, H.2, 271-278.

II
Helmholtz über die mechanischen Grundlagen der Naturwissenschaft

Theoretical and Mathematical Interpretations of Energy Conservation: The Helmholtz-Clausius Debate on Central Forces 1852-54

Fabio Bevilacqua

Summary

An important stage of Helmholtz's struggle with 'energy' is marked by the strong controversy he had with Clausius between 1852 and 1854. The now famous 1847 *Erhaltung* did not enjoy immediate success: experimental physicists criticised the lack of new results. Moreover, in 1852, Clausius started criticising some of the basic physical-mathematical aspects of Helmholtz's essay: Clausius denied the mathematical necessity, but not the physical probability, of central forces depending only on positions to fulfil vis viva conservation. In his answers, Helmholtz attempted to defend his own views and at the same time to qualify as an expert on physical-mathematical grounds, not a minor task for a physiologist. Helmholtz revealed himself to be a great master of mathematical physics and of scientific methodology; through the introduction of a few (partly ad hoc) physical hypotheses he managed not to lose ground. But Helmholtz's successful defence could only mask what was a very serious internal problem in his own research program: he eventually recognised the soundness of Clausius's criticisms thirty years later, but he probably already realised it in the fifties. In fact there is no trace of central forces and of the related dichotomy of potential and kinetic energies in the next stage of his commitment to energy problems, the popularisation of the doctrine.

The difficult emergence of a theoretical approach

Hermann Helmholtz's *Ueber die Erhaltung der Kraft* (Helmholtz 1847b), presented in Berlin on 23 July 1847 to the Physikalische Gesellschaft, did not prompt an enthusiastic reaction, except amongst the society's members. Despite Emil du Bois-Reymond's pressure, the presentation of the work by the Berlin physics professor Gustav Magnus to the editor of the *Annalen der Physik,* Johann Christian Poggendorff, was very cool. Magnus objected to what he supposed was an attempt at unifying physics through mathematics. Poggendorff refused to publish the paper in his *Annalen* because there were no new experimental results and it was too long (Koenigsberger 1906: 38-42, Jungnickel and McCormmach 1986: Vol.1, 157-8). I believe the acceptance of the *Erhaltung* would have been different if Helmholtz, instead of criticising James Joule's experiences of 1843 and 1845, had presented them as corroborating his own views. Helmholtz's criticisms of Joule were probably due to his own mistranslation from British to Continental units of Joule's numerical results (Bevilacqua 1993: 321-326). Georg Reimer in Berlin published the essay, together with the philosophical introduction abandoned for the presentation at the Physikalische Gesellschaft, and Helmholtz, to his surprise, received an honorarium.

Helmholtz's theoretical approach to energy problems was very innovative, both from a methodological and a physical point of view (Bevilacqua 1993: 304-9). On the one hand he contributed to the conscious emergence of a theoretical physics based on the interplay between principles and conceptual models, on the other, with the specific choice of linking the impossibility of perpetual motion with the central Newtonian positional forces, he attempted a unification of contrasting traditions in mechanics. The problem of appreciating the specific characteristics of Helmholtz's innovative theoretical approach was a serious one: Helmholtz claimed recognition among both the experimental and the mathematical physicists. To receive it he had to fight a fierce battle. Only Carl Gustav Jacobi, the mathematician, encouraged him, recognising the links between the *Erhaltung* and the works of the "great mathematicians of the preceding century", despite the contrary opinion of his colleagues Peter Gustav Lejeune Dirichlet and Ferdinand Gotthold Max Eisenstein (Koenigsberger

1906: 43, Helmholtz 1882: 74). In addition to Magnus' and Poggendorff's criticisms, Gustav Karsten, one of the founding members of the Physikalische Gesellschaft, in 1850 showed a basic misunderstanding of Helmholtz's approach to energy conservation: he inserted the report on the *Erhaltung* in the physiological section of the *Fortschritte der Physik im Jahre 1847*, Helmholtz managed to give news of his paper in the physical section of that journal only five years later, in 1855, commenting on the 1850-51 debate on the "Theorie der Wärme", when he started publishing a series of six "Berichte" in the *Fortschritte der Physik* (1855-9) which can now be considered the first history of energy conservation.

Helmholtz, after the publication of the *Erhaltung*, returned to physiology; in 1849 he was appointed extraordinary professor and director of the Physiological Institute at Königsberg: he became professor in 1851. Of course he did not give up his interest in "force" conservation: in Königsberg he struggled to convince Franz Neumann, one of the authorities in the field of mathematical physics, of his own approach (Koenigsberger 1906: 64). Helmholtz knew that Neumann had used an extended interpretation of the vis viva conservation in his research on electromagnetic induction (Helmholtz 1847a: 350), and he managed to get Neumann's help for a paper to appear in the 1851 *Annalen* (Jungnickel and McCormmach 1986: Vol. 1,162; Helmholtz 1851).

In 1851 Helmholtz travelled around Germany (that is, German-speaking states and towns) and met (together with physiologists) the most important physicists. In a letter he remarked that Wilhelm Weber received him in Göttingen with "less apparent cordiality than his brother in Leipzig" (Koenigsberger 1906: 81). For certain, the deep contrast, plainly evident to both, between Weber's force law of 1846 and Weber's potential law of 1848 (Weber 1846, Weber 1848) with Helmholtz's 1847 approach influenced the feelings between the two. Weber's forces, at variance with Helmholtz's, depend not only on positions but also on velocities and accelerations, a radical departure from the standard Newtonian model and from Helmholtz's two basic forms of energy: potential and kinetic. In 1848 Weber had shown that his force law of 1846 could be derived from a potential. This implied that the work done by the force was a complete differential and that the force law did not violate the impossibility of perpetual motion, contra Helmholtz's *Erhaltung*. But, given that the force law included velocities and accelerations, the potential included a kinetic term. The fact that the kinetic and positional part of the potential could not be sharply divided was a serious blow to Helmholtz's approach. Weber's law soon acquired a leading position in electrodynamics, but his potential was not accepted as a viable interpretation of energy conservation till the early seventies.

Helmholtz's fame spread in Britain before it did at home. William Thomson (later Lord Kelvin) read the *Erhaltung* in 1852, and, at du Bois-Reymond's suggestion, John Tyndall translated and published it in 1853, together with Rudolf Clausius's 1852 paper. Helmholtz met Tyndall in 1853 in Berlin, where a lifelong friendship began. When, in the same year, he participated in the meeting of the British Association for the Advancement of Science at Halle, he discovered he was better known there than in Germany (Koenigsberger 1906: 109, 113, 120).

A former friend becomes a competitor

In 1852 Helmholtz added, at least in part through Franz Neumann, Charles Green's theorems to his own mathematical knowledge. These were not well known at the time but were to turn out very useful. In that very year, in fact, Rudolf Clausius started criticising some aspects of the *Erhaltung*, initially very moderately and not without praise for Helmholtz's wide-ranging approach. After Helmholtz's answer in 1853, Clausius's criticisms became sharper; they were based on potential theory and directed at the core of Helmholtz's program, that is, at "the main advance on the investigations of Robert Mayer" (Koenigsberger 1906: 100, 115, 118). Helmholtz received Clausius's criticisms with great concern, given that the latter was already a well-known mathematical physicist. Koenigsberger stresses Helmholtz's concern, which was so great as to make him feel betrayed by a former friend. Rudolf Clausius and Hermann Helmholtz in fact knew each other rather well: Clausius was a participant in Magnus' colloquium from the beginning. Together with Gustav Wiedemann they regularly took their meals together in the winter of 1847 and, since 1848, for a long time, had been in the habit of meeting almost daily (Koenigsberger 1906: 115, Jungnickel and McCormmach 1986: Vol.1, 256).

In 1850 Clausius published a famous paper on the mechanical theory of heat, in which he defined the two fundamental principles of thermodynamics and derived a value for the mechanical equivalent of heat (Clausius 1850: 368-97; 500-24). In 1852 in his first paper on electricity (Clausius 1852: 337-375), which is at the origin of the controversy, he "applied his heat laws to electrical discharge and thermoelectricity" (Jungnickel and McCormmach 1986: Vol.1, 167). He thus united mechanics, heat and electricity in a thorough study based on mechanical principles. At variance with Helmholtz, Clausius's approach derives from his commitment to mathematical potential theory and to his identification of work with the difference of potential. No physical interpretation of potential energy or of the energy concept is given here and conservation is still defined along the lines of the analytical tradition as vis viva conservation. Clausius's 1852 remarks on Helmholtz's results were thus by no means contingent, but deeply rooted in Clausius's research programme. In fact the discussion that followed was to outline a third, more mathematical, formulation of the principle of conservation of energy, after Mayer's correlation of forces (conversion with constant coefficients) and Helmholtz's conservation of "force" (variation of vis viva equals the sum of tension forces). For Clausius the only condition for energy conservation is that the work done by the forces be a complete differential. Clausius in 1859 published the first edition of a famous textbook on the potential and the potential function (Clausius 1859). A French translation of the second German edition of 1866 appeared in 1870 (Clausius 1866, 1870).

The whole controversy of 1852-54 took place in the pages of the *Annalen der Physik*, a clear indication that both Clausius's and Helmholtz's theoretical contributions were now fully accepted by Poggendorff.

Theoretical and Mathematical Physics: A Skilful Duel

The controversy began with some remarks on Helmholtz's *Erhaltung* in the notes of Clausius's 1852 paper on electric discharge. Helmholtz answered in a note of his 1853 paper on the distribution of electrical currents in material conductors (Clausius 1852, Helmholtz 1853: 225). In the same year Clausius fought back, and in 1854 Helmholtz expressed his views at length, in a paper to be reprinted in the energy section in the first volume of his collected works. A short comment of Clausius in 1854 closed the debate (Clausius 1853, Helmholtz 1854, Clausius 1854).

Some commentators have mentioned this debate (Koenigsberger 1906: 115-20, Jungnickel and McCormmach 1986: 163, Heimann 1974: 234-5, Planck 1887: 48-50) but, apart from Koenigsberger's hagiographic approach, without careful analysis. Furthermore, attention has not been paid to the relevance the debate had for the emerging distinction between theoretical and mathematical physics. There were three main points of debate and I will deal with them separately: a) the definition of potential; b) the conceptual model of central forces; c) accusations of lack of precision.

a) The definition of potential

In the second chapter of the *Erhaltung* Helmholtz introduced, in a straightforward fashion, the potential energy (sum of tension forces) and the energy concepts (the variation of vis viva equals the variation of the sum of tension forces, the sum of vis viva and tension forces is a constant) as theoretical entities with a specific physical meaning. Philosophy rather than mathematics played a role in this reinterpretation of the vis viva theorem. A cause-effect relationship of Leibnizian origin between vis viva and the sum of tension forces (the modern version of vis mortua) helped the shift from a theorem to a principle. The peculiarity of Helmholtz's approach is that he shifted from the vis viva theorem to the conservation of "force" principle without entering into a discussion of the concept of potential function or of potential. The concept of *Spannkraft* (potential energy) was introduced without following the now standard pattern which defines the scalar product of force and displacement as work and, if this is a complete differential, identifies it with a difference of potential. In the case of central Newtonian forces assumed by Helmholtz, the expression of the tension forces is indeed a complete differential and thus, from our modern point of view, the two approaches intersect considerably. Through a historical analysis, instead, we can see that the way in which Helmholtz introduced the term potential and related it to the work concept shows the difference between a theoretical and a mathematical approach. In the *Erhaltung* the "potential" was introduced long after the formulation of "force" conservation; only in the fifth chapter, the one dedicated to electrical and magnetic phenomena, and without a clear physical meaning. At the time Helmholtz lacked knowledge of Green's contributions, but he worked "in conformity with Gauss in his magnetical researches", and defined the potential as the sum of tensions consumed by the motion from infinity to *r*, and, equivalently, as the sum of the vis viva produced (Helmholtz 1847b: 38-40). Thus :

> "the increase of vis viva in any movement must be considered equal to the difference of the potential at the end of the trajectory with respect to the potential at the beginning".

That is, the sum of tension forces is equivalent to the difference of potentials:

$$-\int_R^r \varnothing dr = \frac{e_1 e_2}{R} - \frac{e_1 e_2}{r}$$

and, of course, also to the gain in vis viva caused by passing from the distance R to r.

Helmholtz then introduced two more specific concepts: the potential of a body on itself *(W)* and the potential of one body on another *(V)*. In a specific case given as an example, he calculated for two bodies A and B, charged with equal and opposite electricities, the difference of potentials before and after the movement of electricity from one to the other. This difference of potentials was defined as equivalent to the quantity of work done in the process:

$$-(V + (\frac{W_a + W_b}{2}))$$

Clausius followed a different pattern, which here is characterised as "mathematical". In 1852 he criticised Helmholtz's assertion that (in the above expression) the potential W of a body on itself is not equal to the corresponding work done but twice as much (the corresponding work is in fact $W/2$). The accusation of not having understood the deep relations between potential and work was a serious one from Clausius' point of view, who quotes both Carl Friedrich Gauss and Charles Green (Clausius 1852: Engl.:5-7).

Clausius in the same 1852 paper had established a relation between vis viva and potential different from that in Helmholtz's *Erhaltung*. He too started from the vis viva theorem but equated the increase of vis viva to the quantity of mechanical work produced during the same time in the system. Clausius did not accept Helmholtz's "sum of tension forces" (potential energy) and the corresponding interpretation of the conservation principle. For Clausius, if the elementary work dw is a complete differential, its integral w depends only on the initial and final positions, and thus work can be identified with a difference of potential. The potential of an exterior system of masses on a given system is introduced as the function:

$$\frac{\mu m}{\rho},$$

where μ are immovable masses, m the masses of the system and ρ the distance of the two masses from each other.
Similarly:

$$\frac{mm}{\rho}$$

is the potential of the given system of masses upon itself -

> "...the work consists simply in the increasing of these potentials" (Clausius 1852: Engl. 5).

Thus in 1852 Clausius explicitly asserted that the potential is work stored in the system. Work as complete differential and difference of potential are identical concepts. In the physical-mathematical tradition the principle of conservation was often still called, following the 18th century tradition, the vis viva conservation, and the only really important requirement was that work be a complete differential (Sturm 1861, Riemann 1875). This is a statement of the greatest relevance, but very different from Helmholtz's. The more mathematically-oriented interpretation left open the possibility that forces other than the central Newtonian ones could satisfy the conservation principle, if the work done by these forces satisfied the mentioned requirement. It is obvious that Helmholtz felt that an answer to Clausius was due.

In 1853 Helmholtz clarified that the meaning of the term "free tension" he had used in the fifth paragraph of the *Erhaltung* "is identical to what Gauss defined as potential and Green as potential function" (Helmholtz 1853: 224-5). Thus, in 1853, Helmholtz explicitly acknowledged a result of the unifying power of the mathematical potential theory: the concept of electrical tension, formerly with Alessandro Volta and Georg Simon Ohm volume density of the elastic fluid called electricity, had been redefined by Gustav Robert Kirchhoff in 1849, both for static electricity and for currents, as difference of electrical potential (Kirchhoff 1850: 463-8). Electrostatics and galvanism had thus been unified. However, for Helmholtz, this acknowledgement did not signify a shift from a theoretical to a mathematical approach.

In a note on the same page, Helmholtz replied to Clausius's criticisms of 1852, asserting that the problem raised was a minor one: they had utilised two different definitions of potential (Helmholtz 1853: 225). This did not imply any substantial disagreement between them: Helmholtz believed that his own definition of potential could be considered correct and his own development consistent, even if he admitted that, because it did not agree with F. Neumann's, his own was not a good definition. Helmholtz remarked that he had simply assumed for the potential of a body (system of bodies) on itself double the value that Clausius had assumed. That is, he had considered for the potential on itself of each of the two electrically-charged bodies *A* and *B*, the work done to transport to its initial position each element of the electricity with respect to each other element. In so doing, for each body the work relative to each couple of elements is calculated twice (a_m respect to a_n, plus a_n respect to a_m). Notwithstanding this, the work done in the example under discussion had been calculated correctly: in fact, half the potential on itself had been related to the work. It is plainly evident here that Helmholtz considered potential and work two different concepts to be related through a definition.

Clausius could not be satisfied with this answer: he spent six pages in 1853 clarifying that Helmholtz's approach rested on what, from his point of view, was a serious mistake, the conceptual difference between potential and work (Clausius 1853: 568-574). For Clausius, Helmholtz's correct result for the work done in the example at issue is only due to the compensating effect of two opposite mistakes. He first showed that the great importance of the concept of potential of two masses, each respect to the other, is connected with its equivalence "in a specific but frequent case, with that of mechanical work". Work had to be considered as one of the most important quantities for the whole of mechanics and mathematical physics, "and also Helmholtz in his essay utilised it with this meaning".

Thus, if we have to introduce the concept of "potential of a mass on itself", we cannot do that by defining it with twice the value of the corresponding mechanical work; priority should be given to the other concept of potential (that is, to the potential of one body on another which is equivalent to the work). The opposite choice of Helmholtz is a flaw.

A more serious mistake, for Clausius, is to have added together in the same expression a potential corresponding to the work (the potential of one body on another) and one corresponding to twice the work (the potential of a body on itself). In the simple example given by Helmholtz the final result is correct only because the mistake of the double value of the potential of a mass on itself is compensated by the particular choice of the values of the electricities involved.

Helmholtz's answer in 1854 on this point is interesting: he recalled that in that particular example his definition was supposed to hold only under specific hypotheses, that, in general, his definition of potential and work were correct, for example for electrified bodies and for magnets; and asserted that, in the case of the demonstration criticised by Clausius, a wrong work equivalent for the potential was not presupposed; on the contrary the very goal of the demonstration was to find the work equivalent (Helmholtz 1854: 242). In no way could the different methodologies be better clarified.

In a short answer in 1854 Clausius reasserted his point (Clausius 1854: 601-2).

To summarize this point of the controversy: for both Helmholtz and Clausius the difference of potential has to be related to the work done; Helmholtz in the *Erhaltung* faced some difficulties in the definition, partly for his ignorance of Green's papers (he in fact only quoted Gauss) and partly for his theoretical rather than mathematical grasp of the problem. Nonetheless, it is remarkable that his "free tension" was later to be identified with the "potential function" and that, as everywhere else in the *Erhaltung*, he started from the theoretical assumption of the equality between quantity of Spannkraft and variation of vis viva to find the work done in the specific cases. The equivalence between potential and work is not, as in Clausius, a prerequisite, but a consequence of the existence of a potential energy. In the above-quoted case of the electric discharge (i.e. where a variation of the charge distribution exists), Helmholtz knows that the tension forces would then be modified during the discharge, and thus builds up his example to avoid the problem, through considerations of symmetry. Without doubt Clausius was right in pointing out that Helmholtz's definition of potential of a charge on itself was twice the work done, but in the context of Helmholtz's research this was a minor mistake and does not imply that Helmholtz had not understood, despite his lack of knowledge of part of the literature, the relations between work and potential. Moreover, Clausius had an advantage: after Helmholtz's 1847 essay and before Clausius's 1852 paper the relations between potential and work had been clarified by Weber (1848) and Kirchhoff (1849).

b) The conceptual model of central forces

The second point raised in the controversy posed a much more serious problem for Helmholtz: Clausius challenged the necessity of the central force hypothesis, one of the roots of Helmholtz's definition of energy.

Clausius's approach to energy conservation was in fact different: as already stated, he started from the same work-vis viva equation as Helmholtz, but he focused on the work concept rather than on potential energy. He then recognised that in certain cases the work done is a complete differential, and thus the vis viva can be equated to a difference of potential. Clausius did not mention at this stage the potential energy, nor did he think that it should be a positional term. From now on, for Clausius, the principle of conservation of energy was the requirement that the work done by the forces could be expressed as a complete differential. Clausius, of course, believed in the kinetic theory of heat, but he did not apply to it the dichotomy of vis viva and tension forces as Helmholtz had done in the fourth chapter of the *Erhaltung*. Clausius asserted in 1852 that

> "heat consists in a motion of the ultimate particles of bodies and is a measure of the vis viva of this motion" (Clausius 1852: Engl. 5).

Thus the work done, or the difference of potential between the initial and final conditions, includes this effect too:

> "The sum of all the effects produced by an electric discharge is equal to the increase of the potential of the entire electricity upon itself" (Clausius 1852: Engl. 8).

Mechanical work, electricity and heat are unified here, but "potential energy" and "energy" do not appear. Even farther away from Clausius's perspective is the idea of a conservation of the sum of sharply separated kinetic and positional terms. In the paper mentioned, Clausius too relied on the model of central Newtonian forces depending only on distance, but he explicitly introduced it only as a case which is mathematically simple, often occurring in physical situations:

> "The determination of the work may be much simplified in particular cases which very often present themselves" (Clausius 1852: Engl. 4).

Clausius did not believe central Newtonian forces to be a conceptual model necessary for the intelligibility of nature. This criticism of Helmholtz's approach did not appear in the 1852 paper, but, after Helmholtz's remarks of 1853, Clausius in the same year expressed his views on this point rather clearly.

This basic issue is immediately faced, "not for personal reasons, but because it is connected with problems of general scientific interest" (Clausius 1853: 574). Clausius defines the problem in the following way: Helmholtz asserted that the vis viva principle holds only when the acting forces can be decomposed into forces acting on material points, in the directions that join them and whose intensity depends only on the distance, that is only for central forces. But the demonstration given by Helmholtz does not justify the assertion.

In fact, during the demonstration, one of the very statements to be derived, that the intensity of the force is a function of the distance, is assumed. This assumption is part of Helmholtz's conceptual model of Newtonian forces as expressed in the "*Einleitung*" to the *Erhaltung*. According to Clausius the result is that the principle of vis viva is either unnecessary to derive the central forces (if we already presuppose that the intensity is a function of the distance) or is insufficient (if we do not make that assumption) (Clausius 1853: 576). This was not a minor blow to the entire programme of Helmholtz.

Moreover, Clausius asserted that from the vis viva theorem, assuming that the vis viva is a function only of the space coordinates, we can derive that the work is a complete differential and also that the force acting is a function of the space coordinates. But this does not imply that all the forces are central. The only possible implication for Clausius is that, if one of the two conditions holds (direction or intensity depends on the positions) the other does too.

Finally Clausius, while not questioning the "physical probability", denies the "mathematical necessity" of central forces.

Helmholtz, apart from deriving the necessity of central positional forces from the vis viva principle in the second chapter of the *Erhaltung*, assumed in the Introduction that a point acts in different directions with the same force. Clausius did not consider this as evident either: the contrary is not "unthinkable". Helmholtz again in the first chapter of the *Erhaltung*, deriving the principle of virtual velocities from the vis viva principle tried to show that the forces of two points act on the direction joining them. For Clausius this statement, again, was part of the assumptions and cannot be considered a deduction. Helmholtz's whole attempt to demonstrate the mathematical need for central forces is, for Clausius, flawed (Clausius 1853: 576).

The core of Helmholtz's program was seriously shaken. No doubt he was deeply worried by the criticisms and he devoted himself to preparing a twenty-page paper which appeared in 1854. What this paper meant for Helmholtz can be understood by the fact that he reprinted it in 1882 in the energy section of the first volume of the *Wissenschaftliche Abhandlungen*, immediately after the *Erhaltung*.

Helmholtz answered that he could derive from the vis viva principle the need for the forces to be central, without further hypotheses, because he expressed the principle with reference only to the relative and not to the absolute positions of the points. Avoiding "absolute" positions and directions was important for defining "relational" central forces, depending only on the relative distance between the bodies and acting through the direction joining the bodies. In Helmholtz's view this meant taking into account the actual bodies and not "purely imaginary systems of coordinates" (Helmholtz 1854: 247). Helmholtz admitted that he had assumed that the vis viva depends only on the positions of the points and thus only on their distance. Following this version of the principle, Helmholtz could derive, through a massive use of mathematics, both the hypotheses leading to central forces (the intensity depends on the distance and the direction is in the joining line), also for the case in which there is a point moving under the action of a material element. In this case, in principle, the vis viva of the point could depend on the orientation of the distance (and not only on the distance); that is the material element could exert different forces along different directions, as in the case of magnets. Helmholtz managed to achieve his goal showing the possibility of reducing every material element to a set of points acting with central forces. Furthermore, Helmholtz explicitly demonstrated the correctness of having (implicitly) adopted the principle of superposition of the effects for the force (Helmholtz 1854: 254-5).

Clausius's prompt reply was very brief: he denied that in the *Erhaltung* the concern was only with relative positions and reasserted his views on the impossibility of a mathematical deduction of the central forces (Clausius 1854: 603-4).

c) Accusations of lack of precision

Other points were part of the controversy: Clausius in 1852 claimed that some minor parts of the *Erhaltung* were incorrect. He quoted and criticised a proposition that Vorsselman de Heer had published in the *Annalen* and that Helmholtz had accepted ("the total heat which is provoked in the entire circuit by an electric discharge is independent of the nature of the circuit") and Helmholtz's interpretation of an experimental law on the heat of discharge of Leyden jars given by the Berlin physicist Peter Riess (Clausius 1852: 16, 21) .

In 1853 Clausius also criticised Helmholtz's interpretation of Karl Holtzmann's results: the latter, who taught at the technical school in Stuttgart, did not believe in the consumption of heat but in its conservation, that is, not in the kinetic but in the caloric theory of heat (Clausius 1853: 578). This is a point which Helmholtz seems to have misunderstood and which seems to imply a serious flaw in his analysis of the work equivalent of heat in the fourth chapter of the *Erhaltung*. Clausius ended his 1853 paper with an acknowledgement and an appreciation of Helmholtz's achievements in the *Erhaltung*, despite his criticisms.

In 1854 Helmholtz responded, in a detailed and balanced way, to the specific points about de Heer and Riess and admitted his own mistake as far as the interpretation of Holtzmann is concerned (Helmholtz 1854: 256). He added four new points to the electromagnetic research of the *Erhaltung*, based on Poisson's magnetic induction and on his own researches of 1853 on the oscillations of induced currents. He stated that a galvanic current has an electrodynamic potential in itself, proportional to the square of the current intensity. If the circuit is interrupted there is a conversion of this "force equivalent" in heat either in the spark or in the extra current (Helmholtz 1854: 259). Finally, he showed that Franz Neumann's law of induction through magnets or currents is in agreement with conservation of energy, thus correcting some serious inconsistencies of the 1847 energy expressions for electrical circuits.

Clausius in his final remarks of 1854 acknowledged the clarification, but reasserted the disagreement on the problem of central forces.

A wider context

The result of the controversy was remarkable: Helmholtz certainly achieved something, even if he did not enjoy the great success claimed by Koenigsberger (Koenigsberger 1906: 116). He had demonstrated the skills of a mathematical physicist, able to debate with a researcher as clever as Clausius without letting him prevail. In the circumstances, this was a particularly difficult achievement for Helmholtz, who was in a weak position in a debate based on specific demonstrations and applications. His strength, instead, derived from his broad, general methodology, from the great heuristic and justificatory power of his formulation of energy conservation. A fact that, with great chivalry, Clausius himself acknowledged. The 1853 paper was the third Helmholtz published in the *Annalen* and, as already recalled, the entire controversy was published in that journal, which prompted a greater diffusion in Germany of the *Erhaltung*'s ideas. After the debate, Helmholtz's target of being accepted also among mathematical physicists was achieved.

At the same time, the controversy with Clausius carried a lesson for Helmholtz. In fact, there is no further trace of central forces and of the related dichotomy of potential and kinetic energies in the next stage of his commitment to energy problems: the popularisation of the doctrine. In a series of extraordinarily successful and influential talks in the fifties and sixties, Helmholtz spread the doctrine of energy, but he relied on the correlation of forces rather than on the conservation of an energy split into a positional and a kinetic term. Perhaps he started having some doubts about his own approach.

The basic difference in the force model assumed and in the related interpretation of the energy concept was to create a lifelong barrier between these two champions of the mechanical view of nature. Clausius later dedicated great efforts to an electrodynamical theory, based on non-Newtonian forces, which he saw as part of his own mechanical theory of heat. In 1865, in fact, he started republishing his papers on electrodynamics in the second volume of the first edition of *Die Mechanische Wärmetheorie* (Clausius 1865). In 1879 he went much further: for the second edition of the book he rewrote the entire second volume and dedicated it entirely to his own interpretation of electrodynamics; thus the book appeared with the surprising title of *Die Mechanische Wärmetheorie, Bd. 2: Elektrodynamik: Die Mechanische Behandlung der Elektricität* (Clausius 1876, 1879). In this volume, where Clausius "demonstrated" his electrodynamic force law of 1875 (Clausius 1875: 657-60) based on forces depending on positions, velocities and accelerations and on the related kinetic potential, very little room is left for James Clerk Maxwell's and Helmholtz's electrical theories. Clausius's law admitted, as did Weber's, forces depending on velocities and accelerations; but, at variance with Weber, the velocities were 'absolute' (with respect to the ether) and thus Clausius's forces were also non-central. In 1882 Helmholtz still criticised this approach (Helmholtz 1882, 1883, 1895: Vol.1, 69-70, 71). In 1876 Clausius, discussing his own electrodynamic law of the previous year, clarified his interpretation of the conservation principle in a way strongly connected with the approach of his 1852 work on electricity: the only condition needed to fulfil the principle of conservation is that the work done by the forces is a complete differential (Clausius 1876: 489-94). It is certainly a flaw of contemporary historical scholarship that very little effort has been dedicated to the analysis of Clausius's contributions to electrodynamics.

While Helmholtz believed in two sharply-divided forms of energy, positional and kinetic, based on central forces depending only on distance, Clausius admitted (with Weber) forces that were not central and depended also on velocities and accelerations as far as they could be derived from a potential. The fact that in this (kinetic) potential appeared terms including both positions and velocities, was not seen as a problem by Clausius. The existence itself of a potential was considered the fulfilment of the impossibility of perpetual motion, for him the only root of energy conservation.

Helmholtz's interpretation of the energy concept and of the conservation principle played a great role in the electromagnetic debate of the seventies. His discussions with Weber and Clausius cannot be summarised here. It is sufficient to point out that the root of the problem, the acceptance of a privileged role for the central forces, had already been at stake in the Helmholtz-Clausius controversy of the early fifties.

Later on, in the early eighties, this controversy had an unexpected reappraisal: Helmholtz himself indirectly acknowledged that Clausius was right in the specific instances of criticising a) the derivation of the central forces from the principle of vis viva and b) the definition of potential. In fact, in 1882, in the appendix to the reprint of the *Erhaltung* in the first volume of his collected works, Helmholtz admitted, in part and without quoting Clausius, that some of Clausius's former objections were correct (Helmholtz 1882, 1883, 1895: Vol.1, 70, 74-5). He accepted the criticism that his definition of the potential of a charge (body) on itself is less convenient than the one which equates it with the work done. Still, he again denied having given, in 1847, a wrong value for the work: he had asserted in the *Erhaltung* that the work was half the mentioned potential. He also admitted that the demonstration in the second chapter of the *Erhaltung*, that vis viva conservation holds only for central forces, required a restriction, that is: forces that depend also on velocities and accelerations can fulfil the vis viva theorem, and thus do not contradict the impossibility of perpetual motion. Helmholtz asserted that Rudolf Lipschitz clarified this point to him, but he did not mention the debate with Clausius. Nevertheless, he insisted that non-Newtonian forces should not be accepted because they cannot satisfy the principle of action and reaction.

Not long afterwards, Helmholtz, unable to give a satisfactory expression for the field energy along his preferred line, was forced to accept kinetic potentials. At the same time he decided that the principle of least action, rather than energy conservation, should now be seen as the most general regulative principle (Helmholtz 1886: 137-166, 213-222). Under this new perspective he tried once again to give a unified view of physical phenomena. While the methodology he outlined in the *Erhaltung*, a theoretical physics based on the interplay of principles and conceptual models, had enjoyed the widest success, his specific choice of the model and his formulation of the principle underwent major changes. In my view, the controversy with Clausius outlined here, already pointed in the early fifties to some of the problems at the root of Helmholtz's shift in the eighties.

Bibliography

Bevilacqua, Fabio: Helmholtz's *Ueber die Erhaltung der Kraft*. The Emergence of a Theoretical Physicist. In: Cahan, David (ed.) : *Hermann von Helmholtz and the Foundations of Nineteenth-Century Science*. Berkeley/Los Angeles./London: California University Press, 1993, 291-333.

Clausius, Rudolf: Ueber die bewegende Kraft der Wärme und die Gesetze, welche sich daraus für die Wärmelehre selbst ableiten lassen. *Poggendorff's Annalen der Physik*, 79, 1850, 368-97; 500-24.

Clausius, Rudolf: Ueber das mechanische Aequivalent einer elektrischen Entladung und die dabei stattfindende Erwärmung des Leitungsdrahtes. *Poggendorff's Annalen der Physik*, 86, 1852, 337-375. Tr.: On the Mechanical Equivalent of an Electric Discharge, and the Heating of the Conducting Wire which accompanies it. In: Tyndall and Francis (eds.): *Scientific Memoirs on Natural Philosophy*. 1853: 1-32.

Clausius, Rudolf: Ueber einige Stellen der Schrift von Helmholtz "Ueber die Erhaltung der Kraft". *Poggendorff's Annalen der Physik*, 89, 1853, 568-579.

Clausius, Rudolf: Ueber einige Stellen der Schrift von Helmholtz "Ueber die Erhaltung der Kraft", zweite Notiz. *Poggendorff's Annalen der Physik*, 91, 1854, 601-604.

Clausius, Rudolf: *Die Potentialfunction und das Potential, ein Beitrag zur mathematischen Physik*. Leipzig: J.A.Barth, 1859.

Clausius, Rudolf: *Die Mechanische Wärmetheorie*. 2. Braunschweig, 1865.

Clausius, Rudolf: *Die Potentialfunction und das Potential, ein Beitrag zur mathematischen Physik*. 2nd ed. Leipzig: J.A.Barth, 1866.

Clausius, Rudolf: *De la fonction potentielle et du potentiel*. Folie, F. (tr.). Paris: Gauthier-Villars, 1870.

Clausius, Rudolf: Ueber ein neues Grundgesetz der Elektrodynamik. *Poggendorff's Annalen der Physik*, 156, 1875, 657-60. *Philosophical Magazine* 1 (1876): 69-71.

Clausius, Rudolf: Ueber das Verhalten des elektrodynamischen Grundgesetzes zum Prinzip von der Erhaltung der Energie und über eine noch weitere Vereinfachung des ersteren. *Poggendorff's Annalen der Physik*, 157, 1876, 489-94. *Philosophical Magazine* 1 (1876): 218-21.

Clausius, Rudolf: *Die Mechanische Wärmetheorie*. 2 vols., 2nd ed. Braunschweig: 1876, 1879.

Heimann, Peter M.: Helmholtz and Kant: The Metaphysical Foundations of 'Ueber die Erhaltung der Kraft'. *Studies in History and Philosophy of Science*, 5, 1974, 205-38.

Helmholtz, Hermann von: Bericht über "die Theorie der physiologischen Wärmeerscheinungen" betreffende Arbeiten aus dem Jahre 1845. *Die Fortschritte der Physik*, 1, 1847a, 346-55. Reprinted in :*Wissenschaftliche Abhandlungen* 1: 3-11.

Helmholtz, Hermann von: *Ueber die Erhaltung der Kraft. Eine physikalische Abhandlung.* Berlin: G. Reimer, 1847b. Reprinted in: *Wissenschaftliche Abhandlungen* 1:12-75 (including an appendix [68-75] from 1881)

Helmholtz, Hermann von: Ueber die Dauer und den Verlauf der durch Stromesschwankungen inducirten elektrischen Ströme. *Poggendorff's Annalen der Physik*, 83, 1851, 505-40.

Helmholtz, Hermann von: Ueber einige Gesetze der Vertheilung elektrischer Ströme in körperlichen Leitern mit Anwendung auf die thierisch-elektrischen Versuche. *Poggendorff's Annalen der Physik*, 89, 1853, 211-33, 353-77; 224-5. Reprinted in: *Wissenschaftliche Abhandlungen*: 1, 475-519; 487-8.

Helmholtz, Hermann von: Erwiderung auf die Bemerkungen von Hrn. Clausius. *Poggendorff's Annalen der Physik*, 91, 1854, 241-60. Reprinted in: *Wissenschaftliche Abhandlungen* 1: 76-93.

Helmholtz, Hermann von: *Wissenschaftliche Abhandlungen*. 3 vols. Leipzig: Barth, 1882, 1883, 1895.

Helmholtz, Hermann von: Ueber die physikalische Bedeutung des Princips der kleinsten Wirkung. *Crelle's Journal*, 100, 1886, 137-166, 213-222. Reprinted in: *Wissenschaftliche Abhandlungen*. Vol. 3, 203-248.

Jungnickel, Christa and McCormmach, Russell: *Intellectual Mastery of Nature. Theoretical Physics from Ohm to Einstein.* Vol.1: The Torch of Mathematics 1800-70; Vol 2: The Now Mighty Theoretical Physics 1870-1925. 2. Chicago/London: The University of Chicago Press, 1986.

Kirchhoff, Gustav: On a Deduction of Ohm's Laws, in Connexion with the Theory of Electrostatics. *Philosophical Magazine*, 37, 1850, 463-8. Tr. from *Poggendorff's Annalen der Physik* 78, 1849: 506.

Koenigsberger, Leo: *Hermann von Helmholtz*. Welby, Frances A. (tr.). Oxford: Clarendon, 1906.

Planck, Max: *Das Prinzip der Erhaltung der Energie*. Leipzig: Teubner, 1887. 2nd ed. 1908.

Riemann, Bernhard: *Schwere, Elektricität und Magnetismus, nach den Vorlesungen von B. Riemann.* Hannover: 1875.

Sturm, M.: *Cours de Mécanique*. 2 Vols., M.Prohuet (ed.) Paris: Mallet-Bachelier, 1861.

Weber, Wilhelm: *Elektrodynamische Maassbestimmungen über ein allgemeines Grundgesetz der Elektrischen Wirkung*. 1846. Reprinted in: *Werke*, 6 vols. Berlin, 1892-4. Vol.3, 25-214.

Weber, Wilhelm: Elektrodynamische Maassbestimmungen. *Poggendorff's Annalen der Physik*, 73, 1848, 193. *Taylor's Scientific Memoirs* 5 (1852): 489-529.

Weber, Wilhelm: *Werke*. 6 vols. Berlin: 1892, 1894.

Actio, Quantité d'action und Wirkung: Helmholtz' Rezeption dynamischer Grundbegriffe

Hartmut Hecht

In seiner Arbeit *Ueber die physikalische Bedeutung des Princips der kleinsten Wirkung* stellt Hermann von Helmholtz einleitend fest:

> "Wenn ich in diesem Aufsatze vom Princip der kleinsten Wirkung spreche, so möchte ich darunter nicht allein die ursprüngliche 1744 von *P. L. M. de Maupertuis* veröffentlichte Form desselben verstanden wissen, die übrigens erst viel später, wohl durch Lagrange, präcise Bestimmung der Variationsbedingungen und vollständigen Beweis empfangen hat; sondern ich will unter diesem Namen, als dem ältesten und bekanntesten, die verschiedenen transformirten Formen desselben Satzes mitbegreifen, welche später durch *Sir W. Rowan Hamilton* daraus entwickelt [worden] sind" (Helmholtz 1895: 203).

Helmholtz versucht demnach, die Geschichte und die aktuellen Darstellungen eines der fundamentalen Integralprinzipien der Physik des 19. Jahrhunderts mit einem Begriff zu erfassen, der in dieser Verwendung physikalische Bedeutung schlechthin haben soll. Die genannten Forscher, so Helmholtz weiter, "haben das Princip der kleinsten Wirkung zunächst nur auf die Mechanik wägbarer Körper angewendet", während er selbst dessen Gültigkeitsbedingungen und mithin seine Verallgemeinerungsfähigkeit prüfen will (Helmholtz 1895: 204).

Die Einführung des Wirkungsbegriffs in der Helmholtzschen Diktion wird so zum Schlüsselbegriff eines Programms, das die Fortschritte der Naturwissenschaften an dem Ziel mißt, "die allen anderen Veränderungen zugrunde liegenden Bewegungen und deren Triebkräfte zu finden, also sich in Mechanik aufzulösen" (Helmholtz 1971: 164). Es ist ein Programm, das sich als Reflex auf den Entwicklungsstand der zeitgenössischen Wissenschaften hinsichtlich der Physik in die Pflicht genommen sieht, den Status von Disziplinen wie Thermodynamik und Elektrodynamik als Teilgebiete der Physik zu bestimmen, und das hieß für Helmholtz, ihr Verhältnis zur Mechanik zu untersuchen; genauer, mit dem Prinzip der kleinsten Wirkung eine Vereinheitlichung der Physik im Bilde der Mechanik anzustreben. Der Wirkungsbegriff erlangte damit gleichsam paradigmatische Bedeutung und ließ sich also nicht auf seinen physikalischen Gehalt als Größe einschränken.

Helmholtz hat dies zweifellos gesehen, ja er hat eben diesen Sachverhalt zum Ausgangspunkt eigener historischer und methodologischer Untersuchungen gemacht, die er mit den Worten resümierte:

> "Während der letzten Jahre haben sich meine Arbeiten auf die Anwendungen, den Sinn und den Gültigkeitsbereich eines grossen allgemeinen Naturgesetzes bezogen, welches, im vorigen Jahrhundert zuerst gefunden, allmälig an Wichtigkeit und in seinem Geltungsbereiche gewachsen ist, nämlich das sogenannte *Princip der kleinsten Action.* Meine Untersuchungen führten mich dazu, die alte Litteratur durchzusehen, wobei ich eine der seltsamsten Geschichten kennen lernte für die Art, wie eine bedeutende Wahrheit entdeckt werden kann" (Helmholtz 1900: 282).

Doch hat Helmholtz selbstverständlich nicht nur seltsame Geschichten und seiner Zeit bereits phantastisch anmutende Debatten und Probleme gefunden. Er war durchaus auch erfolgreich in eigener Absicht, denn es gelang ihm, sein Anliegen, d.i. die Integration der physikalischen Teildisziplinen seiner Zeit auf der Grundlage des Wirkungsprizips, historisch zu verankern. Helmholtz versicherte sich nämlich nicht nur des universellen Charakters des erstmals von Maupertuis formulierten Prinzips, er gewann auch die Gewißheit, daß dieser nur mit seiner, der Helmholtzschen Fassung des Problems als Prinzip der kleinsten Wirkung einlösbar ist.

Das freilich ist in der Folgezeit nicht unwidersprochen geblieben. Noch in neuerer Zeit liest man bei Arnold Sommerfeld:

> "Der Name ist leider unglücklich gewählt. Wenn wir von Ursache und Wirkung sprechen, so verstehen wir unter Wirkung die Folgeerscheinung oder den Erfolg. Beim Prinzip der kleinsten Wirkung aber ist die Meinung, daß die Natur ihr Ziel auf dem direktesten Wege, also mit dem kleinsten Aufwand an Mitteln, erreicht. Man sollte daher besser sagen: Prinzip des kleinsten Aufwandes bei größter Wirkung. Doch scheint es aussichtslos, nachdem das Wort 'Wirkung' durch *Helmholtz* und *Planck* sanktioniert ist, es durch ein anderes zu ersetzen" (Sommerfeld 1955: 173).

Was in historischen Bezügen dieser Art zum Ausdruck kommt, ist ein bis heute in der Diskussion virulent gebliebenes Bewußtsein darüber, daß sich hinter den verschiedenen Bezeichnungen des in Rede stehenden Prinzips Begriffsverschiebungen verbergen, die im engeren Sinne physikalisch nicht erklärbar sind. Aus diesem Grunde versuchte Sommerfeld, seine Argumentation durch Explikationen philosophischer Kategorien zu stützen. Und wir werden sehen, daß sich der historische Aktionsbegriff bei Helmholtz einer modernen Sinngebung nur durch besondere anschauliche Deutungen öffnet, deren Subtilität und innovative Potenz sich freilich nur dem erschließen, der die Quellen kennt. Sie heißen für Helmholtz Leibniz und Maupertuis.

Ein Exkurs über das Maupertuissche Prinzip

Es ist in der Philosophie- und Wissenschaftsgeschichte unstrittig, daß die erste Formulierung des in Frage stehenden Prinzips auf Maupertuis zurückgeht, der sich dafür einer Größe bediente, die in den Bewegungstheorien des 17. und 18. Jahrhunderts nicht eben selten ist und unter dem Terminus *"actio"* oder *"action"* firmiert. Man findet sie bei Leibniz und Jacob Bernoulli, natürlich bei Christian Wolff, Malebranche und, wie allgemein bekannt, auch bei Newton. Der Begriff selbst ist also geläufig. Er wird freilich gelegentlich unterschiedlich definiert, doch hatte dies kaum weitergehende wissenschaftliche Folgen, denn zum Kristallisationskeim einer der berühmtesten Gelehrtenkontroversen des 18. Jahrhunderts, in die nicht nur ganze Wissenschaftlerkongregationen verwickelt wurden, sondern in der die Angelegenheit schließlich eine Dimension erlangte, die in einem handfesten höfischen Eklat endete, wurde er erst in der Verwendung durch Maupertuis.

Dieser übernimmt nämlich die Leibnizsche Definition, wonach die Aktion als das Produkt aus der Masse des Körpers, seiner Geschwindigkeit und des von ihm durchlaufenen Weges bestimmt ist, stellt sie aber in einen ganz anderen Kontext. Während Leibniz die *actio* als bloße Definition für die apriorische Ableitung seines Kraftmaßes benötigte, d.h. als reine Rechengröße behandelte, wird sie bei Maupertuis mit einem Integralprinzip verknüpft, das von diesem als mathematischer Ausdruck des Wirkens gänzlich anderer natürlicher Entitäten gedeutet wird, als man sie von den Leibnizschen Kräften her kennt. Das *principe de la moindre quantité d'action*, wie es bei Maupertuis heißt, transportiert also von vornherein einen anderen dynamischen Inhalt als Leibniz' Erhaltungssatz für die lebendige Kraft. Es lautet: "Lorsqu'il arrive quelque changement dans la Nature, la quantité d'action employée pour ce changement est toujours la plus petite qu'il foit possible" (Maupertuis 1971: xxii). Sobald sich nur irgendeine Veränderung in der Natur vollzieht, ist also nach Maupertuis die aufgewendete Aktionsmenge immer die kleinstmögliche. Das zum Verständnis der historischen Debatte Entscheidende liegt darin, daß in dem berühmten von Samuel Koenig zitierten Leibniz-Brief, dessen Original übrigens noch immer nicht gefunden wurde, Leibniz zwar mitteilt, daß bei allen Bewegungsänderungen die Aktion gewöhnlich ein Maximum oder Minimum annimmt, er damit aber nicht die Idee eines Maßes der Bewegung verknüpft, worauf es Maupertuis gerade ankommt!

Man muß schon genau hinsehen, um dies zu bemerken, denn bei Maupertuis ist im Unterschied zum heutigen Sprachgebrauch nicht von der Aktion, sondern von der Aktions*menge* die Rede. Nicht die Aktion oder Wirkung schlechthin gehen in seine Formulierung ein, sondern eine Maßbestimmung. Und hinter dieser filigranen Differenzierung verbirgt sich, das ist meine Behauptung, eine fundamentale Verschiebung hinsichtlich der philosophischen Legitimation naturwissenschaftlichen Wissens.

Maupertuis sieht sich in der Tradition von Descartes und Leibniz, aber auch, was bei den ebengenannten Namen nicht selbstverständlich ist, in der Tradition von Newton. So wie Descartes mit der ***quantitas motus*** oder Leibniz mit der ***quantité de la force***, will Maupertuis ein Maß der Bewegung finden, und er will dies nicht jenseits der Newtonschen Theorie ansiedeln. Maupertuis sucht ein allgemeines Bewegungsmaß oder -prinzip, dessen philosophische Grundlegung nicht nur die kontinentalen Variationen der cartesischen Physik erklärt, sondern mit diesen zugleich deren englisches Pendant. Er will, wie er sich in seinem *Essay de Cosmologie* ausdrückt, dieses "monstre métaphysique", das metaphysische Ungeheuer der Attraktion, aus der Welt schaffen, das er als die physikalische Erscheinungsform jenes Skandals der Vernunft ansah, der in der Mitte des 18. Jahrhunderts die Gelehrtenwelt in zwei Lager gespalten hat (Maupertuis 1971: 46). Und als das entscheidende Mittel dafür führt er, selbstbewußt und mit der Würde des Präsidenten der Berliner Akademie der Wissenschaften wohlausgestattet, sein Prinzip der kleinsten Aktions*menge* ins Feld.

Maupertuis' Prinzip ist also nicht als Anwendung einer mathematischen Methode auf die Physik zu verstehen. Eher schon ist es umgekehrt, die Physik soll durch dieses Prinzip überhaupt erst begründet werden! Seine Vorgänger waren, das ist Maupertuis' Überzeugung, an diesem Vorhaben allesamt gescheitert. Descartes konnte unter der Voraussetzung der *res extensa* bestenfalls die Bewegungsgesetze für harte Körper ableiten. Leibniz dynamisierte zwar die Physik, schränkte sie damit jedoch auf elastische Körper ein. Selbst Newton war nicht erfolgreicher, da er die von ihm eingeführten Attraktionskräfte nicht philosophisch rechtfertigen konnte. Was Maupertuis moniert, ist die in jedem der erwähnten Fälle unterstellte Notwendigkeit einer Legitimation der Physik durch Entitäten, die sich entweder nicht in der Erfahrung oder, wie im Falle Newtons, nicht metaphysisch rechtfertigen lassen.

Seine Frage läuft deshalb darauf hinaus, welche Wege denn überhaupt noch offen bleiben, wenn der Geist der Systeme den unvoreingenommenen Blick für die Empirie verbaut und die empirische Forschung keine notwendigen Gesetze liefert, ja, wenn die Sicherheit und Gewißheit der Bewegungsgesetze zudem nicht einmal aus ihrer mathematischen Formulierung resultiert. Die Antwort, und das ist das Faszinierende an seiner Lösung, gibt das Prinzip der kleinsten Aktions*menge*, und zwar nicht als ein mathematisches oder physikalisches, sondern als ein metaphysisches Prinzip. Denn Maupertuis ist sich selbstverständlich darüber im klaren, daß die Begründungsfunktion der Metaphysik hinsichtlich der Naturerkenntnis nicht ersatzlos gestrichen werden kann, wenn die Physik als Wissenschaft gelten soll. Das Gebäude der Physik soll bei ihm vielmehr auf zwei tragenden Säulen ruhen, auf der Mathematik und einer, wie er sagt, *science superieure*.

Maupertuis sieht in dieser Situation nur eine Lösung: So wie Leibniz' Gott in einem einmaligen Akt die Monaden geschaffen hat, die in ihrer Entfaltung das Reich der Phänomene aufspannen, so ist auch Maupertuis' Gott nicht unmittelbar in jedem Phänomen wirksam, doch erschafft dieser Gott keine Substanzen, sondern erhält sich in einem allgemeinen Prinzip, das die Gewißheit der besonderen Phänomene verbürgt. Mit diesem Prinzip, das selbstverständlich kein anderes als das von Maupertuis formulierte sein kann, werden die physikalischen Wirkungen daher durch andere dynamische Entitäten, die Aktionen erklärt. Und es lautet aus diesem Grunde *principe de la moindre quantité d'action*.

Maupertuis führt mit seinem Prinzip also eine neue Dynamik in die Physik ein, eine Dynamik, die weder auf den Newtonschen Bewegungsgleichungen beruht noch auf Leibniz' Kraftbegriff. Ihr Charakteristikum besteht vielmehr darin, daß sie mit dem Aktionsprinzip zugleich Entitäten postuliert, die als natürliche Wirkungen verstanden werden, ohne daß ihnen - wie dem Leibnizschen Kraftbegriff - ein metaphysisches Moment inhärieren müßte. Und sein Beharren auf der Eigenständigkeit seiner wissenschaftlichen Leistung beruht in erster Linie auf dieser Tatsache. Den Streit um die Priorität hinsichtlich des von ihm formulierten Bewegungsprinzips vor Augen, stellt er daher berechtigterweise fest, daß Leibniz, wenn er mit seinem Begriff der Aktion ein Prinzip der Bewegung hätte aussprechen wollen, die Aktion analog zur Kraft hätte a priori einführen müssen. Das aber hat er nicht getan, und Maupertuis sagt auch, warum. Er hätte nämlich an den Grundfesten seines metaphysischen Systems rütteln müssen, weil das Aktions*prinzip*, das heißt die Maupertuissche Verwendung der Leibnizschen *actio*, mit einem Erhaltungsprinzip à la Descartes oder Leibniz nicht zu vereinbaren ist. Leibniz' Begriff der Aktion ist eben eine andere Ausdrucksform für sein Maß der lebendigen Kraft und keinesfalls eine neu entdeckte Natur"kraft" wie bei Maupertuis.

Maupertuis' Leistung beruht also auf einer grundlegenden naturphilosophischen Neubesinnung, die das Reich der Phänomene als durch dynamische Entitäten beherrscht annimmt, die sich von Leibniz' und auch Newtons Kräften signifikant unterscheiden. Das allein wäre jedoch zu wenig, um den erwähnten Skandal der Vernunft aus der Welt zu schaffen. Die entscheidende Frage besteht darin, ob - und wenn ja, wie - Maupertuis mit seiner dynamischen Entität den Gegensatz von Leibniz und Newton, der sich ja in ganz erheblichem Maße an dem Problem der philosophischen Legitimation der von ihnen jeweils unterstellten Kraftbegriffe entfaltete, bewältigt hat.

Darin sehe ich die Bedeutung der kurz vor seinem Tode erst fertiggestellten Schrift *Examen philosophique de la preuve de l'existence de Dieu employée dans l'Essai de Cosmologie*. In ihr führt Maupertuis aus, wie man zu wissenschaftlichen Begriffen gelangen kann, ohne auf metaphysische Wesenheiten wie die Leibnizschen Kräfte zurückgreifen zu müssen. Und er findet den Grund dafür in einer Eigenschaft der Perzeptionen, die er *réplicabilité* nennt. Diese Eigenschaft ist es, auf der die Evidenz und Gewißheit der mathematischen Gegenstände beruht. "Je puis ajouter une étendue à une autre étendue égale, & avoir d'une étendue double, triple &c. une idée aussi claire que je l'avois de la primiere", schreibt er (Maupertuis 1965: 395). Der Vorteil, den die Zahlen und geometrischen Figuren gegenüber den Gegenständen aller anderen Wissenschaften haben, besteht nun darin, daß sie exakt *réplicable* sind. In der Physik gilt dies nach Maupertuis nur in eingeschränktem Maße. Sie habe es nur zum Teil mit Größen zu tun, deren Eigenschaften sich in nichts von denen der mathematischen Objekte unterscheiden. Das sei etwa bei der Geschwindigkeit der Fall. Für andere, wie die Kräfte, könne man dies nur dann voraussetzen, wenn man konstante und homogene Maße angeben könne, "qui se rapportent à l'étendue ou aux nombres" (Maupertuis 1965: 399). Von allen bekannten Kraftmaßen, so Maupertuis, würde das aber nur auf seine Maßbestimmung zutreffen. Newtons Definition der Kraft nämlich sei eine momentane, die sich genau deshalb der Vergleichbarkeit entziehe. Und um Leibniz' lebendige Kraft stehe es nicht besser, denn sie muß stets als Ganzes vorhanden sein, um die Bewegung zu ermöglichen. Weder eine momentan wirkende Kraft noch eine immer als Ganzes vorhandene lassen sich jedoch im Sinne der *réplicabilité* zusammensetzen oder teilen. Da für die Variation der Aktionsgröße aber von vornherein ein Intervall vorauszusetzen ist, erfüllt sie als dynamischer Begriff die Forderungen exakter wissenschaftlicher Begriffsbildung.

Die *réplicabilité* ermöglicht damit eine meßtheoretische Fundierung des physikalischen Wissens, das durch Gott legitimierte Prinzip seine Gewißheit und objektive Bedeutung. Finalprinzipien erscheinen so als Einsicht in die Notwendigkeit eines meßtheoretisch bestimmten Erfahrungsbegriffs. Das wird möglich, weil die Finalität nicht auf Substanzen, sondern auf Phänomene bezogen wird.

Man muß an dieser Stelle freilich hinzufügen, daß Maupertuis' Philosophie eher als ad-hoc-Hypothese denn als durchgearbeitete Theorie anzusehen ist. Dessen ungeachtet hat er mit der *réplicabilité* die entscheidenden erkenntnistheoretischen Fragen der Naturphilosophie seiner Zeit thematisiert und mit seinem Prinzip jene Entwicklung eingeleitet, die zu neuen Formulierungen der Dynamik auf der Grundlage von Extremalprinzipien führte. Mehr noch, *réplicabilité* und Extremalprinzip bedingen sich gegenseitig, weil, wie gesagt, jede Variation einer Größe von einem gegebenen Intervall auszugehen hat. Maupertuis' metaphysisch verstandenes Prinzip macht das mathematische Instrument der Extremalprinzipien dadurch für die Physik praktikabel, daß es den Erfahrungsgehalt physikalischer Aussagen an deren Meßmöglichkeit bindet. Das ist für ihn freilich nur einlösbar, wenn man auf den Begriff der Kraft verzichtet, weil sich das Meßproblem unter der Voraussetzung von Kausalerklärungen nicht stellen läßt. In diesem und nur in diesem Sinne hat Maupertuis eine Lösung gefunden, mit der sich der Gegensatz von Leibnizscher und Newtonscher Dynamik aus der Welt schaffen läßt. Er zeigt, daß die Kontroversen der Newtonianer und Leibnizianer auf der Voraussetzung der metaphysisch legitimierten oder zu legitimierenden Kräfte beruhen. Gibt man diese Voraussetzung auf, so schafft man sich den Freiraum für andere dynamische Entitäten, die physikalisch die gleichen Phänomene beschreiben, ohne dabei dem Gegensatz sich ausschließender kausaler Erklärungsweisen zu verfallen.

Maupertuis' Prinzip resultiert folglich aus einer naturphilosophischen Fragestellung, die aufs engste mit den Begründungsschwierigkeiten der Physik des 18. Jahrhunderts zusammenhängt. Er löst sie so, daß er mit der Aktion ein neues Maß der Bewegung angibt, d.h. die physikalische Größe *action* nicht nur nennt, sondern die Bedingungen der Möglichkeit ihrer Bildung formuliert.

Der Begriff der Wirkung

Sommerfelds weiter oben zitierte Bemerkung zu Helmholtz und Planck zeigt, daß die im Zusammenhang mit Maupertuis' Prinzip diskutierte Problematik in der Physik - wenigstens in der 1. Hälfte des 20. Jahrhunderts - nicht als erledigt angesehen wurde, denn Sommerfeld beruft sich ja explizit auf philosophische Kategorien, um den Gehalt der physikalischen Begriffe zu erfassen. Er wertet genau genommen die Schwierigkeiten der Helmholtzschen Verwendung des Wirkungsbegriffs und namentlich die von ihm konstatierte Begriffskonfusion als ein Philosophie-Defizit, das er meint bei Helmholtz entdecken zu müssen.

Daran ist unstrittig, daß Helmholtz keineswegs die historische Kontinuität des Wirkungsprinzips in gleicher Weise wie Sommerfeld fortschreiben will. Daß Helmholtz die Maupertuissche Lösung nicht mehr genügen konnte, lag freilich nicht nur an den persönlichen und wissenschaftlichen Eigenarten des Akademiepräsidenten, die von ihm ausgiebig referiert werden, und wohl auch nicht daran, daß Maupertuis' Ansichten hinsichtlich der Philosophie von Immanuel Kant längst aufgehoben worden waren. In der Helmholtz-Zeit lagen vielmehr jene wissenschaftlichen Grundbegriffe, deren philosophische Begründung ein wesentliches Anliegen des 18. Jahrhunderts war, bereits als gegeben vor. Und das wissenschaftliche Selbstverständnis dieser Epoche bestand nicht darin, das Reservoir dieser Grundbegriffe zu erweitern. Eher schon hat man es hier mit dem Versuch zu tun, das Interpretationspotential der gegebenen Größen auszuschöpfen. So jedenfalls läßt sich Helmholtz' Anliegen verstehen, die Anwendungen, den Sinn und den Gültigkeitsbereich des Aktionsprinzips als eines allgemeinen Naturgesetzes zu bestimmen. Daß Helmholtz in diesem Zuge das Aktions- oder, wie er nicht konsequent sagt, das Wirkungsprinzip auf die Ebene eines allgemeinen Naturgesetzes hebt, macht deutlich, wie sich dabei die philosophische Fragestellung wandelt. Um sich nämlich von der übergreifenden Allgemeinheit eines zunächst auf besondere Bewegungen eingeschränkten Prinzips zu überzeugen, bedarf es klarerweise anderer Mittel als derjenigen, die für die Begründung physikalischer Größen erforderlich sind. Und Helmholtz hat die Schwierigkeiten dieser philosophisch neuartigen

Situation in der *Einleitung zu den Vorlesungen über Theoretische Physik* auch deutlich artikuliert. Dort nämlich heißt es:

> "Da die Philosophen, bei denen ich mir Rath holen wollte, ihre Untersuchungen meist erst bei dem Wissen anfangen, was bereits in Worten ausgedrückt werden kann, und die davorliegenden Processe des Sammelns von thatsächlicher Erfahrung meistens gar nicht oder doch nur vom Hörensagen kennen, so habe ich mir selbst helfen und mir die Dinge vielfach in eigener Weise zurecht legen müssen" (Helmholtz 1984: 14).

Er legt sich die Dinge nun so zurecht, daß er erst einmal die physikalische Problemlage resümiert. Dafür bezieht sich Helmholtz auf die Hamiltonsche Theorie als der, wie er sagt, für seine Untersuchungen bequemsten Form der Darstellung, die hinsichtlich seiner Vereinheitlichungsbemühungen indessen noch immer auf zu engen Voraussetzungen beruht. Hamilton nämlich hatte den Ausdruck des Wirkungsintegrals mit

$$\int_{t_0}^{t_1} (F - L)dt$$

angegeben, während Helmholtz schreibt:

$$\int_{t_0}^{t_1} \left\{ H + \sum_a [P_a \cdot p_a] \right\} dt$$

Der Unterschied besteht darin, daß die Hamiltonsche Prinzipalfunktion $H = F - L$ mit F als potentieller Energie und L als lebendiger Kraft (kinetische Energie) sich auf mechanische Systeme bezieht, die von der Einwirkung äusserer Kräfte frei sind. Helmholtz läßt demgegenüber mit den P_a Kräfte dieser Art zu und führt darüber hinaus das kinetische Potential H als eine Funktion ein, für deren Variation die explizite Kenntnis der kinetischen Energie gar nicht mehr erforderlich ist. Helmholtz' Verallgemeinerung des Prinzips der kleinsten Aktion zum Wirkungsprinzip manifestiert sich daher physikalisch in der Einführung des Begriffs des kinetischen Potentials, das ihm nun die folgende Formulierung des Prinzips der kleinsten Wirkung erlaubt:

> "Der für gleiche Zeitelemente berechnete negative Mittelwerth des kinetischen Potentials ist auf dem wirklichen Wege des Systems ein Minimum (beziehlich für längere Strecken ein Grenzwerth) im Vergleich mit allen anderen benachbarten Wegen, die in gleicher Zeit aus der Anfangslage in die Endlage führen" (Helmholtz 1895: 205).

Mit dieser Formulierung wird zudem deutlich, daß sich die von Helmholtz vorgenommene Verallgemeinerung keineswegs nur auf das kinetische Potential bezieht. In Termen wie dem des Mittelwerts des kinetischen Potentials verlieren vielmehr die anschaulichen Deutungen physikalischer Größen in Modellen ponderabler Materie ihren Sinn. Und doch ist es genau diese Situation, die ihn hoffen läßt, ein heuristisches Prinzip gefunden zu haben, das mit der mechanischen Darstellung nur eine der möglichen Darstellungsformen bietet. In seinen physikalischen Arbeiten jedenfalls gelangt er zu dem Ergebnis, daß dieses Prinzip nicht nur für die Mechanik gilt, sondern auch für die Elektrodynamik, die reversible Thermodynamik und für die Lichttheorie. "Begrenzt erscheint seine Gültigkeit bisher", stellt Helmholtz fest, "wenigstens für den jetzigen Zustand unserer Kenntnisse, bei den sogenannten irreversiblen Processen, der Wärmeleitung, Wärmeerzeugung durch Reibung, Stoss, elektrischen Widerstand u.s.w." (Helmholtz 1900: 287).

Seine Formulierung des Wirkungsprinzips hat, wie er versichert, zudem den Vorteil, "die sämmtlichen Bedingungen, welche für die untersuchte Klasse von Erscheinungen von Einfluss sind, in den engsten Rahmen einer einzigen Formel zusammenzufassen, und dadurch einen vollständigen Ueberblick über alles Wesentliche zu geben" (Helmholtz 1895: 210). Damit ist gemeint, daß es den allgemeinsten Charakter derjenigen Bewegungen beschreibt, die zugleich mit dem Wirkungsprinzip dem Energieerhaltungssatz genügen. Seine Formulierung des Wirkungsprinzips zieht, um es genauer zu sagen, den Energieerhaltungssatz nach sich. Die Umkehrung gilt jedoch nicht, und daraus folgt, daß Helmholtz' Prinzip der kleinsten Wirkung "noch einen besonderen Charakter der vorhandenen conservativen Naturkräfte ausspricht, der nicht schon durch ihre Bestimmung als conservative Kräfte gegeben ist" (Helmholtz 1895: 222).

Das freilich ist denn auch der Punkt, an dem der Physiker Helmholtz zum Philosophen wird.[1] Die in seiner Darstellung notwendige Verallgemeinerung von Begriffen, wie sie als Mittelwert des kinetischen Potentials in die Formulierung des Wirkungsprinzips eingehen, entziehen sich ja der unmittelbaren Anschauung des mechanischen Paradigmas. Da nun aber, wie Helmholtz durch seine Kantlektüre weiß, physikalische Größen in einer Anschauung gegeben sein müssen, entwickelt er einen Anschauungsbegriff, der den von ihm eingeführten zunächst rein theoretischen Begriffen einen objektiven, d.i. an-

1 Ich lasse hier unberücksichtigt, daß sich Helmholtz' diesbezügliche Ansichten im Laufe seines Lebens verändert haben, und diskutiere im folgenden den gedanklichen Kontext der Arbeiten zur Geschichte des Wirkungsprinzips.

schaulichen Sinn geben soll. Er entwirft dafür ein Modell, das er zwar als mechanisches begreift, jedoch nicht mehr auf ausschließlich ponderable Materie eingeschränkt wissen will. Anschaulichkeit ist für ihn also durch eine mechanische Interpretation zu gewinnen, die in ihren Modellen die besondere physikalische Disziplin der Klassischen Mechanik transzendiert. In diesem Falle wäre dann in der Tat sein Anspruch eingelöst, die gesamte Physik auf Mechanik zurückgeführt zu haben, wobei die Mechanik, von der hier die Rede ist, sich freilich als metatheoretischer Begriff im folgenden Sinne erweist. Helmholtz schreibt:

> "Für die letzten principiellen Fragen der Wissenschaft, der eigentlichen und berechtigten Philosophia naturalis, wie die Physik von den Engländern immer noch genannt wird, ist der Umstand von hoher Wichtigkeit, dass in den beiden allgemeinsten Gesetzen von der Constanz der Energie und von der kleinsten Action nur noch von den beiden Formen der Energie als den den ganzen Verlauf der Naturprocesse bestimmenden Grössen die Rede ist, nicht mehr von Bewegungskräften und Kraftcomponenten, welche letzteren nur noch als abgeleitete Rechnungswerthe erscheinen. Alles Geschehen wird dargestellt durch das Hin- und Herfluthen des ewig unzerstörbaren und unvermehrbaren Energievorraths der Welt, und die Gesetze dieses Fluthens sind vollständig zusammengefasst in dem Satz der kleinsten Action" (Helmholtz 1900: 287).

Damit ist die neue anschauliche Interpretation gewonnen. Als mechanische Agentien treten jetzt nicht mehr Körper und Kräfte auf, die naturphilosophisch zu rechtfertigen wären. Es werden vielmehr die historisch als Maßbestimmungen eingeführten Entitäten als gegeben angesehen, um daraus ein Modell zu konstruieren, d.h. den Übergang vom Begriff zur Anschauung zu vollziehen. Um nun dieses Verfahren auch als philosophisch akzeptabel ansehen zu können, um das, was eine Interpretation über eine Theorie ist, als "eine Gleichartigkeit des Wesens aller Naturkräfte" begreifen zu können, muß sich Helmholtz darüber hinaus den ursprünglichen naturphilosophischen Fragestellungen zuwenden (Helmholtz 1900: 283). Er muß dieses Modell noch philosophisch begründen, um sich der Gültigkeit des Aktionsprinzips für sämtliche physikalischen Gegenstandsbereiche zu versichern. Aus dem Modell allein folgt die heuristische Bedeutung seines Prinzips, das eine Wesensgleichartigkeit aller Naturkräfte unterstellt, nicht. In seiner Begründung dieses Zusammenhangs beginnt Helmholtz mit der Bestimmung des Gegenstandes der Physik.

Er geht davon aus, daß wir es in den physikalischen Wissenschaften mit Veränderungen zu tun haben, die sich an Naturobjekten vollziehen.

> "Wir suchen in den physikalischen Wissenschaften zu ermitteln, welche Veränderungen überhaupt vorkommen können und welche äußeren Einwirkungen und Umstände vorhergehen müssen, um solche Veränderungen hervorzubringen, oder

> auch, was vorhergehen muß, um das Eintreten von solchen Veränderungen zu verhindern" (Helmholtz 1984: 22).

Die Form dieses Zusammenhangs ist das Kausalgesetz. Und Naturgesetze sind für Helmholtz die eigentlichen Ursachen der Erscheinungen. Sie sind es in dem Sinne, daß sie das hinter allem Wechsel ursprünglich Bestehende und Bleibende bezeichnen, eine in jedem Augenblick fortbestehende Wirkungsbereitschaft. Insofern ist ein Gesetz nichts anderes als das "gleichbleibende Verhältnis zwischen veränderlichen Größen" (Helmholtz 1971: 275). Helmholtz muß also naturphilosophisch so etwas wie Kausalität voraussetzen, die er als gesetzmäßig bestimmt, um überhaupt die Gleichartigkeit von Naturprozessen denken zu können. Ohne diese Voraussetzung wäre es für ihn nicht möglich, das Energie- und Wirkungsprinzip als allgemeingültig für ausnahmslos alle Naturvorgänge zu charakterisieren. Das gesetzmäßige Hin- und Herfluten der Energie spricht - so gesehen - eine Forderung an jeden Naturprozeß aus. In Helmholtz' Terminologie wird das Gesetz dadurch zur Bedingung der *Möglichkeit* von Veränderungen in der Natur. Ein Naturvorgang heißt möglich, wenn er dem Energieerhaltungssatz genügt.

Um nun aus der Menge der *möglichen* die wirklich *realisierbaren* Naturvorgänge bestimmen zu können, bedarf es restringierender Bedingungen, und diese zu formulieren, ist die Leistung des Prinzips der kleinsten Wirkung. Das Energieerhaltungsgesetz

> "giebt nämlich nur eine einzige Grösse, das Quantum der Energie, an, die im Verlauf jedes Naturprocesses constant bleiben muss. Darüber, wie übrigens der Process zwischen seinem Anfangs- und Endstadium ablaufen wird, giebt es nichts weiter an. Um das zu ermitteln, musste man bei den Systemen wägbarer Körper auf die allgemeinen Bewegungsgleichungen zurückgehen. Nun ist es gerade dies, was durch das Gesetz der kleinsten Action bestimmt wird" (Helmholtz 1900: 284).

Die Verallgemeinerungsfähigkeit der Mechanik beruht demnach auf der Allgemeingültigkeit dieser beiden grundlegenden physikalischen *Gesetze*, deren eines die Möglichkeit aller Naturvorgänge thematisiert, während das andere die allgemeinsten Bedingungen ihrer Verwirklichung angibt. Oder anders ausgedrückt: Das Aktionsprinzip fixiert die *Wirkungs*bedingungen des als Allgemeinursache gedachten Energieerhaltungssatzes, und es ist dieser Gedanke, den Helmholtz mit dem Begriff der Wirkung und des Wirkungsprinzips verbindet. Die Ursache wird hier nicht als durch Kräfte bestimmt gedacht, und die Wirkung ist daher nicht die Folge einer solchen Aktion.

Helmholtz will mit diesem Begriff eine Größe angeben, die für die Dynamik das gleiche leistet wie das Hineindefinieren einer Metrik in allgemeine räumliche Zusammenhangsverhältnisse. Es handelt sich demnach um eine Ergänzung des zunächst rein funktionalen Bildes der Mechanik durch anschaulich interpretierbare restringierende Bedingungen. Hinsichtlich der physikalischen Größe *Aktion* bedeutet dies, daß Helmholtz sie als das allgemeine Beharrungsvermögen in dem Hin- und Herfluten der Energie interpretiert, wofür das Kausalgesetz in der Gleichung "Kausalität = Kausalgesetz = funktionaler Zusammenhang" als philosophisch grundlegend vorausgesetzt wird. Die Funktion des Aktionsbegriffs in diesem Modell erläutert er an Leibniz. Seine Größe *Wirkung* soll dem entsprechen, was in Leibniz' Dynamik das Wesen der Trägheit ausmacht, eine primitive passive Kraft, d.h. ein durch die Aktivität der Entelechien als der eigentlichen Wirkpotenzen der Monaden selbst erzeugtes restringierendes Moment, auf dem die Trägheitsbewegung beruht.

Der Sinn des auf Helmholtz zurückgehenden Bezeichnungswandels liegt damit klar zutage. Als Wirkung wird bei ihm das durch den Energieaustausch selbst hervorgebrachte passive Moment, ein Beharrungsvermögen, verstanden, und die Forderung ist, daß dieses ein Minimum sein müsse, um Trägheitsbewegungen zu realisieren. Damit freilich ist Sommerfelds Einwand gegenstandslos. Kräfte sind bei Helmholtz abgeleitete Größen, Rechenhilfen, keinesfalls jedoch die Produzenten von Kausalverhältnissen. Es ist ein anderer Wirkungsbegriff, den Helmholtz hier unterstellt, um seine anschauliche Deutung des Wirkungsprinzips zu ermöglichen. Das wird in seiner Auseinandersetzung mit der Leibnizschen Dynamik zum Gegenstand einer besonderen Untersuchung.

Helmholtz über Leibniz

Es besteht kein Zweifel, daß Helmholtz die physikalischen Intentionen der Leibnizschen Verwendung des Begriffs *actio formalis* trifft. Er zitiert jene Passage, in der die *actio* als der Bewegung schlechthin immanent und also nicht von irgendwelchen äußeren Bedingungen des Bewegungsvollzugs herrührend charakterisiert wird. Helmholtz sieht, daß die Trägheitsbewegung bei Leibniz einer dynamischen Erklärung bedarf, und er bemerkt, daß die Leibnizsche Verwendung des Aktionsbegriffs zugeschnitten ist "auf die Verhältnisse der reinen Bewegung, nach Abstraction von allem wahrnehmbaren Stoffe" (Helmholtz 1887: 227).

Das freilich macht zugleich seine Schwierigkeiten mit dem Leibnizschen Begriff aus. Im Kontext einer rein apriorischen Bewegungstheorie ist dieser nämlich nichts weiter als ein anderer Ausdruck für das Maß der lebendigen Kraft. Das aber kann Helmholtz nicht überzeugen, und er schreibt: "Wenn er [Leibniz - H. H.] aber mit dem Begriffe der Action nichts weiter beabsichtigt hat, als den Werth der actuellen Energie zu finden, so muss man dies einen seltsamen Umweg nennen, der kaum geeignet war, auf seine Gegner, namentlich die Anhänger von *Descartes*, Eindruck zu machen" (Helmholtz 1887: 228). Und doch war genau dies der Fall.

Der Grund dafür läßt sich leicht angeben. Für Leibniz gilt: "Physica per Geometriam Arithmeticae, per Dynamicen Metaphysicae subordinatur" (Leibniz 1971: 104). Um eine physikalische Größe einzuführen, bedarf es für ihn mathematischer und metaphysischer Voraussetzungen, die in seiner apriorischen Bewegungstheorie formuliert werden. Ihre erkenntnistheoretische Funktion besteht darin, eine Nominaldefinition der in Rede stehenden Größe zu ermöglichen, die erst dann zum physikalischen Begriff wird, wenn sie durch eine dynamische Konstruktion als Realdefinition wirklich ist. Das Maß der lebendigen Kraft ist also dann und nur dann als physikalische Größe ausgewiesen, wenn man es a priori und a posteriori herleiten kann und darüber hinaus fordert, daß beide Ableitungen zu dem gleichen Ergebnis führen (Hecht 1991: 142 ff.). Für diese metaphysische Begründung des Bewegungsmaßes, oder anders ausgedrückt, die Legitimation der physikalischen Größe als objektiv und allgemeingültig, nimmt die *actio* eine Hilfsfunktion wahr, die über den Status einer bloßen Nominaldefinition nicht hinauskommt (Stammel 1982: 286 ff.).

Für Helmholtz ist diese metaphysische Interpretation nicht mehr interessant. Er geht von bereits gebildeten Größen aus. Wichtig ist für ihn daher nicht so sehr deren genetische, wohl aber ihre funktionale Beziehung im Rahmen einer physikalischen Theorie bzw. ihr Stellenwert für die Physik überhaupt. Leibnizens Umformulierung des Maßes der lebendigen Kraft in die Größe *actio* muß ihm daher als inkonsequent erscheinen. "Die einzige Verwendung, welche der Begriff der Action in der späteren Entwickelung unserer Dynamik gefunden hat", schreibt er, "ist die im Principe der kleinsten Action, und dieses bekommt in der That einen anschaulicheren Sinn, wenn wir *Leibnizens* Begriff der Action hineintragen" (Helmholtz 1887: 228).

Helmholtz schließt also die Leibnizschen Begriffe der lebendigen Kraft und der Aktion dadurch für seine eigene Fragestellung auf, daß er sie im Gegensatz zu Leibniz mit zwei verschiedenen Prinzipien verbindet, einem Erhaltungsprinzip und einem Extremalprinzip. Er will damit der Größe *Aktion* einen eigenständigen dynamischen Sinn und dem Aktionsprinzip eine inhaltliche Interpretation geben. Das aber gelingt nur, wenn man sie aus ihren historischen Kontexten löst und damit deren erkenntnistheoretische Funktion, nämlich den Nachweis, wie Größen erzeugt werden, als geschichtlich erledigt ansieht.

Damit freilich kehrt sich dann die historische Fragestellung um. Es kommt ja nun nicht mehr darauf an, die Notwendigkeit bestimmter Größen oder Maße aufzuzeigen, sondern in der Gesamtheit der bereits gebildeten jene zu entdecken, die von übergreifender Bedeutung sind. Wie weiter oben gezeigt wurde, war dafür eine anschauliche Interpretation erforderlich, die für Helmholtz' Lesart der Leibnizschen Dynamik bestimmend wird. Was bei Leibniz Sinn und Bedeutung nur im Rahmen metaphysischer Erörterungen einer reinen Bewegungslehre besitzt, wird von Helmholtz für sein anschauliches Bild eines Flutens von Energien vereinnahmt. "Die Trägheit, durch das Gesetz der Energie zur Action gezwungen, vermeidet wenigstens überflüssige Action", schreibt er, und es ist zweifellos eine Leibnizsche Denkfigur, derer sich Helmholtz hier bedient (Helmholtz 1887: 228 f.). Nur tritt das bei Leibniz metaphysisch bestimmte Subjekt der Bewegung jetzt als anschaulich interpretierte Energie auf, die in ihrem Hin- und Herfluten ein als Trägheit zu bestimmendes Beharrungsvermögen erzeugt, als dessen Maß die Action oder Wirkung anzusehen ist. Die Bezeichnung "Wirkung" für diesen Sachverhalt ist vollkommen korrekt, denn es ist das *Gesetz* der Energie, das diese Wirkung erzeugt. Man muß also den Helmholtzschen Kausalitätsbegriff unterstellen, um seine Rede von Wirkungen zu verstehen.

Mit der Bestimmung der Wirkung als Maß des Beharrungsvermögens gelingt es ihm, sowohl den Leibnizschen als auch den Maupertuisschen Ansatz in seinem mechanischen Bild unterzubringen, d.h. ihnen einen anschaulichen Sinn zu geben. Damit wird freilich das, was bei seinen Vorgängern eine Begründungsfunktion hatte, in die Konstruktion des Modells umfunktioniert. Von der Philosophie wird nur noch der Nachweis der Gültigkeit des *Kausalgesetzes* verlangt, alles andere fällt in die Kompetenz des anschaulichen Modells. Der Energieerhaltungssatz und das Prinzip der kleinsten Wirkung sind folglich in der hier vorgestellten Form als allgemeinste Naturgesetze legitimierbar.

Stellt man sich konsequent auf diesen Standpunkt, so ist klar, daß Leibniz' metaphysische Begründung des Maßes der Bewegung als ungereimt erscheinen muß, und das gleiche gilt freilich für Maupertuis. Mit dessen Einführung des Aktionsprinzips kann Helmholtz sogar noch weniger anfangen, und das beste, was er über Maupertuis zu sagen weiß, ist noch der Satz:

> "Wenn er auch die Wahrheit nur gerathen hat, so ist es doch immer die Wahrheit, die er gerathen hat. Und sein felsenfester Glaube an die Möglichkeit, ein allge-

meines Gesetz der Natur zu finden, hat seinen letzten Grund doch in dem richtigen Vertrauen auf die Gesetzmäßigkeit der Natur, d.h. auf das Causalgesetz, welches der letzte Grund all unseres Denkens und Handelns ist" (Helmholtz 1900: 294).

Davon kann freilich nicht die Rede sein. Ein Kausal*gesetz* nach Helmholtzschem Muster kommt bei Maupertuis nicht vor. Zwar finden sich bei ihm gelegentlich Verweise auf Kausalbeziehungen, doch ist dann eher eine wahrnehmbare Wirkung gemeint. Als Erklärungsprinzip ist die Kausalität durch Finalursachen ausgeschlossen. Helmholtz muß diese Uminterpretation vollziehen, weil er sonst Maupertuis nicht als Vorläufer in eigener Sache verstehen könnte, was auch bedeutet, daß er deren Begründungsleistung indirekt und aus der Sicht der Metaphysik verfremdet in seine Modellmethode übernimmt. Mit dem Verweis auf das Kausalgesetz wird Maupertuis' Leistung als unvollständige und unvollkommene Suche des eigenen Ergebnisses deutbar. Dabei aber verschwindet dessen eigenständige Leistung. In Helmholtz' Entdeckungsgeschichte des Prinzips der kleinsten Aktion kommt daher streng genommen die historische Problematik gar nicht vor. Es handelt sich um eine Rekonstruktion der Geschichte im eigenen Bilde, so wie sie später im Begriff der Rekonstruktion durch Reichenbach zu einem wissenschaftstheoretischen *terminus technicus* wurde. Dennoch sind beide nicht identisch. Helmholtz kann in seinem Verfahren auf eine anschauliche Darstellung nicht verzichten, in der das objektivierende Moment philosophischer Begründungsleistung zwar verfremdet, doch immerhin aufbewahrt wird. In seinen diesbezüglichen Arbeiten begegnet uns daher eine jener Formen, in denen sich eine neue philosophische Disziplin, die Wissenschaftstheorie, von der Naturphilosophie als dem originären Maupertuisschen Anliegen absetzt.

Es ist zweifellos Helmholtz' Leistung, diesen Weg nicht bloß pragmatisch, sondern mit philosophischem Bewußtsein beschritten zu haben. Daß er damit keine philosophische Letztbegründung anstrebte, hat er selbst deutlich zum Ausdruck gebracht. Die Schwierigkeiten seines Zugangs liegen wohl darin, daß er eine methodologische Rekonstruktion versucht, die das historische Problem der Bildung von Größen als das Problem einer anschaulichen Interpretation durch Modelle thematisiert. Sie zeigen sich physikalisch darin, daß eine Verallgemeinerung der Physik nach seinen Maximen an ein besonderes mechanisches Modell gebunden bleibt, was sich wissenschaftsgeschichtlich nicht durchhalten ließ. Nach der hier vorgestellten Auffassung kommt dies treffend in der Unterscheidung von *Principe de la moindre quantité d'action* und Prinzip der kleinsten Wirkung zum Ausdruck, falls man nur die Formulierungen ernstnimmt.

Literatur

Hecht, Hartmut: Leibniz' Kategorie des Maßes. In: Ders. (Hg.): *Gottfried Wilhelm Leibniz im philosophischen Diskurs über Geometrie und Erfahrung.* Berlin: Akademie Verlag, 1991, 130-161.

Helmholtz, Hermann von: Zur Geschichte des Princips der kleinsten Action. In: *Sitzungsberichte der Königlich Preussischen Akademie der Wissenschaften zu Berlin.* März 1887, I, 225-236.

Helmholtz, Hermann von: Ueber die physikalische Bedeutung des Princips der kleinsten Wirkung. In: Ders.: *Wissenschaftliche Abhandlungen.* Band 3. Hg. v. A. König. Leipzig: Johann Ambrosius Barth, 1895, 203-248.

Helmholtz, Hermann von: Rede über die Entdeckungsgeschichte des Princips der kleinsten Action. In: Adolf von Harnack: *Geschichte der Königlich preußischen Akademie der Wissenschaften zu Berlin.* Band 2. Berlin: Reichsdruckerei, 1900, 282-296.

Helmholtz, Hermann von: Über das Ziel und die Fortschritte der Naturwissenschaft. In. Ders.: *Philosophische Vorträge und Aufsätze.* Hg. v. H. Hörz u. S. Wollgast. Berlin: Akademie-Verlag, 1971, 153-185.

Helmholtz, Hermann von: Die Tatsachen in der Wahrnehmung. In: Ders.: *Philosophische Vorträge und Aufsätze.* Hg. v. H. Hörz u. S. Wollgast. Berlin: Akademie-Verlag, 1971, 247-282.

Helmholtz, Hermann von: Einleitung zu den Vorlesungen über Theoretische Physik. In: R. Rompe u. H.-J. Treder (Hg.): *Zur Grundlegung der theoretischen Physik. Beiträge von H. v. Helmholtz und H. Hertz.* Berlin: Akademie-Verlag, 1984.

Leibniz, Gottfried Wilhelm: *Mathematische Schriften.* Hg. v. C. I. Gerhardt. Band 6. Hildesheim u. New York: Olms, 1971.

Maupertuis, Pierre Louis Moreau de: Examen philosophique de la preuve de l'existence de Dieu employée dans l'Essai de Cosmologie. In: Ders.: *Oeuvres.* Tome 4. Hildesheim u. New York: Olms, 1965.

Maupertuis, Pierre Louis Moreau de: Essai de Cosmologie. In: Ders.: *Oeuvres.* Tome 1. Hildesheim u. New York: Olms, 1971.

Sommerfeld, Arnold: *Vorlesungen über theoretische Physik.* Band 1. Mechanik. Leipzig: Geest & Portig, 1955.

Stammel, Hans: *Der Kraftbegriff in Leibniz' Physik.* Inaugural-Dissertation. Mannheim 1982.

Muscles and Engines: Indicator Diagrams and Helmholtz's Graphical Methods

Robert M. Brain
M. Norton Wise

> "But there is indeed a close connection between both the fundamental questions of engineering and the fundamental questions of physiology with the conservation of force" (Helmholtz 1861: 115).

Introduction

Historians have typically attributed Helmholtz's use of graphic methods in physiology to the example set by Carl Ludwig's kymograph. Helmholtz himself cited the ingenious instrument of his friend and ally as his model and inspiration. But Ludwig ascribed both his and Helmholtz's use of graphic methods to the principles laid out by James Watt in his indicator diagram (fig. 1), a device invented to measure the work performed in the cylinder of a steam engine: "Die mitgetheilte Curve hat Helmholtz unmittelbar durch den Froschmuskel zeichnen lassen; es geschah dieses nach den Grundsätzen des graphischen Verfahrens von Watt" (Ludwig 1852: 333).

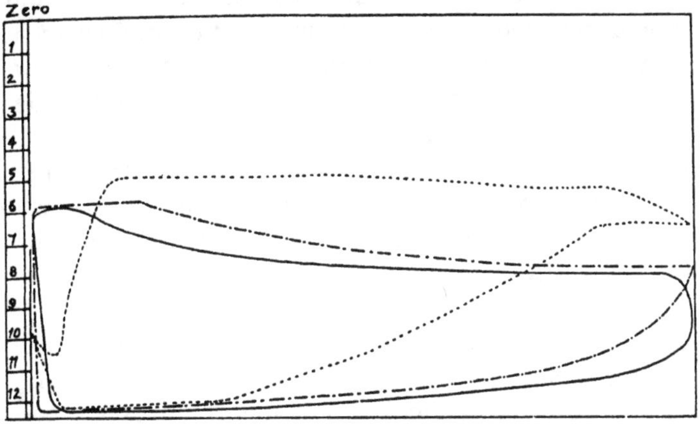

Figure 1. Indicator diagrams with various settings of valves and timing made with Watt's equipment in 1803 (Hills and Pacey 1972: 41; from Portfolio 1381, Boulton and Watt Collection, Birmingham Reference Library).

Indicator diagrams and a variety of self-registering dynamometric instruments found abundant employ in the sixty-five steam engines of mid-1840's Berlin (Mieck 1965: 225-235). Helmholtz, as a regular and "diligent" participant in the Berlin *Physikalische Gesellschaft*, had ample opportunity to ponder them. We will show that he drew heavily upon this the framework for indicator diagrams in the *Erhaltung der Kraft* and used practical modifications of them in his subsequent muscle physiology experiments.

The Young Lions of the Berlin Physical Society

When Helmholtz joined the Berlin Physical Society in 1845, he entered into dialogue with the full range of physical science available in Berlin. The Berlin Physical Society distinguished itself by producing an annual review of the international literature in every branch of pure and applied physics, *Die Fortschritte der Physik*, with individual members contributing reports in their own area of research. "In this manner", wrote an impressed visitor, "the apparently insurmountable task of composing a history of contemporary physics finds itself accomplished effortlessly, with the help of the same principle of division of labor which plays such a great role in industrial enterprises" (Anonymous, 1846: 81).

The membership of the Society reveals the broad base of their coalition for progress. Helmholtz belonged to a small contingent of medical men who had received training in the physiology laboratory of Johannes Müller. At least twelve others of the fifty-four members were engineers; six bore the title "Mechanikus" and another six were listed as military lieutenants, who in all likelihood served as artillery engineers, as did Werner Siemens. The Berlin Physical Society was a club for the young and fiery but not yet established. A number of professors, most notably the club's patron Gustav Magnus, but also others such as Wilhelm Heinrich Dove, participated in meetings but did not become members (Lenoir 1988).

The talented and ambitious activists who filled out the Society's ranks were the offspring of a forceful Prussian industrial policy which for more than twenty-five years had focussed its efforts on educational institutions. Members came not only from the University but from a range of Berlin's more practically oriented scientific, technical, and industrial institutions: the Gewerbeschule, Vereinigte Artillerie- und Ingenieurschule, Bergakademie, and Bauakademie, which had been established in response to Napoleonic power and were modelled on French institutions (Lundgreen 1875: 32; Schubring 1981: 169). The no-nonsense tone of the *Fortschritte* should be ascribed in part to this constituency, which also helps to explain why Helmholtz made the Society his venue and why his *Erhaltung der Kraft* was incubated there (see also Breger 1982: 211-226).

Shopkeeper's Bookkeeping - Liebig's *Thierische Wärme*

Helmholtz's most direct set of problems and resources for the *Erhaltung der Kraft* are set out in his 1845 review, for the first volume of the *Fortschritte der Physik*, of Liebig's newly published "Thierische Wärme", which developed views already published in his *Thierchemie* of 1842. Following a line of interpretation deriving from Lavoisier, Liebig argued that all heat produced by an animal body necessarily derived from respired oxygen, or more specifically, that the heat was produced by combination of the respired oxygen with the elementary carbon and hydrogen in the portions of food absorbed into the bloodstream. Uncertain of where this Stoffwechsel occurred and of the nature of the heat produced, he insisted only that it derived from the interconversion of physical forces (chemical, electrical, mechanical) and rejected any non-physical source, or vital force, whether located in the activity of the nerves or in the contraction of muscles (Kremer 1990: 202-215; Bevilacqua 1993: 297-304).

At the core of his argument, Liebig placed the prohibition of perpetual motion: "Allein aus Nichts kann keine Kraft, keine Thätigkeit entstehen". (Liebig 1843: 28). Elaborating, he insisted that in any chain of conversions of force, the output could never exceed the input, whether in a steam engine, an electric circuit, or a living body. This principle, as principle, served Helmholtz and his physicalist friends as models for their program. Perhaps for that very reason, Helmholtz attacked with vehemence the remaining errors of the Gießen professor. Did the entire bodily heat derive in fact from the combination of carbon and hydrogen with respired oxygen? Liebig's decided support for this view rejected the best measurements available (Liebig 1845: 74-75), while his remarks on the conversion process did not deal consistently with mechanical conceptions of either chemical combination or heat. Implying that the inadequacies of Liebig's position were primitive holdovers from a no longer tenable material theory of heat, Helmholtz insisted on heat as motion, "weil wir Wärme aus mechanischen Kräften ihren Ursprung nehmen sehen" (Helmholtz 1847a: 350).

The motion view, he claimed, implied that only a definite equivalent of heat could derive from either mechanical, electrical, or chemical forces, but little was known either theoretically or experimentally about this "Principle of the Constancy of Force-Equivalents". On the theoretical side, Helmholtz cited the analysis by the French engineers, Sadi Carnot and Emile Clapeyron, of the production of work from the fall of heat in steam engines (also based on the prohibition of perpetual motion) and Franz Neumann's theory of electromagnetic induction of currents. On the experimental side, only studies of chemical heats and of heating by electric currents were available to confirm the principle of constancy.

For the moment, Helmholtz limited his critique to Liebig's respiratory theory, relying on the calorimetric measurements of Dulong and Despretz that Liebig challenged and implicitly on his own newly completed experiments on the Stoffwechsel in muscles. Animal heat did derive from chemical action, he argued, but not simply from burning elementary carbon and hydrogen in respired oxygen. Any equivalence of inputs and outputs would have to account for complex chemical reactions of the foodstuffs, some of which at least, occurred in the muscles, where work too was produced. The whole accounting procedure (Berechnungsart) that Liebig had derived from Lavoisier, Dulong, and Despretz, required reexamination (Helmholtz 1847a: 351-353). Based on an exchange of substances alone, the "Stoffwechsel", it had constituted in effect a balance sheet of material credits and debits, the sort of thing that du Bois-Reymond, with characteristic wit and condescension, called "shopkeeper's bookkeeping" (Cranefield 1982: 3).

This shopkeeper's bookkeeping was one direct target of Helmholtz's *Erhaltung der Kraft*. The problem can be seen in Liebig's formulation of the interconversion principle. He showed great faith in the quantitative interconversion of chemical, electrical, mechanical, and thermal forces, as cause and effect. But he proposed no common analysis of the different forces, only a "substrate", thus effectively a force substance (Liebig 1842: 30-31). Sometimes this force inhered in ponderable matter and was released by a Stoffwechsel as heat, which was not material, since it could be converted into motion, but was also not mechanical, since it could be returned to matter in a Stoffwechsel, which Liebig did not treat as mechanical. In Helmholtz's view, this accounting was literally incoherent. His own solution was to represent both the Stoffwechsel and heat as mechanical (matter moving under the action of forces), but empirical evidence was weak. We will argue elsewhere that he sought a principled way out of this dilemma in the much-discussed Kantian transcendental introduction to the *Erhaltung*, where he sought a priori grounds for the duality of matter and force. In any case, the new bookkeeping would require more than mere Stoffwechsel, more even than exchanges of a substrate labelled force. It required a complex balance sheet for interconversions, based on a common measure of cause and effect, but analyzed in terms of matters, motions, and forces together.

Engineering Mechanics in the *Erhaltung der Kraft*

Helmholtz found his new accounting technology in the immediate circles in which he moved in Berlin, namely in the integrated context for chemistry, physics, and technology provided by Magnus and Dove and the Berlin Physical Society. The character of this community appears in the fact that Magnus, who made his initial reputation as a chemist, taught industrial technology at the University, physics at the Artillery and Engineering School, and chemical technology at the Gewerbeschule. Already in 1830 he gave his trial lecture as Privatdozent on "Die Lehre von den Dämpfen", a theme of later researches on steam as a source of power. As extraordinary professor for technology from 1833 and ordinarius from 1845, Magnus regularly took his class on an excursion to local industrial sites, including in the 1840s the rapidly expanding iron and engine works of August Borsig. And it was to the young group that assembled in Magnus' house that Rudolph Clausius in 1842 reported on the Carnot-Clapeyron theory of steam engines, the source that Helmholtz would choose as his reference (rather than Liebig) in grounding his conservation principle on the prohibition of perpetual motion (Kant, 1977; Wolff, 1993).

A similar sense of the resources immediately available to Helmholtz derives from Dove's activities as professor of physics at the University and teacher at the Gewerbeschule and the Artillery and Engineering School. Characteristic of his style is the *Repertorium der Physik*, which Dove founded in 1836 and edited until 1845. The *Repertorium* emphasized experimental physics: measuring instruments, properties of materials, engineering problems. It contained no speculative theory and little pure mathematical physics. The offerings of Dove's *Repertorium* and of the Magnus circle provide ready material for interpreting the *Erhaltung* both as a bookkeeping principle and as a grounding for Helmholtz's later graphical method in physiology.

As is well known, Helmholtz deployed the Carnot-Clapeyron formulation of the perpetuum mobile principle to prove that the principle of conservation of vis viva, familiar in mechanics, must hold for all natural systems (now necessarily systems of matter and force); that is, that the vis viva of any system of bodies, in passing through a cycle of changes to return to its initial configuration, necessarily returns to its initial value; or, that the total work done by the forces acting on the bodies during the cycle must vanish. Otherwise the cycle could be used to generate work *ad infinitum*. That condition required, Helmholtz argued, that all forces were necessarily central forces acting between material points and dependent only on distance. The principle of conservation of vis viva, now elevated to the "principle of conservation of force", took the form, for a single material point of mass m moving from r to R under the action of a force of intensity φ in the direction of the differential displacement dr,

$$\tfrac{1}{2}mV^2 - \tfrac{1}{2}mv^2 = -\int_r^R \varphi\, dr,$$

where v is the initial velocity and V the final velocity. A system of arbitrarily many material points - meaning any system whatsoever - required a summation over all of them (Helmholtz 1847b: 21-25).

It is for Helmholtz's understanding of this vis viva principle that we seek the local resources. His terminology is already unusual, for he named the work integral on the right the sum of the *Spannkräfte* "welche den Punkt *m* zu bewegen streben, so lange sie eben noch nicht Bewegung bewirkt haben". The term "Spannkraft" had been employed as an equivalent for "force of elasticity", both of steam in the works of Magnus and Dove (Dove 1837; Knochenhauer 1846) and of muscles in those of Eduard Weber (below). Their usages, however, referred not to the amount of work available in the steam or the muscle, but to the pressure of the steam and the tension in the muscle. Helmholtz thus subtly changes the meaning of "Spannkraft" from an *intensity* of Kraft to a *quantity* of Kraft, from the tension immediately present to the work that would be available from it in a differential displacement. That is, Spannkraft becomes *potential* work, whether still available (noch vorhanden) or already used (verbraucht).

Concerning the principle of vis viva itself, its elaboration in terms of "work" derived from engineering mechanics rather than rational mechanics, especially French engineering (Bevilacqua 1993: n. 6). A standard reference was Poisson's *Mécanique*, containing an addition "Relative to the Usage of the Principle of Living Force in the Calculus of Machines in Motion". This usage, wrote Poisson, "forms, so to speak, the point of junction of rational mechanics and industrial mechanics" (Poisson 1833: 747). For detailed discussion, Poisson referred the reader to the yet unpublished works of Navier at the Ecole de Pont-et-Chaussées and to Poncelet, then Colonel of Engineers at the Ecole de l'Artillerie et du Génie. One of the ways in which their works and his entered the Berlin context was through an 1844 review in Dove's *Repertorium* by Ferdinand Minding, professor of applied mathematics in Dorpat, formerly lecturer in Berlin at the Bauschule, Gewerbeschule, and University (Minding 1844: 71-76). Judging from its details, this review was one of Helmholtz's sources. Like Helmholtz, for example, Minding takes vis viva to be $1/2(mv^2)$ rather than the standard mv^2. More significantly, Minding extended Gauss's term "potential" for the integral of what was called a *virtual* moment in statics (force times an imagined infinitesimal displacement), to the work integral correlated with *real* motions in the principle of vis viva, that is, with the integral of the product $\varphi\, dr$. Helmholtz did the same when he came to electrical and magnetic forces, so that the sum of Spannkräfte in a system became its total potential, and Gauss's results for electrostatics and magnetostatics fit right into the sums required for the conservation doctrine (Helmholtz 1847b: 41-43).

For Gauss, the potential apparently had no ontological status in itself. It was a mathematical function from which forces could be expressed as derivatives, greatly simplifying the calculus of static forces. Minding's and Helmholtz's extension into the basic principle of engineering mechanics, which already correlated virtual moment with work, converted Gauss's potential into a real entity, the work content of the system, whether as the work already manifested as vis viva, or the work still available (noch vorhanden) for use. This transformation of rational mechanics into industrial mechanics (in Poisson's felicitous phrase) provided the new accounting system that Helmholtz sought, a new balance sheet for quantities of force.

The same engineering sources typically provided as well a graphic representation of the work integral, a representation not available in rational mechanics, but familiar to engineers and mechanics since the 1820s as Watt's indicator diagram. Minding followed his description of the vis viva principle applied to machines with a description of dynamometers used to measure the work done by moving forces, whether produced by muscles or machines. Citing Arthur Morin, Captain of Artillery at the Metz military school as his primary source, Minding stressed particularly the value of chronometric, self-registering dynamometers, which produced directly a curve of the intensity of force exerted as a function of distance traversed. "Durch Quadratur der gezeichneten Curve erhält man in dem [...] Falle, wo die auf den Ordinaten P senkrechten Abscissen [...] dem durchlaufenen Wege r proportioniert sind, das Integral $\int\varphi\, dr$ oder die gesammte Arbeit der Zugkraft, während der Beobachtung" (Minding 1844: 75; using Helmholtz's r). Helmholtz's description bears comparison with Minding's: "Denken wir uns die Intensitäten von φ [...] durch rechtwinkelig aufgesetzte Ordinaten dargestellt: so würde die genannte Grösse den Flächeninhalt bezeichnen, den die Curve zwischen den zu R und r gehörigen Ordinaten mit der Abscissenaxe einschliesst" (Helmholtz 1847: 22).

Even more directly relevant to the *Erhaltung* than Minding's review is Clapeyron's account of Carnot's theory of the motive power of heat, since Helmholtz stated that he wanted to apply Clapeyron's method to every branch of physics, i.e., to apply the perpetual motion principle to establish that the work produced could not exceed the work expended in any cycle of changes whatever. With respect to the graphical method, Clapeyron referred his entire analysis to indicator diagrams, in the idealized form now commonly called Carnot diagrams, displaying the cycle of pressure versus volume in the cylinder of an engine as a closed series of isothermal and adiabatic curves (fig. 2).

And he described the work integrals associated with the curves in the engineers' graphical language of summing ordinates times differential abscissae under the curve. Clapeyron also drew a direct analogy between the conversion of thermal work (as fall of heat) to mechanical work and the conversion of a raised weight (by falling) to vis viva (Clapeyron 1843: 450, 457). To carry out his own mechanical program for the conservation principle, therefore, Helmholtz had only to show that Clapeyron's basic scheme would be even more coherent if heat were treated as vis viva rather than a substance, that is, if the heat itself, rather than the fall of heat, were converted into mechanical work by a steam engine.

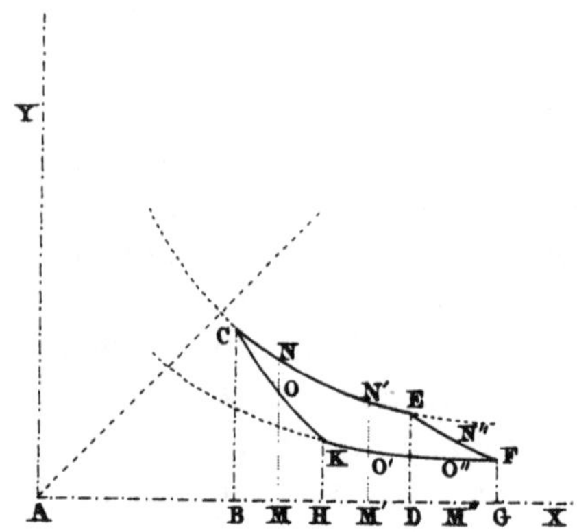

Figure 2. Clapeyron's original Carnot diagram of 1834 (Clapeyron 1843: plate II, no. 13).

As a practical matter, a close relation thus emerged between Helmholtz's conservation problem and the graphical method, because he required accurate measurements for forces varying over time, particularly those in which the work done was converted into heat. Helmholtz complained about the lack of such experiments especially in his sections *V* and *VI* on electrical and magnetic processes, which make up nearly half of his pamphlet. No one had yet devised means for registering rapidly changing currents and voltages, so that calculations of work expended remained gross approximations for many ordinary circuits. Inconstant batteries and inherently time-dependent phenomena like mutual induction, required knowledge of the actual shape of the curve of current, not simply an average of beginning and ending currents, such as Joule had used (Helmholtz 1847b: 52, 55, 65). But experimental physicists had not yet begun to make this kind of measurements. The people who had were engineers, with their dynamometers and indicator diagrams. Luckily for Helmholtz, he was acquainted with such people himself and with others who knew their work.

Accounting for Muscle Contraction

After completing his treatise on the conservation of force, Helmholtz made strategic changes in his ongoing research into the economy of forces at work in the physiology of muscles. He abandoned investigations into chemical and thermal alterations and in 1848 began work on the temporal recording of muscle contraction. In his new research Helmholtz wielded the conceptual core of the *Erhaltung* - the principle of *vis viva* combined with the Clapeyron method - as a *theoretical technology*, a set of conceptual and mathematical tools used to perform intellectual work (Warwick 1992). To this he coupled the *practical technology* of Carl Ludwig's newly invented kymograph, having made the acquaintance of both inventor and instrument while finishing the *Erhaltung*. Ludwig had self-consciously designed the kymograph according to the principles of Watt's indicator, but using a Morin-style chronometric stylus and a Poiseuille hemadynamometer. His sensational instrument realized in practical physiology the theoretical propositions adumbrated in Helmholtz's *Erhaltung*; and vice-versa: Helmholtz's doctrine realized in theoretical physics the principles of Ludwig's kymograph. Thus a theoretical and a practical technology, both the fruits of raids on French engineering, consecrated the alliance of the 1847 group of organic physicists. The following year du Bois-Reymond, the self-appointed ideologist of the group, promulgated the doctrine that the "proper form of physiological representation should be a curve: The dependence of the effect on each condition presents itself in the form of a curve, whose exact law remains, to be sure, unknown, but whose general character one will be able in most cases to trace" (du Bois-Reymond 1848-49: xxvi).

The pairing of the *Erhaltung* and the kymograph bolstered du Bois-Reymond's and Helmholtz's confidence in the necessity of direct methods in physiology, a principle they had insisted upon for some time (Lenoir 1992). In 1845, not long after the two had met, du Bois-Reymond delivered a paper to the Berlin Physical Society which revised Theodor Schwann's famous experiments on the elasticity of muscles by reconstructing his data in the form of curves to show that contracting muscles do not simply compress themselves, but rather they respond to particular forces which act upon them. He cited Helmholtz's experiments on the chemical changes in the active muscle as evidence for these forces, and lauded the direct methods by which he had brought forth the results. Both du Bois-Reymond's graphic methods and Helmholtz's chemical analysis served to ascertain the intermediate processes that

Schwann's "force" approach and Liebig's "energy" approach left concealed. Du Bois-Reymond concluded his discussion by anticipating Helmholtz's attempts with a thermo-multiplier to achieve a "Sichtbarmachung der Nervenwärme" (Kirsten 1986: 84).

The following year Eduard Weber's massive article on "Muskelbewegung" appeared, which according to du Bois-Reymond both summed up and superseded all previous treatments of muscular force (du Bois-Reymond 1848: 95-96). As one of his prominent themes, Weber undertook a review of the long tradition of measuring the mechanical "Nutzeffekt" of human muscular work, to which he himself had made substantial contributions (Weber and Weber 1836). He fastened particularly on a series of experiments by the physiologist Gabriel Valentin, who had sought to improve upon the precision measurement of "the many types of machine-like outward effect of muscles" with an instrument he called the "myodynamometer" (fig. 3) (Valentin 1844: 120).

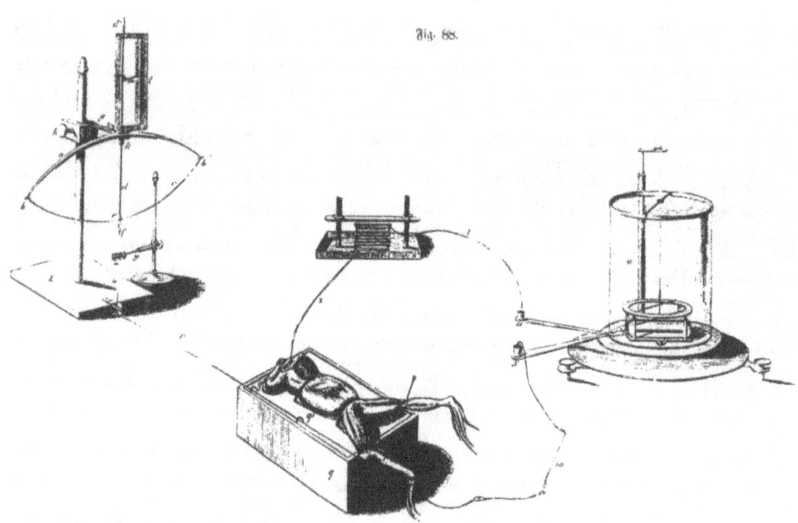

Figure 3. Valentin's Myodynamometer (Valentin, 1844). Cambridge University Library.

Modelled on the standard Regnier spring dynamometer used to measure the traction of wagons and other contrivances, Valentin's myodynamometer measured the "load effect," or the "quantity of force" of the isolated frog muscle in contraction, by a spring mechanism. Weber criticized Valentin's method for its inability to measure the initial force necessary to overcome the tension of the dynamometer, a problem that Valentin sought to circumvent by supposing that the weight of the muscle section was proportional to the muscular force. By contrast Weber argued that the muscle is like an elastic band whose elastic forces are balanced off by a "Spannkraft"; when the muscle passes from the relaxed to the contracted state the elasticity coefficient diminishes. The same weight, in other words, stretches the muscle in its contracted or relaxed state.

Weber contended, moreover, that insofar as Valentin had only measured the force of the muscle in contraction in its normal length and loaded only by its own mass, he had failed to measure Nutzeffect, which he defined as the load multiplied by its increase in height (work). In response Weber himself devised an experimental apparatus, with which one "eine dem Nutzeffecte aequivalente lebendige Kraft gewinnen kann, die sich auf andere Körper durch Maschinen übertragen und dadurch zur Hervorbringung einer bezweckten Bewegung dieser Körper benutzen lässt" (Weber 1846: 91). Using a rotation apparatus to induce muscle contractions of long duration, Weber probed the parameters between the shortening of the muscle, the duration of the contraction, and the weight raised (Weber 1846: 91-100; Olesko and Holmes 1993: 75-76; Kremer 1990: 328-330).

Helmholtz first encountered Weber's "Muskelbewegung" while finishing *Die Erhaltung der Kraft*, and his reformulation of Weber's problem reflected the lessons of his theoretical treatise. It hinged upon the problem of time. Despite Weber's brilliant comparative measurements of muscles at rest and in a prolonged state of contraction, he had assumed that the transition between the two states was instantaneous. Extending his argument from the *Erhaltung* for the necessity of accurate measures of forces varying over time, Helmholtz argued that Weber's method of continuously stimulating muscles "could not answer one of the principal questions of this field, namely that of the mechanical work that it is capable of producing". The continuously stimulated muscle quickly attained a static condition (Gleichgewichtslage), and could not, therefore, produce any work in the sense of mechanics. Helmholtz thus took as his point of departure the principle that: "Um eine Arbeit zu leisten muss der Muskel zwischen Ruhe und Erregung wechseln, und die Grösse seiner Arbeit wird wesentlich von der Geschwindigkeit des Wechsels abhängen" (Helmholtz 1850a: 764).

Helmholtz came well-equipped to the problem of measuring fleeting phenomena. As early as 1845 he had been apprised of a range of new instruments designed to measure high velocities in a paper delivered by Werner Siemens to the Berlin Physical Society, subsequently published in several versions. Siemens' paper served to stake a claim in a raging international priority dispute for work he had conducted with the Prussian Artillery-Testing Commission on techniques used to measure "the velocity of rapidly moving bodies" and "the time-spans in which they traverse specific spaces" (Siemens 1847: 47). Each of the instruments in question measured time intervals in one of two different ways. One method transformed differences of time into differences of space, in a version of graphic representation. The other method measured the "mechanische Wirkung" (work) which, during the time-interval to be determined, brought forth a force of specific intensity, from which one calculated the time. In 1850 Helmholtz reviewed both types of instruments in a lecture intended to provide the background to his research and to show its place within a broader scientific context (Helmholtz 1850b: 862-871).

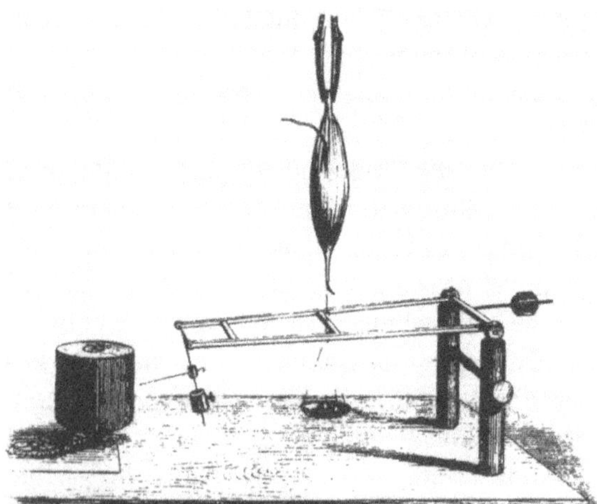

Figure 4. Helmholtz's Myograph, in: Etienne-Jules Marey, *Du mouvement dans les fonctions de la vie* (Paris: Baillière, 1868), 133. Cambridge University Library.

In fact he used both methods to study the velocity of nerve propagation in muscle contraction. For his first attempt he used the graphic method. He designed an instrument, later called a myograph (fig. 4), using the self-registering technique of Ludwig's kymograph. Ludwig had followed James Watt's technique of placing a flotation device in a cylinder to record variations in pressure by inserting a similar device in a Poiseuille hemodynamometer attached to a rod holding a stylus (Ludwig, 1847: 258-267) (fig. 5). To the experimental setup used by Weber, in which an isolated frog muscle was suspended from a frame and stretched by a weight at the lower end, Helmholtz attached a metal rod which moved vertically between two plates. This rod, like the rod in Ludwig's instrument, transmitted the signal to the stylus. Helmholtz's myograph did for the myodynamometer what "Ludwig mit den Höhen des Blutdruckmessers [hemodynamometer] that, die Höhe aufzeichnen, bis zu welcher ein an den Muskel gehängtes Gewicht in den aufeinanderfolgenden Zeitpunkten der Zuckung erhoben wird" (Helmholtz 1850a: 767).

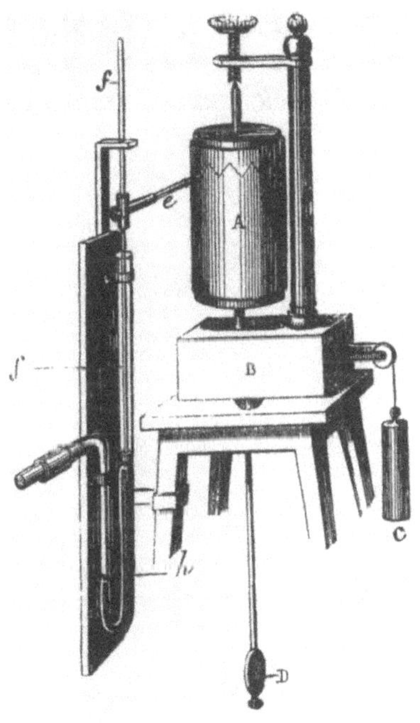

Figure 5. Ludwig's Kymograph, in: Etienne-Jules Marey, *Du mouvement dans les fonctions de la vie* (Paris: Baillière, 1868), 132. Cambridge University Library.

Helmholtz twitched the muscle with a momentary current induced by the opening and closing of a Neefian induction coil. The resulting "Froschcurven" showed the net values of the "energy" of the muscle (work done) along the ordinate, which unfolded in the "time-spaces" represented along the abscissa. He achieved this temporal measure of the total work done by maintaining the force constant with the hanging weight. Despite plaguing problems resulting from friction in the apparatus and the muscle, Helmholtz produced curves which showed a pronounced time-lag before the onset of the muscle contraction (fig. 6). They showed, he wrote, that "die Energie des Muskels nicht im Augenblicke einer instantanen Reizung sich vollständig entwickelt, sondern grösstentheils erst nachdem diese schon aufgehört hat, allmählich ansteigt, ein Maximum erreicht und wieder verschwindet" (Helmholtz 1850a: 770). Helmholtz interpreted the curves as bearing out Weber's findings about the diminishing elasticity coefficient in muscle contraction. The stages of the curve revealed the changing relationship between the contradictory forces which stretched the muscle and those which precipitated its contraction. The convex phases showed the preponderance of the latter, the concavities the preponderance of the former, and the points of inflection the moments of equilibrium between the two forces. (Helmholtz 1850: 769-770; Holmes and Olesko 1994)

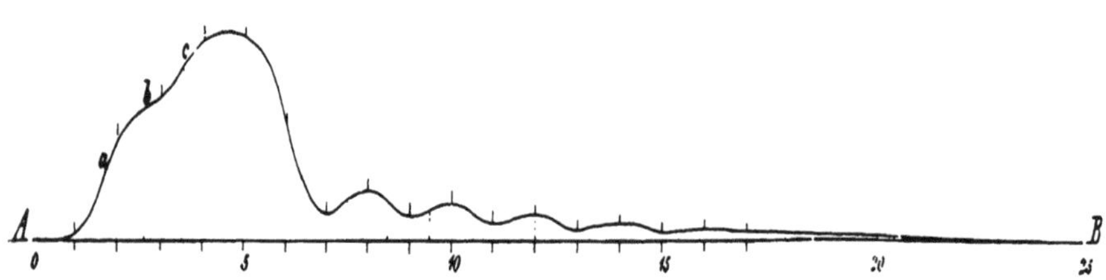

Figure 6: Helmholtz's Myograph curve, in: Helmholtz 1850, Tafel 8, Abb. 3. Cambridge University Library (scale enlarged by a hundred percent along the ordinate).

After his first set of trials with the myograph, Helmholtz took up the second method of measuring time-intervals using the deflection of a ballistic galvanometer, to which the name of C.S.M. Pouillet had, as victor in the priority struggle, been attached. This method availed, by other means, a similar measure of the relation between mechanical effect and the velocity of nerve propagation. Helmholtz found that for purposes of precision the second method held certain advantages, the greatest of which was the near limitless reduction of the time-intervals which could be measured, which encouraged him to work at determining the value of the propagation velocity of the nerve impulse. But this method also proved amenable, as Olesko and Holmes have admirably shown, to a calculus of probable errors in his measurements, a concern to which he attached great importance (Olesko and Holmes 1993: 95-105).

Helmholtz's first communications of these experiments encountered skepticism from different quarters. Du Bois-Reymond lamented the difficulties he had in explaining, even to a most sympathetic colleague such as Volkmann, who "lacked the ABC's" of physics, what was made visible by the curves. "Ich bin zu Ostern bei Volkmann in Halle gewesen", he wrote to Helmholtz, "und ich hätte gute Zeit gehabt, wenn es mir gelungen wäre, Volkmann im Lauf von acht Tagen begreiflich zu machen, daß das Maß der Kraft die Veränderung der Geschwindigkeit sei" (Kirsten 1986: 113).

Ludwig's comprehension rate, by contrast, was instantaneous. While writing his *Lehrbuch der Physiologie* he hosted Helmholtz for an eight day visit, during which time they pondered the frog curves and discussed "all possible physical and physiological topics". In his textbook Ludwig made very explicit the sense in which these studies of the isolated frog muscle resolved old problems of the "work of animate and inanimate motors" (Morin 1838) previously solved by mechanical engineers with dynamometers. "Der Muskel kann begreiflich und bekanntlich die Bewegung, welche seine eignen Theile bei der Formveränderung erleiden, andern Körpern mittheilen, und damit im Sinne der Mechanik Arbeit verrichten. In dieser Beziehung fällt die Betrachtung unseres Apparates mit derjenigen aller arbeitenden Maschinen zusammen" (Ludwig 1852: 342). The newer techniques of engineering could be easily translated into the physiology laboratory because in the problem of measuring and apportioning work the engineering problems had always been both mechanical and physiological.

> "Das Maaß der lebendigen Kräfte, welches für die Leistungen solcher Maschinen in Anwendung gebracht wird, ist wie allbekannt ein Produkt aus den Faktoren der Zeit (*f*), der Gewichte (*g*) und der Hubhöhe (*h*) also = *ghf*. Mit andern Worten dem Techniker ist es wichtig zu wissen, welche Zeit hindurch oder wie oft in gegebener Zeit und wie hoch ein gewisses Gewicht durch eine Maschine gehoben werden kann, weil nach diesen Angaben die Werthung der von einer Maschine gelieferten auf andere Massen uebertragbare Kräfte geschehen kann" (Ludwig 1852: 342).

Helmholtz's method for making the work of the muscles *anschaulich* derived from the spectacular strides of the 1840's in electrotechnology made by Siemens and others, so that by 1850 he could compare the nerve fibers to electric telegraph wires (Helmholtz 1850b: 873). Both transmitted currents at a distance, and as the French telegraph mogul Louis Breguet pointed out, they transmitted information about work being done. In a paper reviewed by both Siemens and Helmholtz (Siemens 1847: 53-58; Helmholtz 1850b: 867), he suggested how his electrification of Morin's dynamometric principle could enable the "Werkführer" to combine shop-floor invigilation and cost-accounting in the same technique. The "instrument may, as I imagine", he wrote, "be employed with advantage in manufactories; for, by means of conductors, leading from the office of the superintendent, and communicating either with the fly-wheel or the cylinder of a steam-engine, he may know at every moment in the day, and without disturbing himself, the velocity of either one or the other" (Breguet 1845).

Conclusion: Titanenarbeit

"Es ist ein Schauspiel für Götter, den Muskel arbeiten zu sehen, wie den Zylinder einer Dampfmaschine" (Kirsten 1986: 123). Du Bois-Reymond's choice of metaphors, to convey to Helmholtz in 1852 his excitement about his latest apparatus for measuring electric current during contraction of a frog muscle, symbolizes our thesis. Muscle physiology in the Berlin Physical Society was the work of a new breed of gods who sought to remake natural science as the philosophical face of the engines and engineers who were remaking the Prussian economy. We have concentrated on the technical use of transformed dynamometers and indicator diagrams to produce graphical methods for analyzing muscles. The diagrams, with their curves of definite shape, carried conceptual content of great significance. They reified organic action as physical work.

But these engine curves *cum* frog curves carried crucial symbolic value as well. They identified the project for progress of the organic physicists with the much larger project for progress of Prussia, promoted at the highest administrative and social levels by a closely linked group of economic modernizers who included the director of the division of trade and industry P.C.W. Beuth, founder and administrator of the Gewerbeschule; his friend and collaborator Karl Friedrich Schinkel, of architectural fame; and their influential associate and royal advisor Alexander von Humboldt, who rescued Helmholtz from his full obligation as a military doctor. In the early 1840s, all of them were members along with Magnus of the Verein zur Beförderung des Gewerbefleisses, which included entrepreneurs like the engine builder Borsig and three out of four of the royal princes, all four of whom had steam engines installed in their private landscape gardens. Helmholtz and his ambitious colleagues moved within the penumbra of this culture-forming elite and envisaged themselves as the successor class.

An immediate sense of how their interests tied into those of the Verein appears in a series of articles on political economy for its *Verhandlungen* by the Oberbaurat von Prittwitz. "Everything which satisfies a human need, or which can be used or enjoyed comes into being through specific forces, and is therefore called a product in the doctrine of political economy" (Prittwitz 1829). Prittwitz sought a general accounting scheme, a means of reckoning the value of all of these productive forces, so as to calculate the national economic capacity, but he limited himself to mechanical forces. Decrying claims to perpetual motion, he reminded his readers that the "mechanical moment" of a moving force was to be measured by the product of resistance overcome and velocity of the motion (rate of working). And he promoted a Prussian effort, analogous to that of French savants, to establish a standard unit of work, the "dynam" or "dynamie", which could serve engineers as a mechanical currency and investors as a guide to cost efficiency. Prittwitz shared some of the same sources and goals as engineers like Poncelet and Morin, whose work figured so prominently in the Berlin Physical Society. To Poncelet, "one of the distinctive characters of mechanical work is that it is the thing that one pays for in the exercise of the force, and that its value, its price in money, increases precisely as its quantity" (Poncelet 1844: vol. 1, 44). Again, citing Navier, Poncelet agreed that units like the "dynam" constituted a kind of "monnaie mécanique", essential in comparing the mechanical value, and simultaneously the economic value, of different motors and machines (Poncelet 1845: vol. 1, 6-8). It was to this context, and to this expression, that Helmholtz returned in both of his lectures for general audiences on conservation of force, in 1854 and 1862, with the remark "Arbeit ist Geld" (Helmholtz 1884: vol. 1, 53, 228).

Not everyone appreciated the cultural imagery of machines and money (Brose 1993: *passim*). Among physiologists, Rudolph Wagner was one of the detractors of the junior sages of Berlin. Writing to du Bois-Reymond in 1852, Helmholtz reported Wagner's displeasure with apparent glee. "Er lässt sich darin über die physikalisch-physiologische Schule los, [calling us] ein junges

Titanengeschlecht, mit ikarischen wächsernen Flügeln." Wagner had at least one of his metaphors right. The image of Titans, if not of Icarus, came readily to mind to observers of powerful engines. And the most impressive symbol of that power had been constructed in 1843 in Helmholtz's home town of Potsdam, housed in its own temple. It is an eighty horsepower steam engine, built by Borsig to drive fourteen pumps simultaneously to supply the fountains of the King's gardens at Sanssouci, the most spectacular of which was a single geyser shooting 130 feet into the air. Promoted by Beuth and housed in a mosque designed by Schinkel's architectural associate, Ludwig Persius, complete with a minaret for a chimney, the magnificent machine and associated fountains were explicitly intended to represent Prussian technological sophistication and growing economic strength to visitors from all over Europe and the wider world.

A visitors guide captures the intended image. "Die lautlose Stille und gleichmässige Bewegung der imposanten Massen und Kräfte dieser hydraulischen Dampfmaschine erwecken eine Regung des Erstaunens über die riesige Macht des Menschengeistes, der sich selbst die Elemente dienstbar zu machen weiss, um wahre Titanenarbeit ohne Anstrengung verrichten zu lassen." And closely following: "Wir erfreuen uns mit Recht dieser Fortschritte deutscher National-Industrie, als eines Beweises, wie immer mehr Deutschland sich von dem bisherigen britischen Industriemonopol emancipiert" (Belani 1843: 67, 68).

Thus the images of Titanenarbeit and Fortschritt move between Borsig's engine and Helmholtz's frog muscles, showing symbolically how those two objects of cultural production belong together in the same city of Berlin and incorporate many of the same values. Analytically speaking, however, the engines and muscles belong together because they shared a common theoretical foundation and a common instrumental foundation. Those two anchors, the principle of vis viva and the kymograph, merged in the indicator diagrams that help to narrate the success of both Helmholtz and Borsig.

References

Anonymous: Progrés des science physique hor de France. *Revue scientifique et industrielle*, 27, 1846, 81-95.

Belani, H.G.R.: *Geschichte und Beschreibung der Fontainenanlagen in Sanssouci*. Potsdam: Janke, 1843.

Bevilacqua, Fabio: Helmholtz's *Ueber die Erhaltung der Kraft*: The Emergence of a Theoretical Physicist. In: Cahan 1993: 291-333.

Breger, Herbert: *Die Natur als arbeitende Maschine: Zur Entstehung des Energiebegriffs in der Physik 1840-1850*. Frankfurt/New York: Campus Verlag, 1982.

Breguet, Louis: Note sur un appareil destiné à mesurer la vitesse d'un projectile dans differents points de sa trajectoire. *Comptes rendus*, 20, 1845, 157; *Annalen der Physik und Chemie*, 59, 1845, 451; *Dingler's Polytechnisches Journal*, 96, 1845.

Brose, Eric Dorn: *The Politics of Technological Change in Prussia. Out of the Shadow of Antiquity, 1809-1848*. Princeton: Princeton Univ. Pr., 1993.

Cahan, David: *Hermann von Helmholtz and the Foundations of Nineteenth-Century Science*. Berkeley/Los Angeles/London: Univ. California Pr., 1993.

Clapeyron, Emile: Ueber die bewegende Kraft der Wärme (1834). *Annalen der Physik und Chemie*, 59, 1843, 446-451, 566-592.

Cranefield, Paul (ed.): *Two Great Scientists of the Nineteenth-Century. Correspondence of Emil Du Bois-Reymond and Carl Ludwig*. Baltimore: John Hopkins Univ. Pr., 1982.

Crump, W. B.: *The Leeds Woollen Industry, 1780-1820*. Leeds: 1931, 216. Quoted in Hills & Pacey 1972: 34.

Dove, H. W.: Analogie zwischen Gasen und Dämpfen. *Repertorium der Physik*, 1, 1836, 58-62.

du Bois-Reymond, Emil: Elasticität fester Körper. *Fortschritte der Physik im Jahre 1846*, 2, 1848, 95-99.

du Bois-Reymond, Emil: *Untersuchungen über die thierische Elektricität*. Berlin: Reimer, 1848-49, 2 vols., vol. 1.

Helmholtz, Hermann von: Physiologische Wärmeerscheinungen. *Fortschritte der Physik im Jahre 1845*, 1, 1847a, 346-355.

Helmholtz, Hermann von: Ueber die Erhaltung der Kraft: Eine physikalische Abhandlung (1847b). In: *WA*, 12-75,

Helmholtz, Hermann von: Messungen über den zeitlichen Verlauf der Zuckung animalischer Muskeln und die Fortpflanzungsgeschwindigkeit der Reizung in den Nerven (1850). In: *Archiv für Anatomie, Physiologie und wissenschaftliche Medicin*, 276-381.

Helmholtz, Hermann von: Messungen über den zeitlichen Verlauf der Zuckung animalischer Muskeln und die Fortpflanzungsgeschwindigkeit der Reizung in den Nerven (1850a). In: *WA*, 764-843.

Helmholtz, Hermann von: Über die Methoden, kleinste Zeittheile zu messen, und ihre Anwendung für physiologische Zwecke (1850b). In: *WA*, 862-880.

Helmholtz, Hermann von: The Application of the Law of the Conservation of Force to Organic Nature (1861). In: Russell Kahl (ed.): *Selected Writings of Hermann von Helmholtz*. Middleton, Conn.: Wesleyan Univ. Pr., 1971.

Helmholtz, Hermann von: *Wissenschaftliche Abhandlungen* (*WA*). Leipzig: Barth, 1882-95, 3 vols., vol. 1.

Helmholtz, Hermann von: *Vorträge und Reden*. Braunschweig: Vieweg, 1884, 4 vols, vol. 1.

Hills, R. L. and Pacey, A. J.: The Measurement of Power in Early Steam-driven Textile Mills. *Technology and Culture*, 13, 1972, 25-43.

Holmes, Frederic L. and Olesko, Kathryn M.: Precision's Images: Helmholtz and Graphical Methods in Physiology. In: M. Norton Wise (ed.): *The Values of Precision*. Princeton: Princeton Univ. Pr., 1994.

Kant, Horst: Entscheidende Impulse für die Entwicklung der Physik in Berlin. Gustav Magnus zum 175. Geburtstag. *Physik in der Schule*, 15, 1977, 187-191.

Kirsten, Christa, et al. (eds.): *Dokumente einer Freundschaft: Briefwechsel zwischen Hermann von Helmholtz und Emil Du Bois-Reymond 1846-1894*. Berlin: Akademie Verlag, 1986.

Knochenhauer: Spannkräfte des Wasserdampfs. *Repertorium der Physik*, 7, 1846, 221-234.

Kremer, Richard L.: *The Thermodynamics of Life and Experimental Physiology, 1770-1880*. New York/London: Garland, 1990.

Lenoir, Timothy: Social Interests and the Organic Physics of 1847. In: Edna Margalit (ed.): *Science in Reflection*. Dordrecht/Boston/London: Kluwer, 1988, 169-91.

Lenoir, Timothy: Die Entwicklung quantitativer Begriffe in den Experimentalwissenschaften: Physiologie, 1845-1860. In: Poser, Hans (ed.): *Beobachtung und Erfahrung*. Berlin: Technische Universitäts-Verlag, 1992, 177-226.

Liebig, Justus: *Die Thierchemie oder die organische Chemie in ihrer Anwendung auf die Physiologie und Pathologie*. Braunschweig: Vieweg, 1843.

Liebig, Justus: Ueber die thierische Wärme. *Annalen der Chemie und Pharmacie*, 53, 1845, 63-77.

Ludwig, Carl: Beiträge zur Kenntnis des Einflusses der Respirationsbewegungen auf den Blutlauf im Aortensysteme. *Müller's Archiv für Anatomie, Physiologie, und wissenschaftliche Medizin*, 1847, 242-301.

Ludwig, Carl: *Physiologie des Menschens*. Heidelberg: Winter, 1852, 2 vols., vol. 1.

Lundgreen, Peter: *Techniker in Preussen während der frühen Industrialisierung*. Berlin: Colloquium Verlag, 1975.

Mieck, Ilya: *Preussische Gewerbepolitik in Berlin, 1806-1844*. Berlin: de Gruyter, 1965.

Minding, Ferdinand: Ueber die Anwendung des Satzes der lebendigen Kräfte in der Maschinenlehre. *Repertorium der Physik*, 5, 1844, 71-76.

Morin, Arthur: *Notice sur divers appareils dynamométriques propres à mesurer l'effort ou le travail dévelopé par les moteurs animés ou inanimés et par les organes de transmission du movement dans les machines*. Metz: Lamort, 1838.

Olesko, Kathryn M. and Holmes, Frederic L.: Experiment, Quantification, and Discovery. Helmholtz's Early Physiological Researches, 1843-50. In: Cahan 1993: 50-109.

Poisson, Simeon-Denis: *Traité de mécanique*, 2nd ed. Paris: Bachelier, 1833, 2 vols., vol. 2.

Prittwitz: Ueber die Anwendung der natürlichen Kräfte für industrielle Zwecke. *Verhandlungen des Vereins zur Beförderung des Gewerbefleisses*, 9, 1829.

Schubring, Gert: Mathematics and teacher training: Plans for a polytechnic in Berlin. *Historical Studies in the Physical Sciences*, 12, 1981, 161-194.

Siemens, Werner: Ueber Geschwindigkeitsmessungen. *Die Fortschritte der Physik im Jahre 1845*, 1, 1847, 47-72.

Valentin, Gabriel: *Lehrbuch der Physiologie des Menschen*. Braunschweig: 1844, 2 vols., vol. 1.

Warwick, Andrew: Cambridge Mathematics and Cavendish Physics: Cunningham, Campbell, and Einstein's Relativity, 1905-1911. *Studies in the History and Philosophy of Science*, 23, 1992, 625-656.

Weber, Eduard: Muskelbewegung. In: Rudolph Wagner (ed.): *Handwörterbuch der Physiologie mit Rücksicht auf physiologische Pathologie*. Braunschweig: Vieweg, 1842-53, 4 vols., vol. 3.

Wolff, Stefan: Gustav Magnus - Sein Weg zum Ordinariat an der Berliner Universität. Unpublished, 1993.

III
Helmholtz, Erkenntnistheoretiker und Naturphilosoph

Die Hypothetisierung des Mechanismus bei Hermann von Helmholtz

Ein Beitrag zum Wandel der Wissenschafts- und Naturauffassung im 19. Jahrhundert

Gregor Schiemann

Die letzten Jahrzehnte des neuzeitlichen Mechanismus begannen in der Mitte des vergangenen Jahrhunderts mit einer großen Renaissance. Die experimentelle Naturforschung schien sich in vielen ihrer spezialisierten Bereiche endlich dem Ziel zu nähern, alle Phänomene durch Zurückführung auf mechanisch bewegte Materie zu erklären. In den 70er Jahren jedoch trat dann insbesondere im Zusammenhang mit weiteren Umsetzungsversuchen des mechanistischen Erklärungsprogramms in der Elektro- und Thermodynamik seine begrenzte Reichweite deutlicher zutage. Die nie gänzlich verstummte Kritik am Mechanismus nahm in den natur- und wissenschaftstheoretisch interessierten Kreisen wieder zu, suchte nach alternativen Erklärungssystemen oder propagierte den vollständigen Verzicht auf einen naturwissenschaftlichen Erklärungsanspruch.[1]

In den Strudel einer letalen Krise gerieten aber nicht nur die Inhalte des mechanistischen Erklärungsprogramms, sondern auch die dabei erhobenen tra-

1 Exemplarisch für die Kritik am mechanistischen Erklärungsprogramm sind Mach 1872, Kirchhoff 1897, Helm 1898.

ditionellen Geltungsansprüche der Naturwissenschaften. Fragwürdig wurde, ob wissenschaftliche Erkenntnis die seit der Antike behauptete alleinige, für immer gültige und vollständige Erkenntnis der Natur sein könne.[2] Während im deutschen Idealismus dieser Wahrheitsanspruch noch einmal mit großem Einfluß auch auf die Naturwissenschaften formuliert wurde und erst mit dem Niedergang der Hegelschen Systemphilosophie an Boden verlor, war seine Kritik bereits vor dem 19. Jahrhundert durch den englischen Empirismus, namentlich durch das wissenschaftskritische Werk von David Hume, vorbereitet worden. Im Kontext der Auseinandersetzung um den Mechanismus verband sich in der zweiten Hälfte des 19. Jahrhunderts die Kritik an absoluten Geltungsansprüchen wissenschaftlicher Erkenntnis zum einen - wie beispielsweise bei Pierre Duhem und Henri Poincaré - mit einer Ablehnung des mechanistischen Erklärungsprogramms.[3] Zum anderen wurde die Relativierung von Geltungsansprüchen aber auch von den Vertretern des Mechanismus selbst betrieben. Neben Emil du Bois-Reymond hat hierfür Hermann von Helmholtz das bedeutendste Beispiel gegeben.[4]

In der ersten Zeit seiner wissenschaftlichen Karriere, d.h. vom Ende der 40er bis etwa zum Ende der 60er Jahre, trägt Helmholtz seine mechanistische Naturauffassung noch mit absolutem Geltungsanspruch vor. Im Verlauf der 70er Jahre wird dann ein Wandlungsprozeß auffällig, der sich am deutlichsten in seinen öffentlichen Reden nachweisen läßt. Die mit dem Mechanismus verbundenen Geltungsansprüche werden einer Hypothetisierung unterworfen, wobei Helmholtz allerdings an den Grundaussagen seiner Naturauffassung im wesentlichen unbeirrt festhält. Er korrigiert zwar seine Aussagen über die elementaren Kräfte und modifiziert auch seine Ansichten über die elementare Struktur der Materie, bleibt aber unverändert seiner Auffassung treu, daß die Naturforschung insgesamt danach streben solle, die Phänomene auf mechanisch bewegte Substanz zurückzuführen. Was die mechanistische Reduktion nicht mehr aufrechterhalten kann, ist der Anspruch, eine endgültige Erklärung aller Phänomene zu erreichen. Sie hat einen unaufhebbar hypothetischen Cha-

2 Zum Wandel der Wissenschaftsauffassung im 19. Jahrhundert vgl. Diemer 1968 und Schnädelbach 1983.

3 Duhem 1978, Poincaré 1974.

4 Du Bois-Reymond 1974. Der Wandel der Mechanikauffassung von C.G.J. Jacobi, den Pulte 1993 untersucht hat, stellt ein weiteres Beispiel der Relativierung von Geltungsansprüchen innerhalb der Tradition des Mechanismus dar.

rakter gewonnen, und ihre vollständige Umsetzung kann lediglich noch als ein Ideal, als ein regulatives Prinzip der Forschung betrachtet werden.[5]

Die Entwicklung von Helmholtz' Mechanismus ist durch diesen Wandel im Geltungsanspruch gekennzeichnet und läßt sich - in einer noch sehr groben Übersicht - in zwei Perioden einteilen. Auf die erste Periode bis etwa zum Ende der 60er Jahre werde ich im ersten Teil meines Beitrages eingehen. Hier rekonstruiere ich umrißhaft die empiristische Begründung, die Helmholtz für den Wahrheitsanspruch seiner Naturauffassung gegeben hat. Im zweiten Teil werde ich dann die wichtigsten Merkmale der im Verlauf der 70er Jahre hervortretenden Hypothetisierungstendenz charakterisieren. Abschliessend will ich in einem dritten Teil zeigen, wie Helmholtz auf der Grundlage seines gewandelten Wissenschaftsverständnisses das Programm einer mechanistischen Naturerklärung mit großem wirkungsgeschichtlichen Einfluß weiter verfolgt hat.

1. Die Begründung des dualen Mechanismus

Mit seinen physiologischen, physikalischen und geometrischen Untersuchungen und mit öffentlichen Reden war Helmholtz maßgeblich daran beteiligt, den letzten Siegeszug des Mechanismus auf den Weg zu bringen.

Die von ihm selbst gegebene Begründung des Mechanismus ist von Anfang an einem Wandlungsprozeß unterworfen, der die grundlegenden Begriffe der Materie und der Kraft, der Ursache und des Gesetzes ebenso betrifft wie das Verhältnis der Geltung von wissenschaftlicher Erkenntnis und lebensweltlichen Wahrnehmungen. Von diesen Veränderungen weitgehend unberührt bleiben bis gegen Ende der 60er Jahre jedoch der (von Helmholtz teilweise

---

5 Veränderungen von Helmholtz' Wissenschafts- und Naturauffassung sind zwar in der Sekundärliteratur wiederholt, jedoch meist nur andeutungsweise, bemerkt worden. Insgesamt überwiegt die Einschätzung, daß den Elementen einer wissenschafts- und naturtheoretischen Veränderung gegenüber einem sich durchhaltenden Standpunkt eine eher untergeordnete Bedeutung zukomme: So z.B. bei Koenigsberger 1895, Conrat 1904, Erdmann 1921, Hörz und Wollgast 1971, Heimann 1974, Elkana 1974, Winters 1985. Nur König 1968 und Heidelberger 1993 messen dem Wandlungsprozeß von Helmholtz' Wissenschafts- und Naturauffassung und der damit verbundenen Hypothetisierungstendenz, die bereits von Erdmann 1921: 10 f. bemerkt wurde, größere Bedeutung bei.

selbst bestrittene)[6] empirische Bezug und die grundlegenden inhaltlichen Bestimmungen der Begründung. Beides geht aus den Texten hervor, in denen sich Helmholtz zu seiner mechanistischen Naturauffassung äußert. Bei diesen Texten handelt es sich einerseits um die Einleitung zu seiner berühmten Schrift *Ueber die Erhaltung der Kraft* von 1847 und andererseits um zahlreiche seiner zu Lebzeiten vielbeachteten Reden, in denen Helmholtz Ergebnisse naturwissenschaftlicher Untersuchungen allgemeinverständlich darstellt, gegen andere Ansätze abgrenzt und allgemein zu den Zielen und Methoden der Wissenschaften Stellung nimmt. Während sein Mechanismus in den meisten Reden im Hintergrund bleibt,[7] nimmt er in der Einleitung und in der Rede vor der Naturforscherversammlung in Innsbruck im Jahr 1869 ("Naturforscherrede") explizit und umfassend zu seiner Naturauffassung Stellung. Diese beiden Texte bilden den Anfangs- und Schlußpunkt der ersten, relativ homogenen Periode in der Entwicklung seines Mechanismus.[8] Um ihre strukturelle Gemeinsamkeit in der Begründung des absoluten Geltungsanspruches zu rekonstruieren, werde ich sie parallel besprechen.

Gemeinsam sind beiden Texten zwei Voraussetzungen, die Helmholtz der Begründung seines Mechanismus voranstellt. Die erste Voraussetzung betrifft das allgemeine Ziel der wissenschaftlichen Erkenntnis, das Helmholtz aus dem "Gesetze der Causalität" ableitet und das er 1847 als das "endliche Ziel der theoretischen Naturwissenschaften" formuliert, "die letzten unveränderlichen Ursachen der Vorgänge in der Natur aufzufinden" (Helmholtz 1882 ff., Bd. 1: 13). 22 Jahre später mißt er dementsprechend den naturwissenschaftlichen Fortschritt "nach dem Maasse, in welchem die [...] Kenntniss eines alle Naturerscheinungen umfassenden ursächlichen Zusammenhanges fortgeschritten ist" (Helmholtz 1903, Bd. 1: 377). In beiden Reden zweifelt er nicht daran, daß die Naturforschung das Endziel objektiver Wahrheit über die Natur tatsächlich erreichen kann. Mit Blick auf die bisherige Entwicklung der neuzeitlichen Wissenschaft, die sich ihm zufolge insgesamt auf die Lösung ihrer einen Aufgabe zubewegt, glaubt Helmholtz in der Naturforscherrede, ein "befriedigendes Facit" ziehen zu können (a.a.O.: 395). Überdies stellt er klar, daß

6 Helmholtz 1856 ff.: 454; Helmholtz 1882 ff., Bd. 1: 68 f.

7 Z.B.: Helmholtz 1903, Bd. 1: 40, 45, 64, 178 und 217 f.; Helmholtz 1882 ff., Bd. 3: 569.

8 Die dieser Periode vorgelagerte Entwicklungsphase, in der Helmholtz erst andeutungsweise mechanistische Positionen vertritt, bleibt hier unberücksichtigt.

der Fortschritt "nicht [...] auf speculativem Wege", sondern nur in Abhängigkeit von der experimentellen Erfahrung vorankommen werde (a.a.O.: 375).

Obwohl Helmholtz wiederholt von der apriorischen Geltung der Kausalität gesprochen hat,[9] gilt der Erfahrungsbezug auch für die Reichweite der kausalen Erforschung der Natur, die auf doppelte Weise empirisch begrenzt ist: Zum einen ist sie beendet, wenn sie, auf Erfahrung gestützt, "letzte Ursachen" findet. Zum anderen ist Helmholtz überzeugt, daß sich geistige Phänomene und willentliche Handlungen nie vollständig kausal erfassen lassen.[10] Seine Naturauffassung ist gleichsam halbiert. Sie bezieht sich nur auf die geistlosen Naturerscheinungen und beansprucht, empirisch den Umfang dieser Erscheinungen und ihre kausale Erklärung aus "letzten Ursachen" angeben zu können.

Die zweite Voraussetzung ist in der Einleitung explizit und in der Naturforscherrede implizit enthalten: Helmholtz setzt die Begriffe der Materie und der Kraft als Grundbegriffe seiner Wissenschafts- und Naturauffassung. "Ihrem bloßen Dasein nach" bezeichne die Wissenschaft die "Gegenstände der Aussenwelt" als Materie; deren "Vermögen Wirkungen auszuüben" bezeichne sie hingegen als Kraft (Helmholtz 1882 ff., Bd. 1: 14). Während Helmholtz in der Einleitung bereits bei dieser Gelegenheit der Materie Massenerhaltung, Trägheit und räumliche Bewegung (als einzig mögliche Veränderung) zuschreibt, ist der Kraftbegriff noch so weit gefaßt, daß jedes Wirkende Kraft heißen kann.[11] Für die Bestimmung des Mechanismus ist damit eine duale Konzeption vorgezeichnet, in der sich weder die Eigenschaften der Materie vollständig auf Kräfte, noch die Eigenschaften der Kräfte vollständig auf Materie reduzieren lassen. Ob Helmholtz diese Vorentscheidung zu Recht nicht als Folgerung des Kausalgesetzes versteht, muß offenbleiben. Strenggenommen müßte er Materie und Kraft gemeinsam als Ursachen des Naturge-

9 Erstmals in Helmholtz 1903, Bd. 1: 116, später in Helmholtz 1856 ff.: 453; Helmholtz 1903, Bd. 2: 243; Helmholtz 1897 ff., Bd. I.1: 17.

10 Helmholtz 1882 ff., Bd. 1: 13. Helmholtz 1903, Bd. 1: 389 f. steht in dieser Hinsicht nicht in Widerspruch zu: Helmholtz 1903, Bd. 1: 37 und 171; Helmholtz 1856 ff.: 454; Helmholtz 1903, Bd. 2: 89 und 419.

11 In den weiteren Abschnitten der Schrift meint Helmholtz mit dem Begriff der Kraft außer der mechanischen Kraft vor allem die heute als Energie bezeichnete physikalische Größe. Zur Vagheit beider Bedeutungen, die gleichwohl gut auseinanderzuhalten sind, vgl. Elkana 1974.

schehens bezeichnen. Er bezieht jedoch lebenslang den Begriff der Ursache meist einseitig auf den Kraftbegriff.[12]

Die Begründung der mechanistischen Naturauffassung setzt nun in beiden Texten am Begriff der Materie an: Helmholtz macht die grundlegende Annahme, daß die (wägbare) Materie in diskrete Elemente zerlegbar sei. Er entwickelt diese Annahme sowohl in der Einleitung als auch in der Naturforscherrede mit ausdrücklichem Bezug auf die experimentelle Forschung, wobei er sich in beiden Texten auf die Erkenntnisse der Chemie stützt. Um sich in der Einleitung "das Weltall zerlegt in Elemente" zu denken, beruft er sich auf die "(chemische[n]) Elemente" (Helmholtz 1882 ff., Bd. 1: 15). In der Naturforscherrede wird dieser rechtfertigende Erfahrungsbezug dann zum zentralen Argument der Begründung. Hier glaubt er, alle Bestimmungen der Materie unmittelbar der Chemie entnehmen zu können. Die Chemie habe "*erwiesen*", erklärt er, daß die von ihr entdeckten Elemente "wirklich unzerstörbar" seien und "alle Masse aus den von ihr gefundenen Elementen zusammengesetzt" sei (Helmholtz 1903, Bd. 1: 378 - Hervorheb. von mir). Elemente sind für ihn also nicht nur durch ihre chemischen Qualitäten und durch die Verhältnisse, in denen sie sich untereinander verbinden, bestimmt, sondern werden darüber hinaus als physikalische Objekte, als faktisch vorhandene Atome verstanden.

Was Helmholtz als erfahrungsfundierte Aussagen der Naturforschung ausgibt, hat für ihn in den 60er Jahren absolute Geltung. Die Naturwissenschaft könne mit dem ihr eigenen induktiven Verfahren zu Begriffen gelangen, in denen das Wesen eines Vorganges "klar vor Augen" liege, und allgemeine Gesetze von "unbedingter Gültigkeit" aufstellen, in denen sich die "Form des logischen Schliessens" vollende (Helmholtz 1903, Bd. 1: 170, 175 und 171). Unter Gesetzen versteht Helmholtz hierbei - knapp gesagt - mathematisch erfaßbare Kausalverhältnisse von Naturveränderungen.[13]

Daß zwischen den Elementen der Materie überhaupt Kräfte wirken müssen, ergibt sich aus dem dualen Charakter seiner Naturauffassung. Aber erst aus dem elementaristischen Aufbau der Materie schließt Helmholtz auf die Eigenschaften dieser Wechselwirkungen: Weil die Elemente allein ihre räumliche Position verändern, können die zwischen ihnen wirksamen Kräfte auch nur "Bewegungskräfte" sein.[14] Ohne bereits auf die Struktur dieser Kräfte ein-

---

12 Helmholtz 1882 ff., Bd. 1: 14 und 68; Helmholtz 1903, Bd. 1: 40 und 377; Helmholtz 1856 ff.: 454 f.

13 Helmholtz 1903, Bd. 1: 169 f. und 374 ff.; Helmholtz 1856 ff.: 454 f. u.ö.

14 Helmholtz 1882 ff., Bd. 1: 15. Entsprechend: Helmholtz 1903, Bd. 1: 379.

zugehen, benennt Helmholtz jetzt erst in beiden Texten die mechanistische Zielsetzung: Zurückführung der "Naturerscheinungen [...] auf Bewegungen von Materien *mit* unveränderlichen Bewegungskräften" (Helmholtz 1882 ff., Bd. 1: 15 - Hervorheb. von mir) oder, gleichbedeutend: "[...] das Endziel der Naturwissenschaft ist, [...] sich in Mechanik aufzulösen." (Helmholtz 1903, Bd. 1: 379).

Die letzte Formulierung macht deutlich, daß er in der physikalischen Disziplin der Mechanik diejenige Wissenschaft sieht, die allein für die nähere Bestimmung der Bewegungskräfte zuständig ist. Die "reinste Form des Ausdruckes der Kraft", heißt es in der Naturforscherrede, sei die "mechanische Kraft", deren Wirkung man in Bereichen von "verschwindend kleinen Massentheilchen" "aus beobachteten Erscheinungen [...] *herauszulesen*" vermöge (Helmholtz 1903, Bd. 1: 376 - Hervorheb. von mir). Nicht nur die Struktur der Materie, sondern auch die der elementaren Kräfte soll mit Gewißheit der induktiv verfahrenden Wissenschaft - in diesem Fall: der Mechanik - entnommen werden. Die elementaren Kräfte sollen deshalb gemäß dem zweiten Axiom von Newtons Mechanik nur Ursache von Bewegungsveränderungen sein. Im Zusammenhang mit seinen Arbeiten zum Energieerhaltungssatz setzt Helmholtz ihre Struktur darüber hinaus mit derjenigen der Gravitationskraft gleich und bestimmt sie als entfernungsabhängige Zentralkräfte, die sich nur in ihrer Stärke voneinander unterscheiden.[15]

Zusammenfassend können als die grundlegenden Bestimmungen von Helmholtz' Mechanismus in ontologischer Hinsicht die Dualität von trägen Elementen und bewegungsverändernden Kräften und in methodischer Hinsicht die Newton-Lagrange-Formulierung der Mechanik angesehen werden. Den Wahrheitsanspruch seiner Naturauffassung begründet Helmholtz nicht vor aller Erfahrung, sondern im Bezug auf Erfahrung, die die einzige von ihm anerkannte Geltungsbasis naturwissenschaftlicher Aussagen ist. Damit kommt die hier angedeutete Rekonstruktion zu einem Ergebnis, das auch für die Bewertung des Verhältnisses von Helmholtz und Kant, welches in der Vergangenheit ebenso häufig wie gegensätzlich beurteilt wurde,[16] nicht folgenlos bleiben

15 Helmholtz 1882 ff., Bd. 1: 15 ff.; in der Rede von 1869 ist die Festlegung auf die Zentralkraftstruktur in der mechanischen Deutung der Energieerhaltung impliziert: Helmholtz 1903, Bd. 1: 379 ff.

16 Einige wenige Titel sind: Krause 1876, Schwertschlager 1883, Cassirer 1973: 92 ff., Heimann 1974 und Fullinwider 1990.

dürfte. Helmholtz' späte Erklärung, "die philosophischen Erörterungen der Einleitung [...] [seien] durch Kant's erkenntnistheoretische Ansichten [...] beeinflußt" (Helmholtz 1882 ff., Bd. 1: 68), findet zumindest durch die im Text eben dieser Einleitung vorgenommene Begründung des Mechanismus keine überzeugende Bestätigung.

Im Unterschied zu Kant vertritt Helmholtz keinen dynamischen (allein auf Kräften aufbauenden), sondern einen dualen Mechanismus, und er gibt dafür in der Hauptsache keine (im Sinne von Kant) metaphysische, sondern eine empiristische Begründung, die jener allerdings im Wahrheitsanspruch keinesfalls nachzustehen gedenkt.[17] Für seinen Empirismus ist es im Zeitraum zwischen dem Ende der 40er und etwa dem Ende der 60er Jahre charakteristisch, daß er die Objektivität der wissenschaftlichen Erkenntnis streng gegen das bloß subjektive Zeugnis der unmittelbaren Sinneswahrnehmung abgrenzt. Das Verhältnis beider thematisiert er im Zusammenhang mit seiner Wahrnehmungstheorie, die er in den 50er und 60er Jahren entwickelt und die weniger für die Begründung des Wahrheitsanspruches des Mechanismus als für seine spätere Relativierung in wissenschaftstheoretischer Hinsicht relevant ist.[18] Anfänglich läßt er zwischen wissenschaftlicher Erkenntnis und dem Zeugnis der Sinneswahrnehmung keinerlei Gemeinsamkeit zu.[19] Erst im *Handbuch der physiologischen Optik* postuliert er 1867 eine Eigenschaft der Sinneswahrnehmung, mit der sie dem Wahrheitsstreben der Wissenschaft uneingeschränkt als empirisches Material zur Verfügung steht: Die Wahrnehmung ereigne sich gleichzeitig mit dem Wahrgenommenen und bilde deshalb die Zeitfolge des Wahrgenommenen ab.[20] Wissenschaft kann jetzt nicht nur, sondern muß auch, wie es später heißt, allein von "beobachtete[n] Thatsachen" (Helmholtz 1903, Bd. 2: 352) ausgehen.

---

17 Zu Kants metaphysischer Grundlegung eines dynamischen Mechanismus vgl. sein *Metaphysische Anfangsgründe der Naturwissenschaft* (1786), in: Kant 1902 ff., Bd. 4.

18 Von der umfangreichen Sekundärliteratur zu Helmholtz' Wahrnehmungstheorie sind zu nennen: Conrat 1904, Hatfield 1990 und Turner 1993.

19 Helmholtz 1882 ff., Bd. 2: 608 f.; Helmholtz 1903, Bd. 1: 40 f. und 171 f.

20 Helmholtz 1856 ff.: 445. Entsprechend nur in: Helmholtz 1903, Bd. 1: 393.

2. Die Hypothetisierung der Wissenschafts- und Naturauffassung

Die in den 70er Jahren hervortretende Relativierung des Geltungsanspruches der mechanistischen Naturauffassung kann als Anwendung bereits vorliegender Bestimmungen der Wahrnehmungstheorie auf die wissenschaftliche Erkenntnis verstanden werden. Zu den kennzeichnenden Merkmalen der Helmholtzschen Wahrnehmungstheorie gehört, daß sie dem qualitativen Inhalt der Sinneswahrnehmung im Gegensatz zum Abbildcharakter der zeitlichen Struktur nur eine relative Geltung zumißt, die von der erfolgreichen Verarbeitung der jeweiligen Wahrnehmung abhängt. Nachdem Helmholtz bereits den Wirklichkeitsbezug wissenschaftlicher Erkenntnis auf Wahrnehmbares eingeschränkt hat, unterwirft er in der Folgezeit auch die wissenschaftliche Zielsetzung und Methodik zunehmend diesen schwankenden Geltungsbedingungen. Die Neuorientierung tangiert alle Bestimmungen seiner Wissenschaftsauffassung. Ihren ersten und auch deutlichsten Ausdruck findet sie aber in der auf wissenschaftlicher Erkenntnis basierenden Gesamtdeutung der Natur, wie Helmholtz sie in seinen öffentlichen Reden vertritt.

Gerade zwei Jahre, nachdem er an prominentester Stelle, auf dem Podium der Naturforscherversammlung, dem Atomismus gehuldigt hat, wendet sich Helmholtz in seiner Rede "Zum Gedächtniss an Gustav Magnus" "gegen das Streben, aus rein hypothetischen Annahmen über [den] Atombau der Naturkörper die Grundlagen der theoretischen Physik herzuleiten" (Helmholtz 1903, Bd. 2: 45). Während Helmholtz hier den Ausdruck "hypothetisch" noch abwertend gebraucht und auf noch nicht gefundene unsichtbare Entitäten bezieht, führt er wenig später den Begriff der Hypothese erstmals mit positivem Sinn in seine Wissenschaftsauffassung ein. "Jede berechtigte Hypothese", schreibt er 1874 in einer Vorrede zu dem von ihm mitübersetzten *Handbuch der theoretischen Physik* der englischen Physiker W. Thomson und P. G. Tait, "ist der Versuch, ein neues allgemeineres Gesetz aufzustellen" (Helmholtz 1903, Bd. 2: 416). Aus der Induktion gehen jetzt nicht mehr unmittelbar Gesetze, sondern Hypothesen hervor, aus denen Gesetze erst durch nachträgliche empirische Verifikation gewonnen werden.[21] 1877, in der Rede "Das Denken in der Medicin", tritt die Verbindung des Wandels der Wissenschaftsauffassung mit

21 Helmholtz 1903, Bd. 2: 414. Vgl. a. Helmholtz 1897 ff., Bd. I.1: 18 f.

der Krise des Mechanismus bestimmter hervor. Einerseits gesteht Helmholtz dort ein, daß man bei einer "ursächliche[n] Erklärung" "nicht immer zurückgehn [könne] auf die Kräfte der Atome" (Helmholtz 1877: 31 f.).[22] Genau diese Reduktion hatte er jedoch bisher für möglich und erforderlich gehalten. Anderseits erklärt er die Verifikation von Hypothesen zu einer Arbeit, die "eigentlich nie" aufhöre und "nie zur unbedingten Wahrheit" führe (Helmholtz 1877: 26 f.). Wenn aber die Verifikation nicht abschließbar ist, erhalten Gesetze selbst hypothetischen Charakter. In diesem Zusammenhang gelangt Helmholtz zu einem Wissenschaftsverständnis, das bereits die Grundüberzeugung des kritischen Rationalismus, wie ihn Karl R. Popper begründen wird,[23] vorwegnimmt: Die von Helmholtz wörtlich so genannte "Infallibilität" wird für ihn zum Kennzeichen einer falschen Wissenschaftlichkeit (Helmholtz 1877: 15).[24]

Bei Helmholtz läuft die Hypothetisierung von Gesetzesaussagen noch nicht der Aufgabe der Wissenschaft zuwider, nach einer allein gültigen Naturerkenntnis zu streben. Diese Aufgabenstellung wird erst durch weitere Schritte des Hypothetisierungsprozesses fragwürdig. Neben geometrischen Überlegungen zur äquivalenten Darstellbarkeit des mathematischen Raumes[25] gehört dazu die Annahme, daß unverzichtbare erkenntnistheoretische Voraussetzungen eine unbegründbare Wahl zwischen entgegengesetzten Positionen implizieren. In diesem Sinn stellt er in seiner wohl bedeutendsten Rede "Ueber die Thatsachen in der Wahrnehmung" von 1878 an zentraler Stelle seiner eigenen realistischen Position die idealistische als ebenso unwiderlegbare und deshalb gleichberechtigte gegenüber und bezeichnet beide als "metaphysische Hypothesen" mit "vollkommene[r] wissenschaftliche[r] Berechtigung" (Helmholtz 1903, Bd. 2: 239).

Von wissenschaftlichen Hypothesen kann bei Helmholtz also in dreifachem Sinn die Rede sein: Im Sinn von noch nicht gefundenen oder fiktiv angenommenen unsichtbaren Entitäten, von nicht hinreichend verifizierten oder nicht vollständig verifizierbaren Gesetzen und schließlich im Sinn von unwiderlegbaren erkenntnistheoretischen Voraussetzungen.

22 Eine "ursächliche Erklärung" geben, heißt an dieser Stelle, ein Gesetz und die Bedingungen seiner Wirksamkeit angeben.

23 Popper 1989: 14 ff.

24 Vgl. entsprechend: Helmholtz 1903, Bd. 2: 239; Helmholtz 1882 ff., Bd. 2: 642.

25 Helmholtz 1903, Bd. 2: 24 ff. und 230 ff.

Folgt man den öffentlichen Reden, so kann man insgesamt von einer Hypothetisierung der Wissenschafts- und Naturauffassung sprechen. Das Spektrum der zu berücksichtigenden Bedingungen und möglichen Ursachen dieses Prozesses ist vielfältiger Art und kann hier nur sehr fragmentarisch angesprochen werden. Zunächst folgt Helmholtz der in der europäischen Wissenschaft der zweiten Hälfte des 19. Jahrhunderts durchgreifenden Tendenz zur Hypothetisierung von Geltungsansprüchen, die als Teil einer allgemeinen kulturhistorischen Umwälzung verständlich gemacht werden kann.[26] Helmholtz scheint hierbei insbesondere durch den englischen Empirismus und durch die Wissenschaftsauffassungen englischer Physiker beeinflußt zu sein. In John Stuart Mills Wissenschaftstheorie findet er das Wesen induktiver Schlüsse "am besten auseinandergesetzt" (Helmholtz 1856 ff.: 447), und den Ansichten von Michael Faraday, James Clerk Maxwell und William Thomson mißt er im Verlauf seiner Karriere wachsende Bedeutung zu.[27] Als ein wesentlicher Faktor im Rahmen seiner eigenen Forschungen ist vor allem der Ausbau seiner Wahrnehmungstheorie zu einer umfassenden Erkenntnistheorie zu berücksichtigen. Nicht auszuschließen ist im weiteren, daß der Wandlungsprozeß der Wissenschaftsauffassung durch die Krise seines Mechanismus maßgeblich angeregt ist. In dieser Krise spiegelt sich vermutlich ebenso eine Reaktion auf die eingangs erwähnte inhaltliche Kritik wie Helmholtz' Bemühen, sich öffentlich von materialistischen Weltanschauungen zu distanzieren.[28] Das Hervortreten der Krise seines Mechanismus fällt ferner beruflich mit seiner stärkeren Hinwendung zu physikalischen Problemen zusammen.[29] Schließlich schwächt der Geltungsverlust der wissenschaftlichen Erkenntnis rückwirkend aber auch den Wahrheitsanspruch des Mechanismus. Dessen gewandelten Stellenwert begründet Helmholtz jetzt wahrnehmungstheoretisch. In ebenso

---

26 Vgl. Anm. 2.

27 Nur die wichtigsten Hinweise können hier gegeben werden: Von seinen seit 1853 unternommenen Reisen nach England berichtet Koenigsberger 1902 f.; 1870 gibt Helmholtz ein Buch von John Tyndall über Faraday heraus, 1871 und 1874 den ersten und zweiten Teil des 1. Bandes des Handbuches von Thomson und Tait (s.o.), zu Maxwell vgl. insb. Helmholtz 1903, Bd. 2: 249 ff. und Helmholtz 1882 ff., Bd. 3: 209. Auf den Einfluß englischer Physiker auf den Wandlungsprozeß von Helmholtz' Wissenschafts- und Naturauffassung hat insbesondere Heidelberger 1993 hingewiesen.

28 Vgl. die erst in die zweite Auflage (1878) der Rede "Das Denken in der Medicin" aufgenommene Distanzierung vom Materialismus: Helmholtz 1903, Bd. 2: 186 f.

29 1871 beendet Helmholtz seine Tätigkeit als Professor für Physiologie in Heidelberg und wird Professor für Physik in Berlin.

pathetischen Worten, wie er ursprünglich die objektive Wahrheit des Mechanismus in Aussicht gestellt hatte, distanziert er sich 1878 grundsätzlich von allen reduktionistischen Absolutheitsansprüchen:

> "Jede Zurückführung der Erscheinungen auf die zu Grunde liegenden Substanzen und Kräfte behauptet etwas Unveränderliches und Abschliessendes gefunden zu haben. Zu einer unbedingten Behauptung dieser Art sind wir nie berechtigt; das erlaubt weder die Lückenhaftigkeit unseres Wissens, noch die Natur der Inductionsschlüsse, auf denen all unsere Wahrnehmung des Wirklichen vom ersten Schritte an beruht." (Helmholtz 1903, Bd. 2: 243)

Das einzige prinzipielle Argument, das Helmholtz hier für die Ablehnung der unbedingten Geltung gibt, "die Natur der Inductionsschlüsse", ist seiner Wahrnehmungstheorie aus den 60er Jahren entnommen, deren Anwendung auf die Wissenschaftsauffassung ich an dieser Stelle exemplarisch vorführen möchte. Mit dem Ausdruck "Inductionsschluss" und seinen Synonymen bezeichnet Helmholtz sehr verschiedene Vorgänge,[30] die jeweils auf ihre Weise zur Relativierung der wissenschaftlichen Erkenntnis beitragen.

Von größter Bedeutung ist der Vorgang, der den grundlegenden Prozeß der Wahrnehmungsentstehung und -verarbeitung in lebenden Organismen betrifft.[31] Nach Helmholtz kommen auf Empfindungen bezogene Wahrnehmungs- und Handlungsweisen in einem naturwüchsigen Lernvorgang induktiv zustande: Ähnliche Empfindungseinwirkungen und ihre erfolgreichen Verarbeitungen werden beständig im Gedächtnis gespeichert und, ohne daß sich das betreffende Individuum in der Regel dessen bewußt wäre, zu Wahrnehmungs- und Reaktionsregeln verallgemeinert. Tritt eine vergleichbare Empfindung wieder auf, bestimmen diese Regeln, nun ausnahmslos unbewußt, die dann entstehenden Wahrnehmungen und Reaktionen. Es handelt sich um die Ausbildung und Anwendung eines psychologischen Schemas, dessen Geltung Helmholtz nur am jeweiligen Erfolg mißt und eben deshalb weder als unveränderlich noch abschließend einschätzt.[32]

Was sich hier am Beispiel der Geltung von Reduktionen zeigt, ließe sich in vergleichbarer Weise auch an den Grundbegriffen von Helmholtz' Wissenschaftsauffassung (insbesondere dem Kraft-, Gesetzes- und Raumbegriff) nachweisen: Elementare Wahrnehmungsprozesse garantieren und begrenzen

30 Conrat 1904: 92 ff. Vgl. Helmholtz 1903, Bd. 1: 112 ff.; Helmholtz 1856 ff.: 430, 449 und 453 f.; Helmholtz 1903, Bd. 2: 226, 233, 243 f. und Anm. 31.

31 Helmholtz 1856 ff.: 447 und 449; Helmholtz 1903, Bd. 1: 358 ff.

32 Helmholtz 1856 ff.: 442 f; Helmholtz 1885 ff: 602 f.

zugleich - fast schon im Sinn eines modernen Lebenswelt-Aprioris - die Geltungsbedingungen der wissenschaftlichen Naturerkenntnis. Weil schon alltägliche Wahrnehmungen nur relativ zu ihrem niemals vollständig gesicherten Erfolg gelten, können die "Substanzen und Kräfte", die Helmholtz den Naturphänomenen unverändert zugrunde legt, nicht mehr mit Unbedingtheit angenommen werden.

Aber welchen Stellenwert hat das Zugrundeliegende, wenn reduktionistische Erklärungen weiterhin zugelassen sein sollen? Helmholtz gibt hierauf, ebenfalls in der Rede "Die Thatsachen in der Wahrnehmung", eine Antwort, in der er sich für die Fortführung eines, allerdings nicht mehr notwendig mechanistischen Reduktionsprogramms ausspricht:

> "Setzen wir [...] voraus, dass das Begreifen [der Naturerscheinungen - G.S.] zu vollenden sein wird, dass wir ein letztes Unveränderliches als *Ursache* der beobachteten Veränderungen werden hinstellen können, so nennen wir das [das - G.S.] regulative Prinzip unseres Denkens, was uns dazu treibt, das *Causalgesetz.* Wir können sagen, es spricht das Vertrauen auf die *vollkommene Begreifbarkeit* der Welt aus. Das Begreifen [...] ist die Methode, mittelst deren unser Denken die Welt sich unterwirft [...] ." (Helmholtz 1903, Bd. 2: 243 - Hervorheb. im Text)

Helmholtz hält am Ziel einer vollständigen Naturerklärung fest, rückt es aber in eine so unerreichbare Ferne, daß das "Begreifen" zu einer gegenüber Inhalten gleichgültigen methodischen Aufgabe geworden ist. Ein regulatives Prinzip kann auch kein Objekt in der Erfahrung bedingen. Es gibt stattdessen eine heuristische Regel vor, die nie zu einer absoluten Grenze führt. Es ist dies das letzte Mal, daß Helmholtz von der "Vollendung" des Begreifens spricht. In den nachfolgenden Reden scheint sich seine Distanzierung von einer endlichen Zielsetzung zu vergrößern, er bleibt aber insgesamt unentschieden.[33]

Für die Äußerungen, mit denen Helmholtz die Hypothetisierung seiner Wissenschafts- und Naturauffassung an den zentralen Stellen seiner Reden in den 70er Jahren vornimmt und von denen ich einige typische angeführt habe, findet sich Vergleichbares im Zeitraum davor nur in seinen wahrnehmungstheoretischen Arbeiten.[34] Diese stellen nicht so sehr schon eine Reflexion auf wis-

33 Vgl. insb. Helmholtz 1903, Bd. 2: 358; Helmholtz 1897 ff., Bd. I.1.: 17 f.

34 So findet der in der ersten Auflage des *Handbuches der physiologischen Optik* auf Wahrnehmungen bezogene relative Wahrheitsbegriff (Helmholtz 1856 ff.: 443) in der zweiten Auflage auch auf die wissenschaftliche Erkenntnis Anwendung (Helmholtz 1885 ff.: 590).

senschaftliche Erkenntnis dar als vielmehr eine nichtreflexive Thematisierung lebensweltlicher Wahrnehmungsvorgänge. Meine Rekonstruktion des Wandlungsprozesses ging davon aus, daß Helmholtz seine Wahrnehmungstheorie aus den 60er Jahren höchstens ansatzweise als Erkenntnistheorie verstanden hat.[35] Er baut sie erst im Verlauf der 70er Jahre zu einer solchen aus und kann mit ihr die mittlerweile eingetretene Geltungsrelativierung begründen.

Trotz des Einschnittes in Helmholtz' öffentlichen Stellungnahmen zu Beginn der 70er Jahre vollzieht sich der wahrnehmungstheoretisch vorbereitete Wandlungsprozeß allmählich und nicht immer kontinuierlich. Altes und Neues stehen teilweise unverbunden nebeneinander.[36] Daß sich die Konturen des Neuen erst langsam abzuzeichnen beginnen, ist Ausdruck der historischen Dimension, die der von Helmholtz mitgetragene Wandel hat: Er bricht mit der bis in die Antike zurückreichenden Tradition der Wahrheitssuche, die sich doch so nah schon an ihrem Ziel wähnte, und bereitet den modernen Prozeß der Wahrheitsrelativierung vor, der sich bis in die Gegenwart hinein fortgesetzt hat.

3. Mechanistische Analogien und mathematische Vereinheitlichung

Nachdem es für Helmholtz fragwürdig geworden ist, ob sich die Materie aus unveränderlichen Atomen zusammensetzt und ob die elementaren Wechselwirkungen nur aus instantan wirkenden mechanischen Zentralkräften bestehen,[37] gibt es für ihn, grob gesprochen, nur noch eine Möglichkeit, an den

35 Helmholtz selbst bewertet seine Wahrnehmungstheorie erst seit etwa dem Ende der 60er Jahre als allgemeine Erkenntnistheorie: Implizit in Helmholtz 1903, Bd. 1: 394 f. und Bd. 2: 222, explizit in Helmholtz 1903, Bd. 1: 16 f. und Bd. 2: 188 und 338.

36 Das gilt insbesondere für Helmholtz' Äußerungen zur Kausalität (Helmholtz 1856 ff.: 453 f.; Helmholtz 1903, Bd. 2: 243 f.). Auch in den Reden, in denen er die Relativierung ehemaliger Geltungsansprüche formuliert oder voraussetzt, betont Helmholtz den nichthypothetischen Charakter wissenschaftlicher Aussagen (z.B. Helmholtz 1903, Bd. 2: 218 und 252) und grenzt streng allgemeingültige Gesetze von Hypothesen ab (Helmholtz 1903, Bd. 2: 184 ff. und 240 u.ö.).

37 1878 ist Helmholtz bereits auf "neue Zerlegungen der jetzt bekannten chemischen Elemente gefaßt" (Helmholtz 1903, Bd. 2: 240), seit Beginn der 80er Jahre problematisiert er ansatzweise die bisher angenommene Struktur der elementaren Wechselwirkungen (Helmholtz 1882 ff., Bd. 1: 68 ff.; Helmholtz 1882 ff., Bd. 2: 1003 ff.).

Grundannahmen seiner bisherigen mechanistischen Naturauffassung und ihrer empirischen Geltungsbasis festzuhalten: Er muß den für sie einst geforderten Wahrheitsanspruch aufgeben und sie als Hypothesen formulieren, die dem allgemeinen Streben der Wissenschaft nach einer vollständigen Naturerklärung nützlich sind.

Mit einer Revision seiner Wissenschaftsauffassung schafft er die hierfür erforderlichen Voraussetzungen. Ihre Liberalisierung läßt jetzt verschiedene, einander möglicherweise widersprechende Hypothesen über einen Gegenstandsbereich zu. Der Mechanismus steht dabei als ein Weltbild neben anderen, prinzipiell gleichberechtigten Naturauffassungen. So behält Helmholtz seine Atomvorstellung als Hypothese bei, entwickelt aus ihr in einem ebenso kühnen wie genialen Gedanken die Idee einer elektrischen Elementarladung und läßt sich in seinen elektro- und thermodynamischen Arbeiten von der Überzeugung leiten, daß bewegte unsichtbare Substanzen und mechanische Kräfte Ursache der beobachtbaren Phänomene seien.[38]
Aber Helmholtz befindet sich in einer Situation, in der die begrenzte Erklärungskraft des dualen Mechanismus allgemein ins Bewußtsein getreten ist. Die größte Schwierigkeit und größte Herausforderung für die mechanistische Erklärung der Natur stellen die elektromagnetischen Erscheinungen dar. Helmholtz vermutet, daß die elektrischen Kräfte mit den chemischen Verwandtschaftskräften "durchaus identisch" seien (Helmholtz 1903, Bd. 2: 289), und er beginnt, sich mit der Thermodynamik chemischer Prozesse zu beschäftigen.[39] Im Zusammenhang mit den sich daran anschließenden Untersuchungen führt er systematisch einen Typ der Phänomendeutung ein, der - am Vorbild der Mechanik orientiert - in jene Stellen einrückt, an denen Erklärungen nicht bzw. noch nicht gegeben werden können: die Analogiebildung zwischen mechanischen Modellen und nicht-mechanischen Phänomenen.

Dabei geht Helmholtz vom mechanischen Prinzip der kleinsten Wirkung aus, aus dem sich durch formale Umwandlungen die empirisch verifizierten Gesetze der Energieerhaltung, der Elektro- und der Thermodynamik ergeben sollen. Sein Vorgehen besteht darin, daß er für ein mechanisches Modell, dessen innere Struktur vollständig bekannt ist und das in seiner konkreten Form nichts mit atomaren Prozessen zu tun haben soll, bestimmte Bedingungen so

38 Zu den mechanistischen Annahmen in Helmholtz' späteren elektrodynamischen Arbeiten vgl. Buchwald 1985: 234 ff. Zu seinen thermodynamischen Arbeiten vgl. Klein 1972 und Bierhalter 1981.

39 Helmholtz 1882 ff, Bd. 2: 958 ff. und 979 ff., Bd. 3: 92 ff.

festlegt, daß sich die empirischen Gesetze ergeben, wenn in den Bewegungsgleichungen des Modells die mechanischen Variablen durch thermo- bzw. elektrodynamische Variablen ersetzt werden.[40] Die Ähnlichkeitsrelation der Analogie muß sich also keinesfalls auf innere Strukturen beziehen. Helmholtz beweist lediglich die (vielleicht nur kontingente) Vergleichbarkeit der mathematischen Form von empirischen Gesetzen mit derjenigen der Bewegungsgleichungen von mechanischen Modellen.

Im Ergebnis glaubt er gezeigt zu haben, daß sich aus dem Prinzip der kleinsten Wirkung alle wesentlichen empirischen Gesetze der Physik ableiten lassen. In der unabhängig von der Annahme bestimmter Ursachen erzielten mathematischen Vereinheitlichung sieht er den entscheidenden Fortschritt seines Verfahrens.[41] Man könnte sagen, daß Helmholtz die auf Substanzen und Kräften basierende frühere Argumentation durch eine stärker phänomenologisch angelegte Betrachtung ersetzt. Die in den Gesetzen ausgedrückten Verhältnisbestimmungen faßt er weiterhin als ein Abbild der - allerdings nur noch hypothetisch angenommenen - Wirklichkeit auf. Ideengeschichtlich besteht zwischen dieser relationalen Widerspiegelungstheorie und Ludwig Wittgensteins späterer Abbildtheorie der Bedeutung eine enge Verwandtschaft.[42]

Die Intentionen, die Helmholtz mit seiner Methode der Analogiebildung verfolgt, beschränken sich allerdings bei weitem nicht auf die formale Vereinheitlichung physikalischer Gesetze. Er selbst rechtfertigt sein Vorgehen unter anderem damit, daß Erklärungen vorläufig noch nicht verfügbar seien, und kann sein Vorgehen nicht erläutern, ohne beständig auf die Hypothese unsichtbarer Bewegungsursachen zurückzugreifen.[43] Die Bedingungen, die er für seine mechanischen Modelle festlegt, sollen gleichermaßen auch für die den Phänomenen zugrundeliegende Struktur gelten.[44] Seine Untersuchungen legen darüber hinaus die Vermutung nahe, die Ursachen selbst könnten rein mechanischer Natur sein. Daß die konkrete Form seiner Modelle nicht als Ursache in Frage kommen soll, schließt keinesfalls die Möglichkeit aus, daß die Ursachen nicht mechanisch sein könnten. Unausgesprochen nutzt Helmholtz vielmehr die Uneindeutigkeit mechanischer Modelle, die darin besteht, daß für eine gegebene mechanische Bewegungsgleichung die Anzahl mechanischer

40 Helmholtz 1882 ff., Bd. 3: 132 ff. und 223 ff.

41 Helmholtz 1882 ff., Bd. 3: 210; Helmholtz 1900: 287.

42 Vgl. Majer 1985.

43 Helmholtz 1882 ff., Bd. 3: 117 f., 135 f., 176 f. und 286; Helmholtz 1900: 286 u. ö.

44 Helmholtz 1882 ff., Bd. 3: 128 ff.

Modelle prinzipiell unbeschränkt ist.[45] Damit eröffnet sich der Analogiebildung ein weites Experimentierfeld, um aus verschiedensten Modellen eine Vielfalt hypothetischer Erklärungen zu entwickeln.

Nicht Helmholtz, sondern sein Schüler Heinrich Hertz wird diesen Aspekt weiterer Relativierung des Mechanismus systematisch ausbauen. Den von Helmholtz noch aufrechterhaltenen Unterschied zwischen Modellbildung und Erklärung wird Hertz in seiner Wissenschaftstheorie ebenso beseitigen wie den Abbildungscharakter der Gesetze. Modelle müssen dann lediglich noch in ihren Voraussagen mit der Wirklichkeit übereinstimmen.[46] Was als leichte Lockerung des ursprünglich selbst verkündeten Wahrheitsanspruches bei Helmholtz begann, wird schon zu seinen Lebzeiten und noch ganz im Rahmen der mechanistischen Naturauffassung bei Hertz zu einem Verlust des Wirklichkeitsbezuges führen, wie er größer in der wissenschaftlichen Erkenntnis bis heute kaum denkbar ist.

Literatur:

Bierhalter, Günther: Zu Hermann von Helmholtzens mechanischer Grundlegung der Wärmelehre aus dem Jahre 1884. *Archives for the History of Exact Sciences*, 25, 1981, 71 - 84.

Buchwald, Jed Z.: *From Maxwell to Microphysics*. Chicago: 1985.

Cahan, David: *Hermann von Helmholtz and the Foundation of Nineteenth-Century Science*. Berkeley/Los Angeles (California)/ London: 1993.

Cassirer, Ernst: *Das Erkenntnisproblem in der Philosophie und Wissenschaft der neueren Zeit*. Vierter Band: Von Hegels Tod bis zur Gegenwart (1832 - 1932). Darmstadt: 1973.

Conrat, Friedrich: *Hermann von Helmholtz' psychologische Anschauungen*. Halle: 1904.

Diemer, Alwin: Die Begründung des Wissenschaftscharakters der Wissenschaft im 19. Jahrhundert - Die Wissenschaftstheorie zwischen klassischer und moderner Konzeption. In: Alwin Diemer (Hg.): *Beiträge zur Entwicklung der Wissenschaftstheorie im 19. Jahrhundert*. Meisenheim am Glan: 1968.

---

45 Vgl. die Helmholtz vermutlich bekannte Bemerkung von Maxwell 1873, Bd. 2: 416, sowie den späteren Beweis Poincarés 1890: IX ff.

46 Hertz 1894: 1 ff.

Du Bois-Reymond, Emil: Ueber die Grenzen des Naturerkennens (1872). In: Ders.: *Vorträge über Philosophie und Gesellschaft*. Hamburg: 1974.

Duhem, Pierre: *Ziel und Struktur der physikalischen Ideen* (franz. EA 1904 f.). Autorisierte Übersetzung von Friedrich Adler. Mit einem Vorwort von Ernst Mach. Nachdr. d. Aufl. von 1908. Hamburg 1978.

Elkana, Yehuda: *The Discovery of the Conservation of Energy*. London: 1974.

Erdmann, Benno: *Die philosophischen Grundlagen von Hermann von Helmholtz' Wahrnehmungstheorie* (Abh. Preuss. Akad., phil.-hist. Klasse, Nr. 1). Berlin: 1921.

Fullinwider, S. P.: Hermann von Helmholtz: The Problem of the Kantian influence. In: *Stud. Hist. Phil. Sci.* 21, 1990, 41 - 55.

Hatfield, Gary: *The Natural und the Normative. Theories of Spatial Perception from Kant to Helmholtz*. Cambridge (Mass.)/London: 1990.

Heidelberger, Michael: Force, Law, and Experiment: The Evolution of Helmholtz's Philosophy of Science. In: Cahan (1993).

Heimann, P. M.: Helmholtz und Kant: The Metaphysical Foundation of "Ueber die Erhaltung der Kraft". In: *Stud. Hist. Phil. Sci.*, 5, 1974, 205 - 38.

Helm, Georg: *Die Energetik nach ihrer geschichtlichen Entwicklung*. Leipzig: 1898.

Helmholtz, Hermann von: *Handbuch der physiologischen Optik*. Leipzig: 1856 ff.

Helmholtz, Hermann von: *Das Denken in der Medicin*. Berlin: 1877.

Helmholtz, Hermann von: *Wissenschaftliche Abhandlungen*. 3 Bde. Leipzig: 1882 ff.

Helmholtz, Hermann von: *Vorträge und Reden* (1. Auflage 1884). 2 Bde. Braunschweig: [5]1903.

Helmholtz, Hermann von: *Handbuch der physiologischen Optik*. 2. Auflage. Hamburg/Leipzig: [2]1885 ff.

Helmholtz, Hermann von: Rede über die Entdeckungsgeschichte des Princips der kleinsten Action. In: Adolf Harnack (Hg.): *Geschichte der königl. preuss. Akademie der Wiss. zu Berlin*. Berlin: 1900, Bd. 2, S. 282 ff.

Helmholtz, Hermann von: *Vorlesungen über theoretische Physik*. Hg. von Arthur König et al. 6 Bde. Leipzig: 1897 ff.

Hertz, Heinrich: *Die Prinzipien der Mechanik*. Leipzig: 1894.

Hörz, Herbert, und Siegfried Wollgast: Einleitung zu: *Hermann von Helmholtz, Philosophische Vorträge und Aufsätze*. Berlin: 1971.

Kant, Immanuel: *Kants gesammelte Schriften*, hrsg. v. der Königlich Preussischen (später: Deutschen) Akademie der Wissenschaften (zu Berlin). Berlin: 1902 ff.

Kirchhoff, Gustav: *Vorlesungen über Mechanik* (1. Auflage 1876). Leipzig: [4]1897.

Klein, Martin J.: Mechanical Explanation at the End of the Nineteenth Century. In: *Centaurus* 17, 1972, S. 58 - 82.

Koenigsberger, Leo: *Hermann von Helmholtz's Untersuchungen über die Grundlagen der Mathematik und der Mechanik*. Heidelberg: 1895.

Koenigsberger, Leo: *Hermann von Helmholtz*. 3 Bde. Braunschweig: 1902 f.

König, Gert: Der Wissenschaftsbegriff bei Helmholtz und Mach. In: Alwin Diemer (Hg.): *Beiträge zur Entwicklung der Wissenschaftstheorie im 19. Jahrhundert*. Meisenheim am Glan: 1968.

Krause, Albrecht: *Kant und Helmholtz über den Ursprung und die Bedeutung der Raumanschauung und der geometrischen Axiome*. Lahr: 1876.

Lenoir, Timothy: *The strategy of life. Teleology and Mechanics in Nineteenth Century German Biology*. Dordrecht/Boston/ London: 1982.

Mach, Ernst: *Die Geschichte und die Wurzel des Satzes von der Erhaltung der Arbeit.* Prag: 1872.

Majer, Ullrich: Hertz, Wittgenstein und der Wiener Kreis. In: H.-J. Dahms (Hg.): *Philosophie, Wissenschaft und Aufklärung*. Berlin: 1985.

Maxwell, James Clerk: *A Treatise on Electricity and Magnetism*, 2. Vols., Oxford: 1873.

Poincaré, Henri: *Electricité et Optique*. Paris: 1890.

Poincaré, Henri: *Wissenschaft und Hypothese.* (franz. EA 1902, dt. Üb. 1904) Nachdruck der dritten, verb. Aufl. von 1914. Darmstadt: 1974.

Popper, Karl: *Die Logik der Forschung* (1. Auflage 1934). Tübingen: ^{9}1989.

Pulte, Helmut: Zum Niedergang des Euklidianismus in der Mechanik des 19. Jahrhunderts. In: *Neue Realitäten. XVI. Dt. Kongress für Philosophie*. Berlin: 1993, 833 ff.

Schnädelbach, Herbert: *Philosophie in Deutschland 1831 - 1933*. Frankfurt: 1983.

Schwertschlager, Joseph: *Kant und Helmholtz erkenntnistheoretisch verglichen.* Freiburg: 1883.

Turner, R. Steven: Consensus and Controversy: Helmholtz on the Visual Perception of Space. In: Cahan (1993).

Winters, Stephen M.: *Hermann von Helmholtz' Discovery of Force Conservation.* Diss. Baltimore: 1985.

Helmholtz' Erkenntnis- und Wissenschaftstheorie im Kontext der Philosophie und Naturwissenschaft des 19. Jahrhunderts

Michael Heidelberger

Durch seinen physiologischen Reduktionismus stellte sich Helmholtz in Gegensatz zu philosophischen und naturwissenschaftlichen Strömungen, die eine Lebenskraft oder spezielle nichtmechanische organische Gesetze für die Erklärung der Organismen annahmen. Da die materialistischen Strömungen des 19. Jahrhunderts diesen Reduktionismus teilten und Helmholtz überdies für seine ablehnende Haltung gegenüber der Philosophie Schellings und Hegels (Helmholtz 1855: 369) aber auch allgemein der Metaphysik (Helmholtz 1877: 183, 186, 188f.) und des Idealismus (Helmholtz 1878: 242, 245.) bekannt ist, kann man sich leicht dazu verführen lassen, Helmholtz' Auffassungen für schlichtweg antimetaphysisch, antiidealistisch oder wenigstens für kryptomaterialistisch zu halten - leicht gemildert durch seine Kantische Erkenntnistheorie, in der die Leistungsfähigkeit des menschlichen Denkens untersucht werden sollte (Helmholtz 1855: 368; Helmholtz 1877: 188f.).

Nun hat Helmholtz sich schon selbst deutlich und unmißverständlich vom Materialismus distanziert (Helmholtz 1877: 183, 186). Ich glaube aber im folgenden noch zusätzlich zeigen zu können, erstens, daß sowohl Helmholtz' Reduktionismus als auch seine Erkenntnistheorie idealistische Wurzeln besitzen, und zweitens, daß er in seiner Wissenschaftstheorie von einer metaphysischen Position ausgegangen ist, die er nie ganz abstreifen konnte. Sein Denken blieb zeit seines Lebens in einen idealistischen und metaphysischen Rahmen eingebettet. Damit soll die Bedeutung seiner epochemachenden Konzeptionen des Mechanismus und des physiologischen Reduktionismus nicht verkleinert oder in Abrede gestellt werden. Wenn man aber tiefere Verbindungen des Helmholtzschen Denkens mit dem weiteren Zusammenhang seiner Zeit aufdecken möchte, dann ist der idealistische und metaphysische Kontext äußerst relevant.

Timothy Lenoir stellte vor nicht allzu langer Zeit die These auf, daß sich in den 50er Jahren des letzten Jahrhunderts in Deutschland eine neue Ideologie entwickelt habe, die bis zur Mitte der siebziger Jahre bestimmend war und die Gesellschaft zusammenschloß - eine Ideologie, die, wie Lenoir schreibt, "auf der Befriedigung materieller Interessen beruhte, alle Formen von Idealismus im politischen und intellektuellen Leben verwarf" und besonderen Wert "auf Macht und Realismus" legte. Er bezeichnete diese Ideologie ausdrücklich als materialistisch und behauptete, daß die "organische Physik von 1847 [also die Gruppe der Physiologen um Helmholtz, die eine physikalische Begründung der Lebenserscheinungen vertraten] eine bedeutende Rolle in der Formulierung dieser materialistischen Ideologie gespielt hat" (Lenoir 1988: 171). Wenn meine Analyse der philosophischen Auffassungen von Helmholtz richtig ist, dann kann diese These Lenoirs, zumindest für die Person von Helmholtz selbst, nicht aufrechterhalten werden. Es trifft zwar zu, daß in der besagten Periode die materiellen Interessen in den Vordergrund drängten; man kann daraus aber nicht auf die Vorherrschaft einer materialistischen Ideologie schließen. Helmholtz' reduktionistische Auffassung förderte zwar durchaus die Indienststellung der Naturwissenschaften für das wirtschaftliche Machtstreben und ihre Emanzipation von der Philosophie. Die Begründungen, die er dafür gab, entspringen aber gerade nicht einer materialistischen Weltanschauung, sondern einer tiefen idealistischen Wurzel seines Denkens. In diesem Sinne arbeitete Helmholtz *gegen* den allgemeinen Trend der Zeit, der die Verbindung der Naturwissenschaften mit den materiellen Interessen der Macht als ein Bündnis "gegen alles, was den Menschen von seinen nächsten Aufgaben abzulenken schien" (Lange 2: 520; vgl. Ziegler 1911: Kap. 11), auffaßte.

Im folgenden möchte ich zuerst Helmholtz' *experimentellen Interaktionismus* behandeln und dessen Wurzeln im Idealismus Fichtes aufweisen. Dieser Teil zeigt, daß Helmholtz' Erkenntnis- und Wissenschaftstheorie auf einer antinaturalistischen Voraussetzung beruht. Diese Voraussetzung ist nach meinem Dafürhalten der Schlüssel, um seine ablehnende Haltung gegen naturalistische Auffassungen zu verstehen - sowohl gegen den Materialismus als auch gegen den Nativismus Ewald Herings und allgemein gegen die Schellingsche Naturphilosophie. Im zweiten Teil gehe ich auf Helmholtz' ***metaphysische Grundlegung der Naturwissenschaften*** ein, die sich aus den Metaphysiken von Immanuel Kant und Johann Friedrich Herbart herleitet. Dem schließt sich im dritten Teil eine Untersuchung des Wandels an, den diese Grundlegung unter dem Einfluß bestimmter physikalischer Auffassungen von Michael Faraday erlitt. Diesen Wandel kann man "Empirisierung" oder "Phänomenalisierung" nennen. Ich lege in meiner Darstellung das Schwergewicht auf systematische und begriffliche Vergleiche.[1]

1. Experimenteller Interaktionismus und seine idealistischen Wurzeln bei Fichte

Helmholtz' philosophisches Grundproblem ist die Frage, wie wir von den Sinnesempfindungen zu einem Wissen über die äußeren objektiven Gegenstände kommen, oder, anders formuliert, wie sich das, was der "eigenen Thätigkeit des Geistes" angehört, von dem trennen läßt, was "aus den Einwirkungen der Körperwelt herrührt" (Helmholtz 1878: 222). Daß hier tatsächlich ein Problem verborgen liegt, sieht Helmholtz durch die Untersuchungen von John Locke, Kant, Herbart und Johannes Müller nachgewiesen. Diese hätten nämlich gezeigt, daß unsere "körperliche und geistige Organisation" Anteil daran hat, wie uns die Dinge erscheinen (Helmholtz 1878: 223). Wenn also die Empfindungsqualitäten bloß subjektive Formen der Anschauung sind, wie ist dann ein objektives Wissen überhaupt möglich?

Der einzige Weg, den Helmholtz für gangbar hält, die gewünschte Trennung von (subjektivem) Denken und (objektiver) Wirklichkeit zustandezubekommen, sieht er im tätigen Handeln. Die Zeichen, die wir von der Natur empfangen, lernen wir lesen, "indem wir sie mit dem Erfolge unserer Bewegungen und den Veränderungen, die wir selbst durch diese in der Aussenwelt hervorbringen, vergleichen" (Helmholtz 1868: 320).

1 In einer längeren Behandlung desselben Themas habe ich noch weitere Indizien für meine Thesen angeführt (Heidelberger 1993b). Vgl. auch Heidelberger 1993a.

> "Durch verhältnismäßig wenige, gut angestellte Versuche bin ich im Stande, die ursächlichen Bedingungen eines Ereignisses mit größerer Sicherheit festzustellen, als durch millionenfache Beobachtung, bei welcher ich die Bedingungen nicht habe beliebig verändern können. [...] Erst indem wir unsere Sinnesorgane nach eigenem Willen in verschiedene Beziehungen zu den Objecten bringen, lernen wir sicher urteilen über die Ursachen unserer Sinnesempfindungen, und solches Experimentieren geschieht von frühester Jugend an" (Helmholtz 1867: 451f. Vgl. auch Helmholtz 1878: 227ff. und Helmholtz 1892: 359f.).

Hier ist *in nuce* der experimentelle Interaktionismus von Helmholtz ausgesprochen - die Auffassung, daß wir nur durch aktiven Eingriff in den Gang der Ereignisse zu einem Verständnis der Erscheinungen, d.h. der Kenntnis ihrer Ursachen, vorstoßen können. Die passive Beobachtung allein reicht nicht aus und kann keine vergleichbare Sicherheit erzeugen.

Indem wir durch aktive Eingriffe den Gang der Ereignisse zu verändern versuchen, lernen wir allmählich die Wirkungen unserer absichtlichen Bewegungen von anderen Ereignissen und Vorgängen zu unterscheiden und können das, was willentlich veränderbar ist, auseinanderhalten von dem, was so nicht verändert werden kann. Dies führt schließlich zur Unterscheidung zwischen bloß Gedachtem und Wirklichem, zwischen der eigenen Tätigkeit des Geistes und der äußeren Wirklichkeit: "Es scheiden sich diejenigen Veränderungen, die wir durch bewußte Willensimpulse hervorbringen und rückgängig machen können, von solchen, die nicht Folge von Willensimpulsen sind und durch solche nicht beseitigt werden können" (Helmholtz 1878: 230f.).

Diese ganze Überlegung von Helmholtz hat zur Voraussetzung, erstens, daß dem Handelnden die Wirklichkeit des eigenen Bewußtseins unmittelbar gegeben und zugänglich ist, und zweitens, daß der Wille des Handelnden frei ist und er sich dieser Freiheit auch bewußt ist. Nur wenn diese beiden Bedingungen gegeben sind, kann der Handelnde wirklich wissen, daß es seine Handlung ist, die gerade dieses Ereignis hervorgerufen hat (Helmholtz 1878: 241). Ein solches Wissen ist aber notwendig, um überhaupt Ursachen in der Welt konstatieren zu können. Würde ich mein Bewußtsein nicht kennen, könnte ich nicht wissen, daß es die *Handlung* ist, die das Ereignis verursacht und nicht irgendetwas anderes an oder von mir. Wäre ich unfrei, wäre auch nicht *ich* die Ursache der hervorgerufenen Wirkung, sondern eine andere Ursache, die ich nicht identifizieren könnte, würde sozusagen durch mich hindurch wirken und sich nur meiner Person bedienen, um die Wirkung hervorzurufen (Helmholtz 1878: 241).

Der direkte Zugang, den wir zu unserem eigenen Geist besitzen, geht uns bei den äußeren Objekten ab. Wenn wir das Gefühl bekommen, mit ihnen vertraut zu sein, so beruht das nur auf der Vertrautheit mit den Zuständen des eigenen Geistes, die wir beim absichtlichen Eingreifen in den Gang der Dinge ständig beurteilen. Wir können "direkt immer nur die Nervenerregungen, also die Wirkungen wahrnehmen [...], niemals die äusseren Objecte". Aus der Beobachtung dieser Wirkungen gewinnen wir durch einen Schluß "die Vorstellung von einer Ursache dieser Wirkung" (Helmholtz 1867: 430).

Nach Helmholtz' Meinung unterscheidet sich das Vorgehen des Physikers, der die Ursachen von Erscheinungen aufsucht, nicht grundsätzlich vom aktiven experimentellen Eingreifen eines Menschen, der die Ursachen seiner Empfindungen finden will. In beiden Fällen ist der äußere Widerstand entscheidend, den wir durch eigene Bewegung im freien, tätigen Handeln erfahren. So heißt es in einem sehr frühen Manuskript von Helmholtz: "Naturwissenschaft hat zum Objecte denjenigen Inhalt unserer Vorstellungen, welcher von uns als nicht durch die Selbstthätigkeit unseres Vorstellungsvermögens erzeugt angeschaut wird, d.h. also das als wirklich wahrgenommene" (Koenigsberger 1902: 126).

Nun haben wir alle Elemente von Helmholtz' experimentellen Interaktionismus beisammen, die eine Entsprechung im subjektiven Idealismus von Fichte besitzen.[2] Auch bei Fichte, oder besser gesagt: erst recht und zuerst bei ihm, ist das freie Handeln des Selbstbewußtseins die Voraussetzung für die Scheidung zwischen Denken und Wirklichkeit, beziehungsweise, wie Fichte selbst es ausdrückt, zwischen "Ich" und "Nicht-Ich". Das Ich selbst ist weder eine Vorstellung noch eine Summe von Vorstellungen, sondern unabhängige Handlung. Im freien tätigen Handeln erfährt der Mensch, was ihm möglich ist und was nicht, und wird sich seiner selbst und der Welt außerhalb seiner selbst als davon unterschiedener Wirklichkeit bewußt.

Fichte nennt das unmittelbare Vertrautsein mit den eigenen Empfindungen das Fühlen und unterscheidet davon die Anschauung, in der beobachtend das eigene Fühlen reflektiert wird. Zum Bewußtsein des äußeren Gegenstandes kommt es durch Anschauung der eigenen Empfindungen und "durch das Denken nach dem Satze des Grundes", wie Fichte es ausdrückt. Man ist genötigt, im Denken der Empfindung einen äußeren Grund hinzuzusetzen, eine fremde

2 Zu den historischen Zeugnissen, die Helmholtz' Fichte-Rezeption nachweisen und auf die Fichte-Verehrung von Helmholtz' Vater eingehen vgl. Helmholtz 1891: 17; Koenigsberger 1902: Bd. 1, 7, 161, 168, 242, 244, 284, 331f.; Kremer 1990: 154; Cahan 1993: 86. Fichtes Sohn, der Philosoph Immanuel Hermann Fichte, war Taufpate von Helmholtz. Siehe auch Köhnke 1986: 151ff.

Kraft. Die Existenz der äußeren Kraft ist nicht direkt gegeben, sondern kann nur aus dem Wissen des eigenen Zustandes gefolgert werden. "Alles Bewußtsein des Gegenstandes außer mir ist durch das klare, genaue Bewußtsein meines eigenen Zustandes bestimmt, und es wird in demselben immer ein Schluß vom Begründeten in mir auf einen Grund außer mir gemacht" (Fichte 1800: 233; vgl. 218). Als empirisches Wesen bin ich mir allerdings nicht bewußt, selbst diesen Schluß zu vollziehen. Das Bewußtsein der objektiven Gegenstände, also ihre Wahrnehmung, ist ein "nicht dafür erkanntes" Bewußtsein der Erzeugung einer Vorstellung von einer Kraft außerhalb des Bewußtseins durch das Bewußtsein selbst. "Die Natur, in welcher ich zu handeln habe, ist [...] durch meine eigenen Denkgesetze gebildet [...]. Sie drückt überall nichts aus als Verhältnisse und Beziehungen meiner selbst zu mir selbst" (Fichte 1800: 258). Alles Bewußtsein von etwas ist bedingt durch das unmittelbare Bewußtsein meiner selbst.

Für Fichte ergibt sich daraus aber keineswegs, wie später für Schelling, eine enge Beziehung zwischen Geist und Natur. Vielmehr bleibt nach seiner Auffassung die Natur durch eine tiefe Kluft vom Reich der Freiheit und des Geistes getrennt. Die äußere Natur als Inbegriff aller vom Ich als fremd gesetzten Gegenstände ist das "System der notwendigen Vorstellungen". Sie verkörpert das dem freien, autonomen Ich verfügbare Material und ist daher etwas an sich Totes, Unfreies, ein blinder Naturmechanismus, ein völlig determinierter, gesetzmäßiger Zusammenhang: "Intelligenz [Geist] und Ding sind also geradezu entgegengesetzt: sie liegen in zwei Welten, zwischen denen es keine Brücke gibt" (Fichte 1794: 435). Die Aufgabe des Menschen ist es, den blinden Naturmechanismus seinen eigenen Zwecken zu unterwerfen. "Die Natur muß allmählich in die Lage eintreten, daß sich auf ihren gleichmäßigen Schritt sicher rechnen und zählen lasse, und daß ihre Kraft unverrückt ein bestimmtes Verhältnis mit der Macht halte, die bestimmt ist, sie zu beherrschen, - mit der menschlichen" (Fichte 1800: 129). Auch und gerade diese Botschaft des Idealismus' Fichtes hat Helmholtz übernommen und in seinen experimentellen Interaktionismus übersetzt.

Es ist gerade Helmholtz' tiefe Bindung an Fichtes Idealismus, die seinen starken Gegensatz zur Naturphilosophie Schellings (bzw. allgemein der "Identitätsphilosophie", wie er sich ausdrückte) erklärt (Heidelberger 1993b: 493f.). Die Natur ist in Helmholtz' Konzeption nicht, wie bei Schelling, Quelle der Produktivität, die den Menschen mit seinem Bewußtsein hervorbringt und ihm im Innersten verwandt ist. Sie ist vielmehr, wie bei Fichte, ein der Freiheit und Unabhängigkeit des Menschen entgegengesetzter, fremder Zusammenhang, den man sich zum Beweis der eigenen Freiheit und zur Ausdehnung seiner Autonomie unterwirft und beherrscht.

Demgemäß sieht Helmholtz den gemeinsamen Zweck der Wissenschaften darin, "den Geist herrschend zu machen über die Welt" und "den Menschen von den auf ihn eindrängenden Nothwendigkeiten mehr und mehr zu befreien". Wie ein Feldherr spricht er von den "Siegen des Gedankens über den widerstrebenden Stoff" (Helmholtz 1862: 142), die errungen werden müssen, und von der "siegreichen Unterwerfung der uns theils fremd, theils feindlich gegenüberstehenden Aussenwelt unter die Kräfte unseres Denkens und unseres Willens" (Helmholtz 1863: 151). Bedeutsamerweise findet sich diese Rhetorik besonders gehäuft in einem Vortrag von 1862, in dem Jahr also, in dem in Deutschland an vielen Orten des 100. Geburtstags von Fichte gedacht wurde (Helmholtz 1862. Vgl. Köhnke 1986: Kap. IV,2).

Was die Auffassung der Stellung des Menschen in der Natur angeht, ist der philosophische Materialismus des 19. Jahrhunderts der Naturphilosophie weitaus näher als dem Helmholtzschen Fichteanismus. Beide Positionen, sowohl Naturphilosophie als auch Materialismus, vertreten einen Naturalismus für den Bereich der ganzen Wirklichkeit. Bei Helmholtz ist der Naturalismus *per definitionem* auf den Bereich der Materie, der äußeren Erscheinungen, eingeschränkt. Daß der Materialismus, ähnlich wie Helmholtz, die Beherrschung der Naturkräfte durch den Menschen propagiert, reicht nicht zur Rechtfertigung aus, in Helmholtz einen Parteigänger des Materialismus zu sehen.

Der antinaturalistische Zug, der sich durch den Idealismus Fichtes in Helmholtz' Auffassungen einnistet, zeigt sich nirgends so deutlich wie in seiner Auffassung der Psychologie und allgemein der Geisteswissenschaften. Deren Aufgabengebiet ist derjenige Anteil der Empfindungen und Vorstellungen, der nicht auf äußere Ursachen, sondern auf die eigene, innere Tätigkeit des Menschen zurückgeht. Da wir einen freien Willen und ein spontanes Vorstellungsvermögen besitzen, wie Helmholtz immer wieder betont, ist der entscheidende Teil der psychischen Vorgänge nicht auf strenge Gesetze zurückführbar (Helmholtz 1862: 131, 135). Strenge Gesetzlichkeit ist gerade dasjenige Merkmal, das die äußere Wirklichkeit *per definitionem* von der inneren unterscheidet. Die Wissenschaft muß von der Begreiflichkeit der Natur nach dem Kausalgesetz ausgehen. Man kann nach Helmholtz nicht schlimmer gegen die wissenschaftliche Methode verstoßen, als wenn man wie die Identitätsphilosophie die kausalen Prinzipien der äußeren Wirklichkeit mit der Spontaneität der inneren Welt zusammenbringt und vermengt. Eine solche Vermengung kann sich nur in einer Abschwächung des Kausalgesetzes ausdrücken. In der wahren Wissenschaft sind jedoch keine Spuren von dem zu finden, was das Innere des Menschen ausmacht.

> "Der Natur gegenüber besteht kein Zweifel, dass wir es mit einem ganz strengen Causalnexus zu thun haben, der keine Ausnahmen zulässt. Deshalb ergeht an uns auch die Forderung fortzuarbeiten, bis wir ausnahmslose Gesetze gefunden haben; eher dürfen wir uns nicht beruhigen, erst in dieser Form erhalten unsere Kenntnisse die siegende Kraft über Raum und Zeit und Naturgewalt" (Helmholtz 1862: 138).

> "Die Gesetzlichkeit der Natur wird als causaler Zusammenhang aufgefasst, sobald wir die Unabhängigkeit derselben von unserem Denken und unserem Willen anerkennen" (Helmholtz 1869: 343).

2. Materie und Kraft als metaphysische Grundbegriffe der Naturwissenschaft

Welche Konsequenzen ergeben sich nun aus dem experimentellen Interaktionismus für die wissenschaftliche Untersuchung der Natur? Durch das Experiment - sei es nun das Handeln des wahrnehmenden Menschen allgemein oder speziell des Naturwissenschaftlers im Labor - können wir die Ursachen der uns gegebenen Erscheinungen herausfinden. Die theoretische Naturwissenschaft setzt sich das Ziel, die allgemeine Beschaffenheit der fremden Kraft, die als Ursache von außen auf uns einwirkt, zu bestimmen. Zu welcher Schlußfolgerung wird sie geführt? Welcher Art sind die Ursachen, die hinter den uns gegebenen Erscheinungen tätig sind?

Eine Untersuchung dieser Frage ist in einem doppelten Sinne eine metaphysische: erstens, da sie sich auf ein Sein richtet, das uns nicht in der Erfahrung gegeben ist, sondern hinter dem Schein der Empfindungen, die es verursacht, seine Tätigkeit ausübt, und zweitens, da sie die Begriffe untersucht, mit denen dieses Sein vor aller Erfahrung zu bestimmen ist - die Begriffe, die, wie Helmholtz es ausdrückte, die "allgemeine und nothwendige Form der Naturanschauung" sind (Koenigsberger 1902: Bd. 1, 127).

Dieses Programm nimmt methodische und zum Teil auch inhaltliche Anleihen auf sowohl bei den *Metaphysischen Anfangsgründen der Naturwissenschaft* von Kant als auch bei der Metaphysik des Fichte-Schülers Johann Friedrich Herbart. Herbarts Philosophie galt zu ihrer Zeit als eine der wenigen philosophischen Alternativen, die für Physiker und Mathematiker in Frage kamen. Helmholtz bezog sich später selbst auf Herbart und wurde von seinen

Zeitgenossen mit Herbart in Zusammenhang gebracht (Helmholtz 1868: 288; Hering 1878: Einltg.). Herbart sah es als Aufgabe der Physik an, das Reale, was dem gegebenen Schein zu Grunde liegt, begrifflich so zu bestimmen, "wie in der Astronomie aus den scheinbaren Bewegungen der Himmelskörper die wahren gefunden werden" (Herbart 1813: 227, § 132). Auch nach der Auffassung von Helmholtz "müssen wir in jeder Erklärung von Naturerscheinungen das Gebiet der Sinnlichkeit verlassen und zu unwahrnehmbaren, nur durch Begriffe bestimmten Dingen übergehen", wenn wir "zu den Ursachen der Naturerscheinungen" aufsteigen wollen (Helmholtz 1853: 18). Die ganze Rhetorik um das Gegensatzpaar: Wirklichkeit und Schein bzw. Erscheinung, die bei Helmholtz immer wieder vorkommt, spielt in der Philosophie von Herbart eine prominente Rolle.

Der Bezug zu Kant ergibt sich durch Helmholtz' Grundlegung der Naturwissenschaft in den Begriffen der Materie und der Kraft. Kant machte in der Vorrede seiner *Metaphysischen Anfangsgründe* klar, daß nur diejenige Erkenntnis, "die ihren Gegenstand gänzlich nach Prinzipien a priori" behandelt, im eigentlichen Sinn den Namen 'Wissenschaft' verdient (Kant 1786: A V). Die Naturwissenschaft als "Körperlehre" muß eine "vollständige Zergliederung des Begriffs von einer Materie überhaupt zum Grunde" legen, bevor sie an ihr eigentliches Geschäft gehen kann. Dabei darf sie sich aber keiner besonderen Erfahrung bedienen, "sondern nur dessen, was sie im [von speziellen Erfahrungen] [...] abgesonderten Begriffe" der Materie in Bezug "auf die reinen Anschauungen im Raume und in der Zeit" antreffe. Die Physik kann ihre Gegenstände nur dann mathematisch behandeln, wenn sie "Prinzipien der *Konstruktion* der Begriffe, welche zur Möglichkeit der Materie überhaupt gehören", besitzt (Kant 1786: XII). Diese "Prinzipien der Notwendigkeit dessen, was zum *Dasein* eines Dings gehört", werden in der "Metaphysik der Natur" gewonnen. Sie handelt "von den Gesetzen, die den Begriff einer Natur überhaupt möglich machen", und zwar "ohne Beziehung auf irgend ein bestimmtes Erfahrungsobjekt, mithin unbestimmt in Ansehung der Natur dieses oder jenes Dinges der Sinnenwelt" (Kant 1786: A VIII).

Ganz ähnlich muß auch für Helmholtz die Naturwissenschaft nach den Begriffen suchen, "aus welchen sich die einzelnen bestimmten empirischen Wahrnehmungen ableiten lassen", wenn sie mehr als bloß "eine geordnete Übersicht alles Empirischen" liefern wolle. Diese "allgemeinen Naturbegriffe" liegen "aller Naturanschauung zum Grunde" und sind aus ihrer Möglichkeit hergeleitet. Sie bilden die "allgemeine und nothwendige Form", mit der die Naturwissenschaft erst denkbar wird. "Die Gewißheit" der allgemeinen oder reinen Naturwissenschaften ist daher "eine absolute". Es läßt sich aber aus ihnen "kein empirisches Factum oder Gesetz" gewinnen, sondern "sie können uns nur eine Norm für unsere Erklärungen geben" (Koenigsberger 1902: Bd. 1, 127).

Die Ausführung dieses Programms von Helmholtz findet sich in der "Einleitung" zu seiner Schrift von 1847, *Über die Erhaltung der Kraft*. Als erstes behandelt Helmholtz dort den Begriff der Materie. Er argumentiert, daß die Ursachen der Erscheinungen für uns materieller Natur sind, wenn wir sie unter Absehung ihrer Wirkungen und Veränderungen nur im Hinblick auf ihr Dasein als Gegenstände auffassen. Die einzigen Eigenschaften, die wir der Materie dann vor aller Erfahrung zuschreiben können, sind ihr Ort und ihre "Quantität (Masse)". Da aber die Materie nicht wirkungslos ist und sich die Erscheinungen verändern, müssen wir auch die Wirkungen berücksichtigen, die das Sein hinter den Erscheinungen auf uns ausübt. Wir müssen der Materie also auch noch Kräfte zuerteilen.

Als nächstes versucht Helmholtz, die Natur dieser Kräfte zu klären. Sein Argumentationsgang verläuft folgendermaßen: Jede Veränderung beruht auf dem Wirken einer veränderlichen oder einer unveränderlichen Kraft. Die grundlegenden Kräfte, auf denen die Erscheinungen letztlich beruhen, müssen aber unveränderlich sein, d.h. unter gleichen Umständen zu gleichen Wirkungen führen. Wären sie selbst veränderlich, so müßten als Ursache ihrer Veränderungen wiederum andere Kräfte angenommen werden, die noch grundlegender wären. Folglich muß die Materie in letzter Hinsicht als aus Teilen bestehend aufgefaßt werden, die sich in Bezug auf ihre Wirkung nur hinsichtlich der unzerstörbaren und unveränderlichen Kräfte, mit denen sie ausgestattet sind, voneinander unterscheiden.

Wenn nun aber, so folgert Helmholtz weiter, die wirkenden Kräfte unveränderlich sind, so reduziert sich jede Veränderung der Materie auf Ortsveränderung, also auf Bewegung. Folglich kann die Kraft, die ein Materieteil auf ein anderes ausübt, keine andere Wirkung haben als die Veränderung der Entfernung der Materieteile voneinander. Die Kräfte wirken also entweder anziehend oder abstoßend. Damit ist für Helmholtz der Beweis abgeschlossen, daß alle Erscheinungen auf Bewegungen materieller Gegenstände beruhen, die durch entfernungsabhängige Zentralkräfte aufeinander einwirken. Es kann jetzt genauer formuliert werden, was es heißt, die Natur zu begreifen: Es heißt, die Erscheinungen "zurückzuführen auf unveränderliche, anziehende und abstoßende Kräfte, deren Intensität von der Entfernung abhängt" (Helmholtz 1847: 16). Einige Jahre später faßt Helmholtz dieses Ergebnis noch schärfer: "Eine Naturerscheinung ist physikalisch erst dann vollständig erklärt, wenn man sie bis auf die letzten ihr zu Grunde liegenden und in ihr wirksamen Naturkräfte zurückgeführt hat" (Helmholtz 1853: 18). Wenn gezeigt ist, daß diese Zurückführung vollständig und die einzig mögliche ist, dann ist sie auch als "nothwendige Begriffsform der Naturauffassung" erwiesen (Helmholtz 1847: 7). Bei alledem macht Helmholtz unmißverständlich klar, daß er weder Materie noch (unveränderliche) Kraft als bloße begriffliche Hilfsmittel oder Fiktionen auffaßt, denen "nichts Wirkliches entspräche" (Helmholtz 1847: 14). Vielmehr sind Kräfte etwas tatsächlich Existierendes, das real den Erscheinungen zugrundeliegt.

In Gegensatz zu seiner metaphysischen Deutung der Grundlagen der Naturwissenschaft stellte Helmholtz die Auffassung Goethes, daß sich die Veränderungen in der Welt ohne Bezug auf die Wirklichkeit hinter den Erscheinungen erklären lassen. Für Goethe, meint Helmholtz, haben es die Naturwissenschaften ausschließlich mit den Erscheinungen und ihrem gegenseitigen Zusammenhang zu tun. "Er fordert daher für die Untersuchung physikalischer Gegenstände eine solche Anordnung der Beobachtungen, daß eine Tatsache immer die andere erkläre, damit man zur Einsicht in den Zusammenhang komme, ohne das Gebiet der sinnlichen Wahrnehmung zu verlassen. Diese Forderung", fährt Helmholtz fort, "hat einen bestechenden Schein für sich, ist aber ihrem Wesen nach grundfalsch". Der Physiker darf nicht bei der unmittelbaren sinnlichen Anschauung stehenbleiben, wie es sich für den Dichter ziemt, sondern er muß hinüberführen "in eine Welt unsichtbarer Atome, Bewegungen, anziehender und abstoßender Kräfte, die, in zwar gesetzmäßigem, aber kaum zu übersehendem Gewirre, durcheinander arbeiten. Ihm ist der sinnliche Eindruck keine unumstößliche Autorität" (Helmholtz 1853: 18).

Die Feststellung regelmäßiger Zusammenhänge auf der bloßen Erscheinungsebene hat zwar für Helmholtz ihren guten Sinn im experimentellen Teil der Naturwissenschaft. Eine wirkliche Erklärung der Erscheinungen wird aber erst im "theoretischen Teil" der Naturwissenschaften geleistet, wenn die "unbekannten Ursachen der Vorgänge aus ihren sichtbaren Wirkungen" gefunden sind (Helmholtz 1847: 4). Der experimentelle Teil liefert also nur vorläufige Beschreibungen von Verhältnissen auf der Ebene der Erscheinungen, die dann in der Theorie als Auswirkung fundamentaler, aber verborgener Ursachen dargestellt werden.

3. Die Empirisierung des Kraftbegriffs und der Metaphysik

Helmholtz hielt an dieser metaphysischen Begründung der Naturwissenschaften noch lange Zeit fest. Im Jahre 1855, als William Rankine die Physik als eine "Science of Energetics" aufbauen wollte, die ohne die "bisherigen hypothetischen Suppositionen von Atomen, Kräften usw." auskommt, machte Helmholtz unmißverständlich klar, daß er "die philosophischen Grundanschauungen, von denen Hr. Rankine ausgeht, nicht teilt" (Helmholtz 1858: 365). Auch auf der Naturforscherversammlung in Innsbruck im Jahre 1869 formulierte Helmholtz sein metaphysisches Programm noch in gleicher Form: Wenn alle elementaren Substanzen sich nur durch ihren Bezug auf den Raum ändern können, dann kommt jede Veränderung in der Welt durch Bewegung zustande.

> "Ist aber Bewegung die Urveränderung, welche allen anderen Veränderungen in der Welt zu Grunde liegt, so sind alle elementaren Kräfte Bewegungskräfte, und das Endziel der Naturwissenschaften ist, die allen anderen Veränderungen zu Grunde liegenden Bewegungen und Triebkräfte zu finden, also sich in Mechanik aufzulösen" (Helmholtz 1869: 379).

In derselben Rede beginnt sich jedoch ein tiefgreifender Wandel in Helmholtz' Auffassung des Kraftbegriffs abzuzeichnen. Die grundlegende Rolle, die der Kraftbegriff bisher für ihn gespielt hat, wird nun vom Gesetzesbegriff übernommen. Der Kraftbegriff drückt keine Wirklichkeit *hinter* den Erscheinungen mehr aus, sondern eine Beziehung *zwischen* den Erscheinungen. Mit dieser Veränderung leitete Helmholtz eine grundlegende Wende in seinem Denken ein: weg von der unbeobachtbaren Wirklichkeit der nur erschlossenen Ursachen hinter dem sinnlichen Schein der Erscheinungen, hin zu den Erscheinungen selbst und ihren gesetzlichen Beziehungen untereinander; weg von der metaphysischen Grundlegung der Naturwissenschaften, hin zu einer empiristisch-positivistischen Fassung des Kraftbegriffs. Helmholtz hat aber diese Wende nie vollständig vollzogen, obwohl er sie zu vollziehen wünschte, sondern kämpfte sein ganzes Leben mit metaphysischen Restbeständen, die sich von seinem Kraftbegriff nicht lösen wollten.

Nach der neuen Auffassung ist der Kraftbegriff ein Abstraktum, mit dem wir die Beschreibung naturgesetzlicher Beziehungen zwischen den beobachtbaren Objekten zusammenfassen. Kräfte erscheinen nun als tautologische Umformungen von gesetzlichen Tatsachenbeziehungen und werden aufgelöst in Relationen von Erscheinungen. Mit dieser Empirisierung des Kraftbegriffs ist Helmholtz natürlich auch gezwungen, seine Ansicht von der Notwendigkeit,

Allgemeingültigkeit und absoluten Gewißheit der Begriffsform der Naturwissenschaften aufzugeben.

Wenn wir in der Mechanik von der Wirkung einer Kraft auf eine Masse sprechen, dann beziehen wir uns, sagt Helmholtz jetzt, nicht auf ein verborgenes Reich von Kräften hinter den Erscheinungen, sondern meinen damit nur, daß unter gegebenen beobachtbaren Umständen eine bestimmte beobachtbare Beschleunigung eintritt. Wenn wir von der Lichtbrechungskraft sprechen, beziehen wir uns auf das Gesetz der Lichtbrechung, das für durchsichtige Körper gilt; bei der elektromotorischen Kraft einer Batterie beziehen wir uns auf die Stärke des elektrischen Stromes, die bei bestimmten Zusammensetzungen von Metallen und Flüssigkeiten auftritt, usw. auch bei den anderen Kräften. Darin besteht der "thatsächliche Sinn", den unser Kraftbegriff hat. Wir bezeichnen mit einer Kraft "in der That weiter nichts, wenigstens nichts, was noch einen factischen Sinn hat, als was auch in der blossen Beschreibung des Phänomens enthalten ist" (Helmholtz 1903: 11). Mit dieser Formulierung bekennt sich Helmholtz zu einer empiristisch-phänomenalistischen Interpretation des Kraftbegriffs, die schon 1855 von Fechner, 1872 von Mach und 1876 von Kirchhoff vorgeschlagen worden war (Fechner 1855: 90, 107, 113, 179f.; Mach 1872: 37, 45f., 53f.; Kirchhoff 1876: 1. Vgl. König 1968).

Trotzdem läßt Helmholtz nicht vollständig von seinen früheren metaphysischen Überzeugungen ab. Der Begriff einer Wirklichkeit hinter den Erscheinungen fällt nicht, wie bei den eben genannten Empiristen, mit dem Bezug auf die Erscheinungen ersatzlos weg, sondern bekommt einen Ersatz in Form des Gesetzesbegriffs: "Das erste Produkt des denkenden Begreifens der Erscheinung ist das Gesetzliche". Je allgemeiner die regelmäßigen Beziehungen sind, die wir zwischen den Erscheinungen gefunden haben und je mehr wir von ihrer Ausnahmslosigkeit überzeugt sind, desto berechtigter sind wir, sie "Ursache" zu nennen. "Nur in diesem Sinne", sagt Helmholtz, "ist meiner Meinung nach die Anwendung des Wortes gerechtfertigt". Wir können das Gesetz auch "Kraft" nennen, da wir es "als ein unsere Wahrnehmung und den Ablauf der Naturprocesse Zwingendes, als eine unserem Willen gleichwerthige Macht anerkennen" (Helmholtz 1878: 245). Solange wir nur unvollständig und ungenau die Bedingungen der Vorgänge erfaßt haben, solange ist uns auch die eigentliche Realität des Gesetzes und damit der Kraft verborgen, und wir sind auf bloße Hypothesen angewiesen. Es gibt also für Helmholtz weiterhin "den verborgenen dauernd bestehend bleibenden Grund der Erscheinungen", der "hinter dem Wechsel der Erscheinungen stehend auf uns einwirkt" (Helmholtz 1903: 16; Helmholtz 1878: 245), nur daß dieser Grund jetzt nicht mehr durch a priori bestimmte Begriffe von Kraft und Materie zu fassen ist, sondern im empirischen Gesetzesbegriff begründet liegt.

Helmholtz' Abkehr (oder Versuch der Abkehr) von der Metaphysik geht vor allem auf den Einfluß von Michael Faraday, aber auch auf den anderer britischer Physiker zurück: "Die mathematische Physik empfing den Anstoß zu dem besprochenen Fortschritt [...] [weg von den hypothetischen, metaphysischen Annahmen, hin zu den direkten Erfahrungen] hauptsächlich von Faraday" (Helmholtz 1892: 352. Vgl. Heidelberger 1993b: 481). An Faradays Theorien konnte Helmholtz sehen, daß eine erfolgreiche Physik möglich ist, ohne daß man sich bei der Erklärung der Erscheinungen des Begriffs einer anziehenden oder abstoßenden Kraft bedienen muß. Faradays Bestreben "war in der That auf einen wesentlichen Fortschritt in den Principien wissenschaftlicher Methodik hingerichtet, dessen Ziel es ist, die Naturwissenschaft von den letzten Ueberbleibseln der Metaphysik zu befreien" (Helmholtz 1881: 276). "Es war nun Faraday das Hypothetische, was ihm in der Einschiebung der Kräfte zu liegen schien, sehr zuwider. [...] Es gelang ihm in der That, [...] für die magnetischen und elektrischen Kräfte, eine Deutung zu finden [...], welche die bis dahin angenommenen Fernkräfte für die magnetischen und elektrischen und elektro-magnetischen Erscheinungen fallen ließ" (Helmholtz 1903: 12f.). Diese Deutung bestand bekanntlich in Faradays Annahme der Kraftlinien, aus denen der Feldbegriff erwachsen ist. Faraday konnte also die Erscheinungen mit ganz anderen und zudem auch noch experimentell entwickelten "Abstraktionen" erklären als die früheren Auffassungen, die mit verborgenen anziehenden und abstoßenden Kräften operierten. Die Kraft- und Materiebegriffe der früheren Physik waren, wie Helmholtz nun zugab, im Vergleich zu den neuen Begriffen Faradays "übersinnliche, unausdenkbare Abstraktionen" "anschauungsleerer Begriffe" (Helmholtz 1892: 351).

Helmholtz muß durch die Erfahrung einer in diesem Sinne metaphysikfreien Physik bei Faraday stark beeindruckt und zum Umdenken gebracht worden sein. Wenn mehrere physikalisch gleichberechtigte, aber tiefgreifend unterschiedliche metaphysische Voraussetzungen einer physikalischen Theorie möglich sind, dann kann keine von ihnen Notwendigkeit oder Allgemeinheit für sich in Anspruch nehmen. Sie sinken herab auf den Status von Hypothesen, die sich, wie andere Hypothesen auch, an den beobachtbaren Tatsachen messen lassen müssen. "Man hat begriffen, dass auch die mathematische Physik eine reine Erfahrungswissenschaft ist; dass sie keine anderen Principien zu befolgen hat, als die experimentelle Physik" (Helmholtz 1871: 47). Noch in der *Erhaltung der Kraft* hatte Helmholtz darauf bestanden, daß im theoretischen

Teil der Physik die Eigenschaften der unbeobachtbaren Ursachen in einer metaphysischen Abstraktion zu bestimmen sind und nur der experimentelle Teil sich auf die regelmäßigen Zusammenhänge der Erscheinungen bezieht.

Auf seinem Weg von der Auffassung der Kraft als metaphysisch notwendiger Voraussetzung der theoretischen Physik zu ihrer Einschätzung als deskriptive, gesetzliche Zusammenfassung empirischer Erscheinungen war Helmholtz in gewisser Weise schon sein Freund Emil du Bois-Reymond im Jahre 1848 vorausgegangen. Du Bois-Reymond hielt es schon damals für einen Anthropomorphismus, von der Kraft als der Ursache von Bewegung zu sprechen. Es sei nichts damit gewonnen, wenn man sich das Bild gleichsam von einer Hand mache, "welche die träge Materie leise vor sich herschiebt" oder von Polypenarmen, "mit welchen die Stoffteilchen sich umklammern, sich gegenseitig an sich zu reißen suchen" (du Bois-Reymond 1848: 14). Kräfte als Ursachen anzusehen, sei nur zulässig, wenn man sich darüber klar ist, daß der Kraft eigentlich "nichts Wirkliches zugrundeliegt", ihr "keine Wirklichkeit zukommt". Folgerichtig hält du Bois-Reymond die Existenz von Kräften für eine "Fiktion" (du Bois-Reymond 1848: 6, 13). Diese Kritik zielte genau ins Herz der Helmholtzschen Metaphysik und schlug schon damals als Heilmittel die von Helmholtz später übernommene Empirisierung und Phänomenalisierung vor. Als Antwort auf die Frage, "was denn übrig bleibe, wenn weder Kräfte, noch Materie Wirklichkeit besitzen", empfahl du Bois-Reymond, sich auf die Erscheinungen zu beschränken: "Wir ziehen daher vor, statt uns im Kreise unfruchtbarer Spekulationen zu drehen oder mit dem Schwerte der Selbsttäuschung den Knoten zu zerhauen, uns zu halten an die Anschauung der Dinge, wie sie sind", sprich, an die Erscheinungen, die wir von den Dingen haben (du Bois-Reymond 1848: 14). Die Aufgabe der Naturwissenschaften ist es daher nicht, die Ursachen der Bewegungen zu erkennen, sondern ihre Gesetze. "Die Kraft ist uns das Maß, nicht die Ursache der Bewegung" (du Bois-Reymond 1848: 15). Folglich ist bei du Bois-Reymond auch nicht mehr von Zurückführung der Erscheinungen auf ihre Ursachen die Rede, sondern vom "ursächlichen Zusammenhang der natürlichen Erscheinungen", den wir uns "unter dem mathematischen Bilde der Abhängigkeit, der Funktion" vorzustellen haben (du Bois-Reymond 1848: 6).

Seit seiner metaphysikkritischen Umbildung des Kraftbegriffs (aber eben bezeichnenderweise nicht früher) ging Helmholtz nun auch mit der Metaphysik scharf ins Gericht und empfahl der Philosophie, sich überhaupt nur auf Erkenntnistheorie zu beschränken. Zur selben Zeit begann er auch mit seinen Untersuchungen zu den Grundlagen der Geometrie. Auch hier kann man vermuten, daß die Fragestellung, die er verfolgte, aus seiner neugewonnenen Metaphysikkritik erwachsen ist. In der *Erhaltung der Kraft* gehörte die "räumliche Vertheilung", mithin der Raum, noch zu den Eigenschaften, die wir mit der Materie "setzen", also zu ihren metaphysischen Bestimmungen rechnen (Helmholtz 1847: 14).[3] Seit den späten 60er Jahren prüfte er dann die Möglichkeit, ob nicht auch die Gesetzlichkeit des von uns angeschauten Raumes eine empirische Gesetzlichkeit der erfahrbaren Erscheinungen und keine Denknotwendigkeit darstellt. Aber auch hier bleibt ein metaphysischer Restbestand. Die reine Form, das "inhaltsleere Schema" der Raumanschauung, bleibt trotz der neugewonnenen empiristischen Interpretation der Geometrie transzendental gegeben.

Literatur

Abkürzungen:

VR3: Helmholtz, Hermann von: *Vorträge und Reden*. 3. Aufl., 2 Bde., Braunschweig: Vieweg, 1884.

VR5: Helmholtz, Hermann von: *Vorträge und Reden*. 5. Aufl., 2 Bde., Braunschweig: Vieweg, 1903.

Cahan, David (Hg.): *Letters of Hermann von Helmholtz to his parents: the medical education of a German scientist 1837-1846*, Stuttgart: Steiner, 1993.

du Bois-Reymond, Emil: Über die Lebenskraft. Aus der Vorrede zu den 'Untersuchungen über tierische Elektrizität' vom März 1848. In: Estelle du Bois-Reymond (Hg.), *Reden von Emil du Bois-Reymond*, in zwei Bänden, 4. Aufl., Bd. 1, Leipzig: Veit, 1912, 1-26.

Fechner, Gustav Theodor: *Ueber die physikalische und philosophische Atomenlehre*. Leipzig: Hermann Mendelssohn, 1855.

Fichte, Johann Gottlieb: *Erste Einleitung in die Wissenschaftslehre*, 1794. In: Immanuel Hermann Fichte (Hg.), *Johann Gottlieb Fichte's sämmtliche Werke*, Erste Abtheilung, erster Band, Berlin: Veit, 1845, repr. Berlin: de Gruyter, 1965.

3 Dies ist übrigens eine Auffassung, die Herbart abgelehnt hat. Für ihn kommen dem Sein hinter den Erscheinungen weder räumliche noch zeitliche Bestimmungen zu, genausowenig wie anziehende oder abstoßende Kräfte. Vgl. Herbart 1813: 234, § 137; 326, § 160.

Fichte, Johann Gottlieb: *Die Bestimmung des Menschen.* Berlin: Voss, 1800. Zitiert nach der Ausgabe Stuttgart: Reclam, 1981.

Heidelberger, Michael: Räumliches Sehen bei Helmholtz und Hering. *Philosophia Naturalis*, 30, 1993, 1-28. (1993a).

Heidelberger, Michael: Force, Law, and Experiment: The Evolution of Helmholtz's Philosophy of Science. In: David Cahan (Hg.): *Hermann von Helmholtz and the Foundations of Nineteenth-Century Science.* Berkeley/ Los Angeles/ London: University of California Press, 1993, 461-497. (1993b).

Helmholtz, Hermann von: Ueber die Erhaltung der Kraft. Eine physikalische Abhandlung. Berlin: G. Reimer, 1847. Zitiert nach: Helmholtz, Hermann von: *Wissenschaftliche Abhandlungen.* Leipzig: J. A. Barth, 1882-95, 3 Bde., 1: 12-68.

Helmholtz, Hermann von: Ueber Goethe's naturwissenschaftliche Arbeiten (1853). In: *VR³* 1: 1-23.

Helmholtz, Hermann von: Ueber das Sehen des Menschen (1855). In: *VR³* 1: 365-396.

Helmholtz, Hermann von: [Referat zu] W.J.M. Rankine, Outline of the science of energetics. *Fortschritte der Physik im Jahre 1855*, XI, 1858, 365.

Helmholtz, Hermann von: Ueber das Verhältniss der Naturwissenschaften zur Gesammtheit der Wissenschaften. Akademische Festrede (1862). In: *VR³* 1: 117-145.

Helmholtz, Hermann von: Ueber die Erhaltung der Kraft. Einleitung eines Cyclus von Vorlesungen, gehalten in Carlsruhe während des Winters 1862 auf 1863. In: *VR³* 1: 147-189.

Helmholtz, Hermann von: *Physiologische Optik.* 1. Aufl., Leipzig: Voss 1867.

Helmholtz, Hermann von: Die neueren Fortschritte in der Theorie des Sehens (1868). In: *VR³* 1: 233-331.

Helmholtz, Hermann von: Ueber das Ziel und Fortschritte der Naturwissenschaft. Eröffnungsrede für die Naturforscherversammlung zu Innsbruck (1869). In: *VR⁵* 1: 367-398.

Helmholtz, Hermann von: Zum Gedächtniss an Gustav Magnus (1871). In: *VR³* 2: 35-53.

Helmholtz, Hermann von: Das Denken in der Medicin (1877). In: *VR³* 2: 165-193.

Helmholtz, Hermann von: Die Thatsachen in der Wahrnehmung (1878). In: *VR³* 2: 217-271.

Helmholtz, Hermann von: Die neuere Entwickelung von Faraday's Ideen über Elektricität (1881). In: *VR³* 2: 272-337.

Helmholtz, Hermann von: Goethe's Vorahnungen kommender naturwissenschaftlicher Ideen (1892). In: *VR⁵* 2: 335-61.

Helmholtz, Hermann von: *Einleitung zu den Vorlesungen über theoretische Physik.* Arthur König und Carl Runge (Hg.), Leipzig: J. A. Barth, 1903 (= Band I, Abtheilung 1 der *Vorlesungen über theoretische Physik*, hg. Arthur König et al., 6 Bde., Bd. 1-4 und 6: Leipzig: Barth, Bd. 5: Hamburg und Leipzig: Voss, 1897-1907).

Herbart, Johann Friedrich: *Lehrbuch zur Einleitung in die Philosophie.* Königsberg, 1813. Zitiert nach dem Nachdruck der 4. Auflage von 1837, Leipzig: Meiner, 1912.

Hering, Ewald: *Zur Lehre vom Lichtsinne.* Wien: Gerold, 1878.

Kant, Immanuel: Metaphysische Anfangsgründe der Naturwissenschaft. Riga: Hartknoch, 1786. Zitiert nach: *Werke in zehn Bänden*, Wilhelm Weischedel (Hg.), Bd. 8, Darmstadt: Wissenschaftliche Buchgesellschaft 1975.

Kirchhoff, Gustav Robert: *Vorlesungen über mathematische Physik. Mechanik.* Leipzig: Teubner, 1876.

Köhnke, Klaus Christian: *Entstehung und Aufstieg des Neukantianismus*. Frankfurt a. M.: Suhrkamp, 1986.

Koenigsberger, Leo: *Hermann von Helmholtz*. 3 Bde., Braunschweig: Vieweg, 1902-03.

König, Gert: Der Wissenschaftsbegriff bei Helmholtz und Mach. In: Alwin Diemer (Hg.): *Beiträge zur Entwicklung der Wissenschaftstheorie im 19. Jahrhundert*. Meisenheim: Hain, 1968, 90-114.

Kremer, Richard L. (Hg.): *Letters of Hermann von Helmholtz to his wife: 1847-1859*. Stuttgart: Steiner 1990.

Lange, Friedrich Albert: *Geschichte des Materialismus und Kritik seiner Bedeutung in der Gegenwart*. 2 Bde., 2. Aufl. Iserlohn: Baedeker, 1875. Zitiert nach der Ausgabe Frankfurt: Suhrkamp, 1974, 2 Bde.

Lenoir, Timothy: Social interests and the organic physics of 1847. In: Edna Ullmann-Margalit (Hg.): *Science in Reflection*. Dordrecht: Kluwer, 1988, 169-191.

Mach, Ernst: *Die Geschichte und die Wurzel des Satzes von der Erhaltung der Arbeit*. Prag: Calve, 1872, photomechanischer Nachdruck des zweiten Abdrucks von 1909, hrsg. von Joachim Thiele, Amsterdam: Bonset, 1969.

Ziegler, Theobald: *Die geistigen und sozialen Strömungen des 19. Jahrhunderts*. Berlin: G. Bondi, 1911.

Ontologische und erkenntnistheoretische Dimensionen des Gesetzesproblems in den Helmholtzschen Reflexionen über Naturgesetze

Ulrich Röseberg

1. Helmholtz über den Stellenwert von Gesetzeserkenntnis in den Wissenschaften

Gesetzeserkenntnis scheint für Hermann von Helmholtz ein zentrales Problem aller Wissenschaften zu sein. Er sucht den Unterschied zweier von ihm vorgefundenen disjunkten Wissenschaftsklassen über Unterschiede in deren Gegenstandsbereichen festzumachen, wobei er die vom deutschen Idealismus geprägten Bezeichnungen 'Natur-' und 'Geisteswissenschaften' als adäquate Fixierung der jeweils behandelten spezifischen Gegenstände ansieht. Ungeachtet aller philosophischen Debatten über die Gegenstandskonstitution von Wissenschaften ist der Gegenstand der Naturwissenschaften für ihn (ebenso wie für viele Naturforscherkollegen seiner Zeit) schließlich die geistlose Natur selbst, während es die Geisteswissenschaften, wie ja schon ihr Name sagt, im weitesten Sinne mit dem geistigen Leben der Menschen zu tun haben. Natur- und Geisteswissenschaften werden dabei selbstverständlich auch als geistige Tätigkeiten unterstellt. Zugleich aber wird hervorgehoben, daß nur die letzteren Geist thematisieren, wie er sich im Denken und in allen Formen des historischen Wandlungen unterliegenden menschlichen Zusammenlebens zeigt, die durch Politik, Wirtschaft, Sprache, Sitten, Recht, Moral u.a. geregelt sind.

Der Naturforscher Helmholtz arbeitet in einigen seiner populärwissenschaftlichen Vorträge Ähnlichkeiten zwischen der Tätigkeit von Künstlern und Wissenschaftlern heraus und stößt dabei schon sehr früh zu einem Kulturverständnis vor, das sich von manchen der damals weitverbreiteten idealistischen Einseitigkeiten freizumachen vermag, ohne deshalb etwa naiven materialistischen Gegenthesen aufzusitzen. Helmholtz' engagierter Einsatz für die Entwicklung von Medizin, Naturwissenschaften und Technik sowie der entsprechenden Bildung folgt aufklärerischen Einsichten in den Zusammenhang von Weltgeschichte und zivilisatorischem Fortschritt. - Mit diesen wenigen, über das Gesetzesthema hinausführenden Andeutungen soll hier lediglich Mißverständnissen vorgebeugt werden, die heute entstehen können, wenn man liest, daß für einen Wissenschaftler der Gegenstand der Naturwissenschaften selbstverständlich Natur und der der Geisteswissenschaften eben Geist ist.

In dieser begrifflichen Fixierung seines Wissenschaftsverständnisses läßt sich der in Auseinandersetzung mit spekulativer Naturphilosophie den Empirismus experimenteller und mathematisch-theoretischer Naturforschung propagierende Helmholtz auch nicht beirren, als in der zweiten Hälfte des 19. Jahrhunderts die Zweckmäßigkeit des Begriffs 'Geisteswissenschaften' zunehmend hinterfragt wird, der Begriff 'Kulturwissenschaften' stärker in die Diskussion kommt, mit den Sozialwissenschaften ein neuer Wissenschaftstyp entsteht und die methodischen Gegensätze von Natur- und Geisteswissenschaften innerhalb von Methodologien selbst zum Untersuchungsgegenstand werden.[1] Helmholtz jedenfalls hält bis zu seinem Tode an der einmal gewählten Bezeichnung 'Geisteswissenschaften' fest, konzediert diesen allerdings, daß sie es wohl erheblich schwerer als die Naturwissenschaften haben, Gesetze ihres Gegenstandsbereichs zu finden. Wenn er davon spricht, daß die Geisteswissenschaften "es überwiegend mit Urtheilen nach psychologischem Tactgefühl zu thun" haben (Helmholtz 1862: 172), so lese ich daraus zugleich

1 An die kulturalistische Umdeutung der Naturwissenschaften oder die programmatische These, daß alle Wissenschaft Geisteswissenschaft sei (Janich, Mittelstrass), ist vor reichlich hundert Jahren noch nicht im entferntesten zu denken, obgleich eine sich in diesen Interpretationskontext begebende Lektüreweise einschlägiger Texte keine Mühe hat, das letzte Drittel des 19. Jahrhunderts und den Beginn des 20. Jahrhunderts retrospektiv als entscheidenden Umschlagpunkt auszumachen und dabei dann auch Helmholtz anders zu lesen (Chevalley, Sandkühler).

die Hoffnung, auch diese Wissenschaften mögen dereinst eine Entwicklungshöhe erreichen, die ihnen mehr erlaubt. Wie sonst sollte er gleichzeitig sagen können: "Vollständige Kenntniss und vollständiges Verständniss des Waltens der Natur- und Geisteskräfte ist es allein, was die Wissenschaft erstreben kann" (Helmholtz 1862: 182). Hinsichtlich der Existenz von Gesetzen in jenen Wirklichkeitsbereichen, für die die Geisteswissenschaften zuständig zeichnen, drückt sich Helmholtz, wenn überhaupt, sehr vorsichtig aus. - Deshalb soll es im folgenden nur noch um Naturgesetze bei Helmholtz gehen.[2]

Die Naturwissenschaft, so jedenfalls steht es für Helmholtz schließlich fest, sucht nach den Gesetzen in "der Welt der Wirklichkeit" (Helmholtz 1878: 218). In seiner Eröffnungsrede für die Naturforscherversammlung zu Innsbruck hatte er bereits betont, daß alle naturwissenschaftliche Forschungsarbeit letztlich auf Kenntnis der Naturgesetze gerichtet sei (Helmholtz 1869: 374). Was aber ist nun ein Naturgesetz für Helmholtz?

2 Diese Vorsicht ist nicht zuletzt dem während der Konferenz zunehmend verfestigten Eindruck geschuldet, daß die bislang publizierten Helmholtz-Texte nicht hinreichen, um daraus auch noch eine einigermaßen scharf umrissene Philosophie der Geisteswissenschaften zu destillieren. Alle Reden und Vorträge mit einschlägigen Passagen sind aus konkreten Anlässen gehalten worden. Dabei unterlagen wohl insbesondere die universitären Festreden dem Zwang, auch dem Kosmos der selbst nicht verfolgten Wissenschaften Rechnung zu tragen. Hier ist daher die Gefahr groß, daß der spätere Leser mehr Klarheit in den Formulierungen eines Autors vermutet, als dieser selbst besaß. Sieht man von einer derartigen Kontextabhängigkeit der Helmholtzschen Äußerungen zu Gesetzen ab, lassen sich einige seiner Bemerkungen sehr wohl in jene philosophische Traditionslinie einordnen, die die Existenz objektiver Gesetze in Natur, Gesellschaft und im Denken voraussetzt und als eine der wesentlichsten Funktionen von Wissenschaft das Erkennen eben dieser Gesetze bestimmt (Hörz; Hörz/Wollgast; Röseberg 1975). Auf Inkonsistenzen der einschlägigen Helmholtzschen Äußerungen und andere Interpretationsmöglichkeiten verweist vor allem der Beitrag von W. Küttler in diesem Band. Mir scheint es lohnenswert, angesichts der in der Wissenschaftsphilosophie sich gegenwärtig wieder intensivierenden Diskussionen zu Gesetz und Kausalität (Cartwright, Fraassen, Salmon) die wissenschafts- und philosophiehistorischen Entwicklungslinien detailgetreu und unter Verzicht auf vorschnelle Generalisierungen nachzuzeichnen. Ohne im Rahmen dieses Beitrags der Wirkungsgeschichte der Helmholtzschen Position zum Naturgesetz die gebührende Aufmerksamkeit schenken zu können, geht es mir erst einmal um einen Versuch der Rekonstruktion dieser Position selbst. Im Mittelpunkt steht dabei eines der zentralen Probleme, nämlich das des Verhältnisses von Erkenntnistheorie und Ontologie. Dabei stößt man auch auf die Frage nach der inneren Konsistenz der Helmholtzschen Erkenntnistheorie.

2. Zum Begriff des Naturgesetzes bei Helmholtz

Eine relativ geschlossene Ausarbeitung der eigenen erkenntnistheoretischen Anschauungen gibt Helmholtz bekanntlich erst im Anschluß an seine umfangreichen sinnesphysiologischen Untersuchungen. Was er dabei schließlich zu dem für ihn so zentralen Gesetzesproblem ausführt, scheint jedoch Überlegungen zu explizieren, die Schritt für Schritt parallel zur naturwissenschaftlichen Forschungsarbeit entstanden sind. Uns fällt heute besonders auf, daß der Mediziner, Physiologe und Psychologe hinsichtlich des Verhältnisses der ontologischen und erkenntnistheoretischen Dimensionen des Problems wohl nicht gravierend anders denkt als der Physiker. Seine Äußerungen über Existenz und Erkenntnis von Naturgesetzen werden zwar in dem Maße ausdifferenziert, in dem der erfahrene Naturforscher das für die Naturforschung als charakteristisch angesehene methodische Vorgehen reflektiert. Es kommt dabei auch zu Akzentverschiebungen hinsichtlich des Gebrauchs solcher Begriffe wie *Ursache*, *Kraft* und *Macht*.[3]

Entscheidende Aussagen zum Gesetzesproblem kommen bereits in der berühmten Einleitung zu seiner Arbeit *Über die Erhaltung der Kraft* vor. Helmholtz sieht schon hier die Aufgabe der physikalischen Naturwissenschaften darin, "die Gesetze zu suchen, durch welche die einzelnen Vorgänge in der Natur auf allgemeine Regeln zurückgeleitet, und aus den letzteren wieder bestimmt werden können" (Helmholtz 1847: 15). Diese Regeln gelten ihm als allgemeine Gattungsbegriffe, die in der experimentellen Forschung festzustellen seien und sämtliche Erscheinungen des entsprechenden Bereichs zu umfassen haben. Der theoretische Teil der Forschung suche dagegen, "die unbekannten Ursachen der Vorgänge aus ihren sichtbaren Wirkungen zu finden; er sucht dieselben zu begreifen nach dem Gesetze der Kausalität" (Helmholtz 1847: 15). Dieses Unterfangen sieht Helmholtz durch den Grundsatz gedeckt, "daß jede Veränderung in der Natur eine zureichende Ursache haben *müsse*" (Helmholtz 1847: 15). Stößt man in der Forschung zunächst nur auf veränderliche Ursachen einer Naturerscheinung, so nötigt eben dieser Grundsatz, weiter zu fragen, bis schließlich die letzten Ursachen gefunden sind, "welche nach einem unveränderlichen Gesetz wirken, welche folglich zu jeder Zeit, unter denselben äußeren Verhältnissen dieselbe Wirkung hervorbringen"

3 Auf diese wird in anderen Beiträgen des vorliegenden Bandes (M. Heidelberger, L. Krüger, G. Schiemann u.a.) ausführlicher eingegangen.

(Helmholtz 1847: 15). Ziel der theoretischen Naturwissenschaften ist es mithin, "letzte unveränderliche Ursachen der Naturvorgänge" im Gesetz zu finden. Das ist die Sicht von 1847.

In der Heidelberger Rektoratsrede von 1862 wird das Verhältnis von *Gattungsbegriff* und *Gesetz* präzisiert. Helmholtz betont, daß Wissenschaft erst entstehe, wenn sie die Ursachen und Gesetze der Tatsachen enthülle. Sie habe das empirische Material zu verarbeiten, indem sie Ähnliches zusammenschließe und dafür dann einen umfassenden allgemeinen Begriff bilde, der die Einzelheiten im Denken vertreten könne. Diesen nennt er *Gattungsbegriff*, wenn er eine Menge existierender Dinge umfaßt. Er nennt ihn *Gesetz*, wenn er eine Reihe von Vorgängen bzw. Ereignissen umfaßt (Helmholtz 1862: 169). Damit wird das Wissen nicht nur in eine leichter zu handhabende Form gebracht, sondern zugleich auch erweitert, weil die so gefundenen Regeln und Gesetze auch auf zukünftige Fälle ausgedehnt werden können.

Naturgesetze sind also für Helmholtz weder bloße Hilfsmittel zur besseren Ordnung empirischer Befunde, noch will er sie als spekulative Konstruktionen des forschenden Geistes deuten lassen. Verbindlich für alle Naturforscher erklärt er: "Wir müssen sie vielmehr in den Thatsachen *entdecken*" (Helmholtz 1869: 375). Wie man dabei gemäß dem Programm von 1847 auf "letzte unveränderliche Ursachen der Naturvorgänge" stoßen soll, bleibt nun allerdings sein Geheimnis. Der erbarmungslose Kritiker metaphysischer Spekulationen in der Philosophie kann sich offenkundig selbst nur schrittweise aus den Klauen eben solcher Spekulationen befreien.

Nachdem Helmholtz das - wie wir heute sagen - Energieerhaltungsgesetz formuliert hat und ihm auch in der Physiologie und der Psychologie der Sinneswahrnehmungen Gesetzeserkenntnisse gelungen sind, gibt er 1878 schließlich eine funktionale Begriffsbestimmung für *Gesetz*, in der von "letzten unveränderlichen Ursachen der Naturvorgänge" explizit nicht mehr die Rede ist. In Übereinstimmung mit dem Wortgebrauch in der mathematisierten Physik nennt er nun "das gleichbleibende Verhältniss zwischen veränderlichen Grössen: das sie verbindende *Gesetz*" (Helmholtz 1878: 240).

Er weiß inzwischen, daß Wissenschaft der Erörterung von erkenntnistheoretischen Hypothesen nicht entraten kann, und fordert deshalb, alle zulässigen Hypothesen zu erörtern, "um eine vollständige Uebersicht über die möglichen Erklärungsversuche zu behalten" (Helmholtz 1878: 239). Dabei allerdings versäumt er es, ein Kriterium anzugeben, wie wohl eine zulässige von einer unzulässigen Hypothese unterschieden werden solle.

Obgleich es eines wissenschaftlich sein wollenden Denkens unwürdig sei, den hypothetischen Ursprung seiner Sätze zu vergessen, behauptet Helmholtz nun, daß der Wissenschaftler das Gesetzliche in der Erscheinung "unzweideutig und als Thatsache ohne hypothetische Unterschiebung finden" könne (Helmholtz 1878: 240).

Gesetzeserkenntnis hat für Helmholtz einen anderen erkenntnistheoretischen Status als Sinnesempfindungen. Gemäß seiner Wahrnehmungstheorie sind Empfindungen Wirkungen, die über Reize in unseren Sinnesorganen hervorgebracht werden und deshalb auch von den letzteren selbst abhängen. Die Qualität unserer Empfindungen gibt Nachricht über die die Sinne erregenden äußeren Einwirkungen; sie sind Zeichen derselben, nicht aber deren Abbilder. Die wahrnehmungstheoretisch begründete Absage an die Wahrheit der von den Sinnen vermittelten Bilder läßt nur noch Ähnlichkeitsbeziehungen zu. Diese aber sollen erkenntnistheoretisch noch "die Abbildung der Gesetzmässigkeit in den Vorgängen der wirklichen Welt" (Helmholtz 1878: 222) sichern.

Wie ist es möglich, daß unsere Sinne die Wirklichkeit nicht abbilden, wie sie ist, sondern uns nur Zeichen liefern, die richtig zu deuten wir lernen müssen, während wir mit den Gesetzesaussagen empirisch vielfach überprüfter Theorien die Naturgesetze selbst kennenlernen? Helmholtz sieht keine Schwierigkeit, auf der Grundlage seiner Zeichentheorie der Empfindungen die Möglichkeit zur Abbildung der Gesetzmäßigkeit in den Vorgängen der wirklichen Welt zu erklären:

> "Jedes Naturgesetz sagt aus, dass auf Vorbedingungen, die in gewisser Beziehung gleich sind, immer Folgen eintreten, die in gewisser anderer Beziehung gleich sind. Da Gleiches in unserer Empfindungswelt durch gleiche Zeichen angezeigt wird, so wird der naturgesetzlichen Folge gleicher Wirkungen auf gleiche Ursachen, auch eine ebenso regelmässige Folge im Gebiete unserer Empfindungen entsprechen" (Helmholtz 1878: 222).

Helmholtz behauptet also, daß aus seiner Zeichentheorie der Sinnesempfindungen die Abbildtheorie der Gesetzeserkenntnis folgt. Das aber kann er freilich nur, wenn er die durchgängige Gültigkeit des Kausalgesetzes in der Natur voraussetzt und damit gleichsam die erkenntnistheoretische Hypothese der Existenz eines fundamentalen und elementaren Naturgesetzes ontologisiert. Im Zuge eines denkmöglichen (nicht aber praktisch auch notwendigerweise zu verwirklichenden) Reduktionsprogramms lassen sich die nur noch als funktionale Abhängigkeiten gedeuteten Naturgesetze dann prinzipiell immer auf ontologische Gegebenheiten zurückführen, sind also Abbildungen und mithin mehr als nur Zeichen. Die implizit unterstellte Annahme geht ersichtlich weit über das hinaus, was Helmholtz im Rahmen seiner sinnesphysiologisch abgestützten Wahrnehmungstheorie noch zu behaupten wagt. Dort stellt er fest, daß Zeichen für etwas stehen, was dem wahrnehmenden Subjekt so nicht gegeben ist. Hinsichtlich der Gesetzeserkenntnis scheint ihm solche erkenntnistheoretische Vorsicht nicht mehr geboten. Es gibt Naturgesetze, und wir vermögen sie mit den Methoden experimentierender und rechnender Naturwissenschaft schließlich so kennenzulernen, wie sie sind, indem wir sie abbilden. Da die dafür gegebene Begründung an den Apriorismus des Kausalitätsgesetzes gebunden ist, wird klar, daß in diesem Punkte dem Anspruch des Empirismus nicht Genüge getan werden kann.[4]

Helmholtz weiß inzwischen, daß sich Gesetzeserkenntnis nicht im streng empiristischen Sinne als Tatsachenschau mißverstehen läßt, weil sie immer denkendes Begreifen voraussetzt, wozu - wie auch schon bei der einfachen Sinnesempfindung - Aktivitäten des erkennenden Subjekts erforderlich sind:

> "Das erste Product des denkenden Begreifens der Erscheinung ist das *Gesetzliche*. Haben wir es so weit rein ausgeschieden, seine Bedingungen so vollständig und sicher abgegrenzt und zugleich so allgemein gefasst, dass für alle möglicherweise eintretenden Fälle der Erfolg eindeutig bestimmt ist, und wir gleichzeitig die Überzeugung gewinnen, es habe sich bewährt und werde sich bewähren in aller Zeit und in allen Fällen: dann erkennen wir es als ein unabhängig von unserem Vorstellen Bestehendes an und nennen es die *Ursache*, d.h. das hinter dem Wechsel ursprünglich Bleibende und Bestehende; nur in diesem Sinne ist meiner Meinung nach die Anwendung des Wortes gerechtfertigt, wenn auch der gemeine Sprachgebrauch es in sehr verwaschener Weise überhaupt für Antecedens oder Veranlassung anwendet. Insofern wir dann das Gesetz als ein unsere Wahrnehmung und den Ablauf der Naturprozesse Zwingendes, als eine unserem Willen gleichwertige Macht anerkennen, nennen wir es '*Kraft*'. Dieser Begriff der uns entgegentretenden Macht ist unmittelbar durch die Art und Weise bedingt, wie unsere einfachsten Wahrnehmungen zustande kommen" (Helmholtz 1878: 240 f.).

4 Der erkenntnistheoretische Bruch zwischen Zeichentheorie der Sinnesempfindungen und Abbildtheorie der Gesetzeserkenntnis kann mittels der starken metaphysischen Annahmen des Kausalgesetzes (unabhängig von seiner Begründung) vermieden werden. Helmholtz ist diesen Weg gegangen, hat seine Annahme allerdings nicht expliziert, weil sie ihm vermutlich in diesem Zusammenhang als die wohl einfachste und natürlichste erschien. Wer diese Annahme nun nicht akzeptieren kann oder will, kommt m.E. nicht umhin zu konstatieren, daß zwei wesentliche Stücke der Helmholtzschen Erkenntnistheorie nicht zusammenpassen. Dann gibt es im Unterschied zu Helmholtz' gegenteiliger Behauptung einen Bruch zwischen seiner Zeichentheorie der Sinnesempfindungen und der Abbildtheorie der Gesetzeserkenntnis.

Das erkenntnistheoretische Problem, nämlich die Beziehung von Gedachtem und Wirklichem, entsteht - wie Helmholtz betont - erst, "wenn wir die Scheidung dessen, was das Ich ändern und nicht ändern kann, zu vollführen wissen. Diese wird aber erst möglich, wenn wir erkennen, welche gesetzmässigen Folgen die Willensimpulse zur Zeit haben. Das Gesetzmässige ist daher die wesentliche Voraussetzung für den Charakter des Wirklichen" (Helmholtz 1878: 242). Der hier von ihm gebrauchte Begriff des Wirklichen hebt auf das Wirken ab und läßt das Bestehen unveränderlicher Substanz hinter dem Wechsel der Erscheinungen außen vor. Helmholtz lehnt es in Übereinstimmung mit Kant ab, ein "Ding an sich" in positiven Bestimmungen vorstellen zu wollen, ohne es in die Form des Vorstellens aufzunehmen. "Was wir aber erreichen können, ist die Kenntniss der gesetzlichen Ordnung im Reiche des Wirklichen, diese freilich nur dargestellt in dem Zeichensystem unserer Sinneseindrücke" (Helmholtz 1878: 242). Dabei sind Erkenntnisfortschritte über adäquatere Formulierungen der Gesetzesaussagen oder die Präzisierung jener Bedingungen denkmöglich, unter denen die jeweiligen Gesetzesaussagen notwendig gelten.

In Auseinandersetzung mit Kirchhoffs Formulierung, die Aufgabe der Mechanik darin zu sehen, die in der Natur vorkommenden Bewegungen vollständig und auf die einfachste Weise zu beschreiben, hält Helmholtz an der Orientierung an Gesetzeserkenntnis fest, betont nun aber, daß Gesetze nicht direkt beobachtet werden, sondern über Hypothesen gefunden werden. "Jede richtig gebildete Hypothese stellt ihrem thatsächlichen Sinne nach ein allgemeineres Gesetz der Erscheinungen hin, als wir bisher unmittelbar beobachtet haben; sie ist ein Versuch, zu immer allgemeinerer und umfassenderer Gesetzlichkeit aufzusteigen" (Helmholtz 1878: 242). Dabei hat man sich des Induktionsschlusses zu bedienen, der "ein Vertrauen auf die Gesetzmässigkeit alles Geschehens" voraussetzt (Helmholtz 1878: 243), die als Bedingung der Begreifbarkeit der Natur gilt.[5] Das in der Argumentation zum Abbildcharakter der Gesetzeserkenntnis vorausgesetzte Kausalgesetz wird jetzt nur noch als regulatives Prinzip unseres Denkens apostrophiert (Helmholtz 1878: 243). In der II. Beilage zum Vortrag über "Die Tatsachen in der Wahrnehmung" hinterfragt Helmholtz seine vorher aufgestellte These, derzufolge Gesetzeserkenntnis schließlich zu apodiktischer Gewißheit führe. Jetzt sind alle Naturgesetze nur noch durch Induktion approximativ zu erweisen. Es heißt nun: "Aber der Wunsch, exacte Gesetze kennen zu wollen, ist noch kein Beweis dafür, dass es

5 Siehe hierzu den Beitrag von L. Krüger in diesem Band.

solche gibt" (Helmholtz 1878: 393).[6] Damit aber sind die ontologisch als Naturgegebenheiten unterstellten Gesetze auf der Reflexionsebene streng genommen Hypothesen. Zu dem radikalen Schritt seines Schülers H. Hertz, den Abbildcharakter der in mathematisierten wissenschaftlichen Theorien formulierten Gesetze aufzugeben (Hertz: 1894: 1f.), ist Helmholtz aber nicht bereit. Zwar läßt sich auch für ihn zukünftig durchaus ein anderer Aufbau der Mechanik denken. Er vermag jedoch keinerlei Notwendigkeit zu einer prinzipiell neuen Mechanik zu sehen, wie sie etwa Boltzmann und Gibbs zumindest als vage Denkmöglichkeit in Betracht ziehen. Die Mechanik gilt Helmholtz nach wie vor als dauerhaft und gut gesichert, wenngleich er seine mechanistische Naturauffassung nur noch mit einem reduzierten Geltungsanspruch vertritt.

Er hatte 1862 davon gesprochen, daß in der mathematisierten Physik allgemeine Gesetze "von unbedingter Gültigkeit und ausserordentlich umfassendem Umfange" vorliegen (Helmholtz 1862: 175), ohne diesen Gedanken später je wieder zu dementieren. Das Newtonsche Gravitationsgesetz, der Energieerhaltungssatz und schließlich das Prinzip der kleinsten Wirkung sind für ihn Muster jener "allgemeinen Gesetze von unbedingter Gültigkeit und außerordentlich umfassendem Umfange", denen andere Wissenschaften (noch?) nichts Gleichwertiges an die Seite zu setzen haben.

Der Vorrede zur ersten Auflage der Helmholtzschen populären Vorlesungen entnehmen wir, daß ihr Autor darin den Gedanken realisiert sieht, "das Wesen und die Tragweite der Naturgesetze und ihre Beziehungen zu den geistigen Tätigkeiten des Menschen anschaulich zu machen" (Helmholtz 1865: VI). Nahezu drei Jahrzehnte später heißt es, das Naturgesetz sei nicht nur für unseren beobachtenden Verstand Leitfaden:

> "[...] es beherrscht auch den Ablauf aller Vorgänge in der Natur, ohne dass wir darauf zu achten, es zu wünschen oder zu wollen brauchen, ja leider oft genug auch gegen unser Wünschen und Wollen. Wir müssen es also als Aeusserungsweise einer Macht anerkennen, die in jedem Augenblicke, wo die Bedingungen für ihre Wirksamkeit eintreten, zu wirken bereit ist. In diesem Sinne bezeichnen wir es als *Kraft*, und da diese *Kraft* eben in jedem Augenblicke als wirkungsbereit und wirkungsfähig sich bewährt, schreiben wir ihr dauernde Existenz zu. Darauf beruht meines Erachtens auch die Beziehung der Kraft als *Ursache* der Veränderungen, die unter ihrem Einfluss vorgehen; sie ist das hinter dem Wechsel der Erscheinungen verborgene Bleibende" (Helmholtz 1892: 353 f.).

---

6 G. Schiemanns Lesart der Helmholtzschen *Vorträge und Reden* hat mich überzeugt, dem Prozeßcharakter der dahinterstehenden Reflexionen größte Aufmerksamkeit zu widmen, auch und gerade wenn es um eine der Invarianten in diesen Wandlungen geht.

Diese Formulierungen belegen, daß Naturgesetze für Helmholtz ebenso existieren wie Dinge, die zu Gegenständen der Erfahrung werden können, sie gehören also zu den Fakten der Natur selbst. Er setzt mithin eine Ontologie voraus, in der neben Dingen und Prozessen auch Gesetze ihren Platz haben. Das methodische Denken des Naturwissenschaftlers, der sich zeitbedingt als Antimetaphysiker profiliert, setzt selbst - wie wir gesehen haben - metaphysische Elemente voraus, ohne sich darüber in jedem Falle Rechenschaft abzulegen.

3. Metaphysische Elemente im Denken des Antimetaphysikers Helmholtz

Obgleich die kritische Distanz der sich emanzipierenden Naturwissenschaften zur Philosophie seit dem 17. Jahrhundert in dem Maße wächst, in dem mit der Anwendung spezifischer Forschungsmethoden (Experimente, messende Beobachtungen und mathematische Berechnungen) nachweisbar Erkenntnisfortschritte erzielt werden, weisen spätere methodologische Reflexionen des naturwissenschaftlichen Methodengebrauchs immer wieder darauf hin, daß die Naturwissenschaften bis heute explizit oder implizit Voraussetzungen als gültig unterstellen, die traditionell in der Philosophie thematisiert werden. Der unter Naturwissenschaftlern besonders im 19. Jahrhundert zunehmende Einfluß des Empirismus reflektiert den stürmischen Erkenntniszuwachs und will zugleich die wiederaufblühende spekulative Metaphysik abwehren. Dabei bemerken viele der als Antimetaphysiker auftretenden Naturwissenschaftler nicht, in wie starkem Maße ihr eigenes philosophisches Denken von unreflektierten Voraussetzungen ausgeht. Zu diesen Voraussetzungen gehört nicht selten - Helmholtz ist hier keine Ausnahme - die unerschütterliche Überzeugung forschender Naturwissenschaftler, daß die Natur selbst festen, unveränderlichen Regeln und Gesetzen unterliegt, denen zufolge alle beobachtbaren Erscheinungen und Prozesse notwendige Folgen der dem Augenschein verborgenen, dem Verstand aber einsehbaren Ursachen darstellen. Aufgabe der Naturwissenschaft ist es dann, die mit der Natur selbst ontologisch gegebenen Regeln und Gesetze zu erkennen.

Die erkenntnistheoretische Dimension des Gesetzesproblems besteht also unter dieser Voraussetzung nur noch darin, die Beziehung zwischen den theoretisch formulierten Gesetzesaussagen und deren ontologischen Entsprechungen zu klären, oder anders ausgedrückt, den ontologischen Gehalt der in bewährten Theorien formulierten Gesetze zu bestimmen. Welchen Sinn in diesem Zusammenhang die Redeweise von einer ontologischen Entsprechung oder dem ontologischen Gehalt hat, ist so überhaupt nicht mehr zu hinterfragen. Die Sinnhaftigkeit einer solchen Denkweise wird unterstellt. Helmholtz' diesbezügliche Auffassungen treffen sich mit denen der von ihm kritisierten Identitätsphilosophen und mechanischen Materialisten.

Helmholtz hingegen sieht sich selbst eher in der Tradition von Kants Erkenntniskritik. Dazu gibt es eine umfangreiche Literatur, auf die ich hier natürlich nicht näher eingehen kann. - Für Kant ist Naturwissenschaft in den sicheren Gang einer Wissenschaft eingemündet, als die Naturforscher begriffen, "daß die Vernunft nur das einsieht, was sie selbst nach ihrem Entwurfe hervorbringt" (KrV, B XIII). Die Helmholtzsche These von der Gesetzmäßigkeit der Natur als Bedingung ihrer Begreifbarkeit scheint mir damit noch kompatibel zu sein, nicht aber die Versuche, wissenschaftliche Gesetzesaussagen zu ontologisieren. Naturgesetze sind für den nachkritischen Kant nicht länger ontologisch als Momente in einer Welt "an sich seiender Dinge" zu verorten. Die Helmholtzsche Abbildtheorie der Gesetzeserkenntnis steht auf dem Standpunkt des vorkritischen Kant.

In den naturwissenschaftlichen Texten, in denen Helmholtz von Naturgesetzen spricht, gelten die in der Theorie formulierten Gesetzesaussagen als adäquate Abbildungen der unabhängig vom erkennenden Subjekt existierenden Zusammenhänge. Die Gesetzesaussagen wissenschaftlicher Theorien werden (völlig unbeeindruckt von der Hypothetisierung der Gesetzeserkenntnis auf der Reflexionsebene) praktisch ontologisiert.[7]

Während also die Helmholtzsche Zeichentheorie für die Sinnesempfindungen in Übereinstimmung mit der Kantischen Erkenntniskritik keine ontologische Deutung der von den Sinnen wahrgenommenen Zeichen zuläßt, mündet seine Abbildtheorie der Gesetzeserkenntnis in eine bestenfalls mit dem vorkritischen Kant noch vereinbare Ontologie, in der die wissenschaftlich erkannten Gesetze als in der Natur unter bestimmten Bedingungen wirkende Mächte angesehen werden. Akzeptiert man den Helmholtzschen Versuch nicht, die Abbildtheorie der Gesetzeserkenntnis aus der Zeichentheorie der Sinnesempfindungen zu erklären, weil darin eine schwerwiegende metaphysi-

7 Helmholtz selbst benutzt allerdings das metaphysisch belastete Wort 'ontologisieren' nicht. Es soll hier sein an naturwissenschaftlichen Texten nachweisbares Vorgehen adäquat wiedergeben.

sche Unterstellung enthalten ist, sieht man sich unweigerlich mit der Inkonsistenz dieser Erkenntnistheorie konfrontiert und muß entscheiden, welcher der beiden Teile als der problematisierungsbedürftigere anzusehen ist. Nach den bisherigen Ausführungen dürfte es jetzt nicht mehr verwundern, wenn ich der Abbildtheorie der Gesetzeserkenntnis skeptischer gegenüberstehe als der Zeichentheorie der Sinnesempfindungen.

Der Physiker Helmholtz steht in jener Traditionslinie neuzeitlicher Naturwissenschaft, die nach den mathematisch formulierbaren Zusammenhängen von Naturerscheinungen fragt, dabei schon frühzeitig die ursprünglich normativ gemeinte Gesetzesmetaphorik deskriptiv gewendet hat und dann dazu übergegangen ist, die mathematisch formulierten Gesetzesaussagen zu ontologisieren, also mathematische Strukturen mit wirklichen Zusammenhängen zu identifizieren. Wie beispielsweise aus den Helmholtzschen Vorlesungen über theoretische Physik ersichtlich, ergeben sich daraus für den Aufbau und das Verständnis der klassischen Physik keine naturwissenschaftlich inakzeptablen Folgerungen. Auf diese stoßen wir erst, wenn wir seine Sicht auf das Gesetzesproblem auch in der nichtklassischen Physik beibehalten wollen.

4. Grenzen des Helmholtzschen Konzepts der Gesetzeserkenntnis in den Naturwissenschaften

Helmholtz reflektiert das Gesetzesproblem mit der Erfahrung aus der Blütezeit der klassischen Physik. Seine Sicht auf die erkenntnistheoretischen und ontologischen Dimensionen des Gesetzesproblems ist - wie wir heute wissen - nur eine der mit der klassischen Physik verträglichen Interpretationen. Die Einsicht, daß sie nicht die einzig mögliche Interpretation ist, verdanken wir weniger der philosophischen Kritik als vielmehr der im Anschluß an die Theorienentwicklung der Physik des 20. Jahrhunderts angestellten Reinterpretationen der damit klassisch gewordenen, aber eben nicht deshalb überholten Theorien.

Diesen Gedanken will ich nun noch kurz im Zusammenhang mit der nichtrelativistischen Quantenmechanik zu illustrieren versuchen.

Eine der Darstellungen dieser Theorie, nämlich die Feynmansche, läßt sehr schön erkennen, warum der von der klassischen Mechanik vorausgesetzte Bahnbegriff aufgegeben werden muß. Will man bei der Darstellung quantenmechanisch zu beschreibender Prozesse weiterhin mit Bahnen in Raum und Zeit operieren, muß die Mannigfaltigkeit aller unter den jeweiligen makrophysikalischen Beobachtungsbedingungen geometrisch möglichen Kurven in Rechnung gestellt werden. Gleichzeitig wird die Frage sinnlos, welche von diesen Bahnkurven die der tatsächlichen Bewegung ist. Alle geometrisch möglichen Kurven liefern über Phasenintegrale einen Beitrag zur Wahrscheinlichkeitsvoraussage eines tatsächlich beobachtbaren Versuchsausgangs. Aus der Theorie folgt jetzt, daß, mit Ausnahme einer Menge vom Maße Null, die überall stetigen Bahnkurven in keinem Punkte differenzierbar sind. Wir sehen also, wie weitblickend es von Helmholtz war, für die in der Newtonschen Mechanik unterstellten Bahnkurven nicht nur Stetigkeit in jedem Punkt des Raum-Zeit-Kontinuums zu verlangen, sondern überdies auch die zweifache Differenzierbarkeit des Weges nach der Zeit (Helmholtz 1893: 8). Diese Forderung ist also nicht automatisch für eine überall stetige Funktion erfüllt. Sie gehört zu den explizit zu benennenden Bedingungen dafür, daß wir mit den Größen Geschwindigkeit und Beschleunigung rechnen können. Für Helmholtz ergab sich die Notwendigkeit zur Formulierung dieser Bedingung aus der mathematischen Funktionentheorie, die zu seiner Zeit schon viele Beispiele für überall stetige, aber nirgends differenzierbare Funktionen kannte. Inzwischen gibt es mit der nichtrelativistischen Quantenmechanik eine physikalische Theorie, die zeigt, daß auch diese von der Mathematik eröffnete Möglichkeit nicht ohne empirische Relevanz ist. Damit aber wird die von Helmholtz vorausgesetzte Interpretation klassisch mechanischer Bewegungsgrößen problematisch. Um mit physikalischen Größen arbeiten zu können, die als mathematische Differentialquotienten eingeführt werden, müssen die Bedingungen ihrer Meßbarkeit geklärt werden. Wir wissen seit der Feynmanschen Darstellung der Quantenmechanik, daß und warum sie nicht als ontologische Gegebenheiten unterstellt werden dürfen.

Die Quantenmechanik aber geht hinsichtlich der Frage der mathematischen Darstellung gesetzmäßiger Zusammenhänge in der Wirklichkeit noch einen Schritt weiter. Während es für die Newtonsche Mechanik physikalisch folgenlos bleibt, wenn die in Größengleichungen mathematisch beschriebenen Beziehungen mit entsprechenden Verhältnissen in der Wirklichkeit identifiziert werden, führt das bei quantenmechanischen Bewegungsgleichungen zu unauflösbaren Paradoxien. Die quantenmechanischen Zustandsgrößen sind in einem mathematischen Darstellungsraum, dem Hilbert-Raum, formuliert, der den dreidimensionalen Meßraum nicht als Unterraum enthält. Wir können also die mathematischen Strukturen der physikalischen Theorie nicht mehr als Abbildungen realer Verhältnisse ansehen. Jetzt werden auch die innerhalb mathematisierter Theorien vorliegenden Gesetzesformulierungen zu Zeichen (oder, wie Bohr es ausgedrückt hat, zu 'Symbolen'), die für reale Verhältnisse stehen. Damit aber verschwindet zumindest für diese Gesetzeserkenntnisse der bei Helmholtz ausgemachte erkenntnistheoretische Bruch.

So wie wir es im Laufe unserer Individualentwicklung verstehen, die mit unseren Sinnen wahrgenommenen Zeichen mit Blick auf praktisches Handeln richtig zu deuten, haben es die Physiker gelernt, die Mitte der 20er Jahre gefundenen quantenmechanischen Gesetzesformulierungen richtig zu deuten. Dafür spricht die erfolgreiche Laborpraxis der Physiker und ein nicht unerheblicher Teil der zeitgenössischen industriellen Praxis. Zugleich aber verändert sich damit unser theoretisches Weltverständnis. Wir vermögen nämlich nicht mehr problemlos zu sagen, was eigentlich der ontologische Gehalt eines quantenmechanischen Gesetzes ist.

Philosophisch betrachtet, haben wir in dieser Situation zwei Möglichkeiten. Wir können die Quantenmechanik als einen esoterischen Sonderfall behandeln. Dann ist es nicht ausgeschlossen, hinsichtlich der Gesetzeserkenntnis auch weiterhin den Helmholtzschen Standpunkt der Abbildtheorie beizubehalten. Wenn wir uns aber entschließen, die zugegebenermaßen auf den ersten Blick etwas esoterisch anmutende Quantenmechanik als einen Fingerzeig zu werten, auch die Verhältnisse in der klassischen Physik noch einmal kritisch zu durchdenken und der Frage nachzugehen, warum eigentlich die Helmholtzsche Position so lange als die einzig mögliche gegolten hat, dann gewinnen zugleich philosophische Denksysteme der Vergangenheit wieder an Interesse, die eine konsistente Zeichentheorie der Erkenntnis favorisiert haben.

Mir will scheinen, daß mittels solcher Zeichentheorien der Erkenntnis ein wissenschaftlicher Realismus verteidigbar wird, der den rationalen Kern konstruktivistischer Kritik zu bewahren weiß, trotzdem aber nicht Gefahr läuft, in radikal konstruktivistischer Wendung schließlich nicht mehr zu verstehen, inwiefern Naturwissenschaft eine Natur zu erkennen sucht, die selbst dann noch existieren würde, wenn es weder diese Wissenschaft noch Erkenntnissubjekte mit ihren spezifischen Erkenntnisinteressen gäbe.

Ich hoffe, mit diesem Schluß nicht allzu weit vom Gegenstand dieses Bands, dem historischen Helmholtz, weggeführt zu haben. Noch lieber aber sähe ich es, wenn er dazu angetan wäre, meine These vom erkenntnistheoretischen Bruch zwischen Sinnesempfindungen und Gesetzeserkenntnis bei Helmholtz ins Bewußtsein zu heben.

Ulrich Röseberg hat diesen Beitrag noch selbst für den Druck bearbeiten können. Der Text gibt einen Eindruck von dem Weg, den Röseberg im Ausgang von seinen Büchern zu neuen Studien über das Verhältnis von Naturwissenschaft, Erkenntnistheorie und Metaphysik der Natur beschritten hat, bis sein Tod am 24. März 1994 seiner Arbeit ein jähes und vorzeitiges Ende setzte.

Literatur

Helmholtz, Hermann von: *Ueber die Erhaltung der Kraft* (1847). Hg. Christa Kirsten und Kurt Zeisler. Berlin: Akademie - Verlag, 1982.

Helmholtz, Hermann von: *Ueber das Verhältnis der Naturwissenschaften zur Gesamtheit der Wissenschaft* (1862). In: Ders.: *Vorträge und Reden.* Erster Band. Braunschweig: Verlag von Friedrich Vieweg und Sohn, 1896.

Helmholtz, Hermann von: Vorrede zur Ersten Auflage (1865). In: Ders.: *Vorträge und Reden.* Erster Band. Braunschweig: Verlag von Friedrich Vieweg und Sohn, 1896.

Helmholtz, Hermann von: Ueber das Ziel und die Fortschritte der Naturwissenschaft (1869) In: Ders.: *Vorträge und Reden.* Erster Band. Braunschweig: Verlag von Friedrich Vieweg und Sohn, 1896.

Helmholtz, Hermann von: Die Thatsachen in der Wahrnehmung (1878). In: Ders.: *Vorträge und Reden.* Zweiter Band. Braunschweig: Verlag von Friedrich Vieweg und Sohn, 1896.

Helmholtz, Hermann von: Goethe's Vorahnungen kommender naturwissenschaftlicher Ideen (1892) In: Ders.: *Vorträge und Reden.* Zweiter Band. Braunschweig: Verlag von Friedrich Vieweg und Sohn, 1896.

Helmholtz, Hermann von: Vorlesungen über die Dynamik discreter Massenpunkte (1893). In: Ders.: *Vorlesungen über theoretische Physik.* Band I, Abteilung 2. Leipzig: Verlag von Johann Ambrosius Barth, 1911.

Hertz, Heinrich: *Die Prinzipien der Mechanik.* Leipzig 1894.

Hörz, Herbert: *Der dialektische Determinismus in Natur und Gesellschaft.* Berlin: Akademie-Verlag.

Hörz, Herbert; Wollgast, Siegfried: Einleitung zu: Hermann von Helmholtz: *Philosophische Vorträge und Aufsätze.* Berlin: Akademie-Verlag 1971.

Kant, Immanuel: *Kritik der reinen Vernunft*, Akademieausgabe Bd. III.

Kirchhoff, Gustav: *Vorlesungen über Mechanik.* Leipzig 1876.

Röseberg, Ulrich: *Determinismus und Physik.* Berlin: Akademie-Verlag, 1975.

Helmholtz über die Begreiflichkeit der Natur

Lorenz Krüger

I

Zunächst möchte ich Helmholtz' abstrakten Begriff der Begreiflichkeit darstellen und einige in ihm angelegte Probleme analysieren. Dabei ist es nützlich, von der Frage auszugehen, welchen Stellenwert dieser Begriff in Helmholtz' Verständnis von Wissenschaft innehat. Hier gibt uns der Vortrag Auskunft, den er 1869 vor der Versammlung der Naturforscher in Innsbruck gehalten hat und der den Titel trägt "Über das Ziel und die Fortschritte der Naturwissenschaft". Als dieses Ziel und damit zugleich den Maßstab zur Beurteilung des wissenschaftlichen Fortschritts bestimmt er dort die unlösbare Verbindung zweier Teilziele: (1) "geistig faßbarer Zusammenhang der Kenntnisse", und (2) "Herrschaft des Menschen über die Naturmächte" (1869, VR I, 374). Die Zusammengehörigkeit beider Ziele ergibt sich unter anderem daraus, daß sie eine gemeinsame notwendige Bedingung haben: "die Kenntnis des Gesetzes der Erscheinungen" (ibid.), so daß die Aufgabe (im Singular!) der Wissenschaft geradezu bestimmt werden kann als diejenige, "die Gesetze der Thatsachen zu finden" (1871, VR II, 47). Eben hierauf beruht es, daß die Wissenschaft "das Leben der modernen Menschheit so eingreifend umgestaltet hat, sowohl durch ihren geistigen Einfluss wie durch die Unterwerfung der Naturkräfte unter die Zwecke des Geistes" (ibid., 50). An diese doppelte Zielsetzung nun knüpft Helmholtz den Begriff der Begreiflichkeit (1847, WA I, 13) oder Begreifbarkeit (1878, VR II, 243) der Natur. Er sagt: "Die Gesetzmäßigkeit [...] ist die Bedingung der Begreifbarkeit." Und weiter: "Das Begreifen [...] ist die Methode, mittels deren unser Denken die Welt sich unterwirft, die Thatsachen ordnet, die Zukunft vorausbestimmt" (1878, VR II, 243). Kurz zusammengefaßt: "Das *Gesetz* der Erscheinungen finden, heisst sie ***begreifen***" (1869, VR I, 375).

Dieser Bestimmung zufolge ist Begreifen als der denkende und handelnde Zugriff auf die Natur das Ziel der Naturwissenschaft und das Erfassen der Gesetzmäßigkeit der Natur die Bedingung, unter der wir das Ziel verfolgen oder erreichen können. Um den Begriff der Begreiflichkeit zu verstehen, müssen wir also fragen, was Gesetzmäßigkeit bedeutet und unter welchen Bedingungen sie erfaßt werden kann.[1]

Zwecks Klärung dieser Fragen hält sich Helmholtz zunächst an die Etymologie von "begreifen": er bestimmt das Gesetz als allgemeinen *Begriff*, unter den sich gleichartige Naturgegebenheiten zusammenfassen lassen (1869, VR I, 375; 1862, VR I, 169). Hier unterscheidet er sogleich zwei Fälle: existierende Dinge, die unter einen "Gattungsbegriff", und Vorgänge oder Ereignisse, die unter ein "Gesetz" gebracht werden können (1862, VR I, 169); Beispiele sind "Säugetier" und "Brechungsgesetz" (1869, VR I, 375). Solche Begriffe aufzusuchen, ist "das Geschäft des experimentellen Theils unserer Wissenschaften" (1847, WA I, 13), in dem sichtbare Erscheinungen in Zusammenhänge geordnet werden. Zum eigentlichen Begreifen der Natur sind indes zusätzlich tiefere Zusammenhänge erforderlich, nämlich solche zwischen unbekannten Ursachen und ihren sichtbaren Wirkungen. Derartige Zusammenhänge zu entdecken, ist die Aufgabe des theoretischen Teils der Wissenschaften; hier geht es darum, "zu begreifen nach dem Gesetze der Causalität" (1847, WA I, 13).

Ferner folgt Helmholtz der mit den Begriffen Gesetz und Kausalität damals (und nicht selten heute noch) wie selbstverständlich verbundenen Meinung, daß sie keine Ausnahmen dulden: "Wo wir ein Naturgesetz vollständig kennen, müssen wir auch die *Ausnahmslosigkeit seiner Geltung* fordern und diese zum Kennzeichen seiner Richtigkeit machen" (1869, VR I, 375), denn - so meint Helmholtz - "[d]er Natur gegenüber besteht kein Zweifel, daß wir es mit einem ganz strengen Causalnexus zu tun haben, der keine Ausnahmen zulässt" (1862, VR I, 178). Eben darum müssen veränderliche Ursachen, deren Wirk-

---

1 Zur Einbettung des Begriffs in den weiteren Kontext von Helmholtz' Naturphilosophie vergleiche man auch den Beitrag von Heidelberger, zum Gesetzesbegriff den von Röseberg, beide in diesem Band.

samkeit noch von weiteren Bedingungen abhängt, schrittweise auf ihre jeweiligen Ursachen zurückverfolgt werden, "bis wir zuletzt zu letzten Ursachen gekommen sind, welche nach einem unveränderlichen Gesetz wirken" (1847, WA I, 13). Nach dem üblichen Verständnis liegt es in der Natur des Begriffs, daß er als "Begriff einer in der Zeit wechselnden Reihe von Erscheinungen das zusammenzufassen suchen muss, was in allen ihren Stadien gleich bleibt". Und hier zitiert Helmholtz Schiller:

> "Sucht das vertraute Gesetz in des Zufalls grausenden Wundern,
> suchet den ruhenden Pol in der Erscheinungen Flucht".

Als ruhende Pole nennt er, der schon genannten Einteilung der Begriffe folgend, "die *Substanz*" und "das gleichbleibende Verhältnis zwischen veränderlichen Größen: das sie verbindende Gesetz" (1878, VR II, 240), in welcher Unterscheidung das schon in der philosophischen Einleitung zur Arbeit über den Energiesatz verwendete Begriffspaar *Materie* und *Kraft* (1847, WA I, 14) wiederkehrt. Darauf wie auf das im Schiller-Zitat anklingende Bestreben, allen Zufall loszuwerden, komme ich noch zurück.

Zunächst muß eine weitere Implikation der Gesetzmäßigkeit hervorgehoben werden, die ihr allererst den Charakter einer äußeren Realität verschafft. Wir müssen ihren "Erfolg eintreten sehen ohne Willkür, ohne Wahl, ohne unser Zuthun, mit einer die Dinge der Aussenwelt ebenso so gut, wie unser Wahrnehmen, zwingenden Notwendigkeit". So gesehen, werden uns Gesetze zur "objectiven Macht", benennen wir sie mit dem Wort "Kraft" (1869, VR I, 376) und, wenn die Kraft in der Zeit andauert, mit dem Wort "Ursache" (1892, VR II, 354). Der abstrakte Begriff der Kraft bringt nach Helmholtz nur das eine zusätzlich zum Ausdruck: "dass dieses Gesetz nicht willkürlich erfunden, sondern dass es ein zwingendes Gesetz der Erscheinungen sei" (1869, VR I, 376). Den Hintergrund dieser Begriffsbildung liefert seine Erkenntnistheorie, derzufolge wir schon beim Wahrnehmen experimentieren: "Wir haben [...] nicht nur wechselnde Sinneseindrücke, die über uns kommen, ohne dass wir etwas dazu thun, sondern wir beobachten unter forthdauernder eigener Tätigkeit, und gelangen dadurch zur Kenntnis des *Bestehens* eines gesetzlichen Verhältnisses [...]. Jede unserer willkürlichen Bewegungen, durch die wir die Erscheinungen der Objecte abändern, ist als ein Experiment zu betrachten [...]" (1878, VR II, 237). Nur scheinbar paradox zusammengefaßt: nur dank unseres Zutuns können wir etwas erfassen, was ohne unser Zutun besteht, das Objekt (vgl. ibid., 226). Zwar nicht die Existenz, wohl aber das Erkennen natürlicher Gesetzmäßigkeit hat zur Voraussetzung, daß wir ihr nicht (oder jedenfalls nicht völlig) unterworfen sind. Du Bois-Reymond's zweite Grenze des Naturerkennens, das Rätsel des Bewußtseins, ist als Konsequenz des Helmholtzschen Begriffs der Begreiflichkeit der Natur angelegt.

Natürlich ist damit noch nicht entschieden, ob es nicht auch Gesetze des Geistes geben könne; nur der Dualismus ist besiegelt: diese Gesetze können nicht selbst einen Teil der Gesetze der Körperwelt bilden. Wie Helmholtz ausführt, hängt die Kraft der Überzeugung davon, daß wir in einem Experiment etwas Objektives erfaßt haben, an einer, wie er sagt, "wesentlichen Voraussetzung", nämlich "dass unser Willensimpuls weder selbst durch physische Ursachen, die gleichzeitig auch den physischen Process bestimmten, schon mit beeinflusst worden sei, noch seinerseits psychisch die darauf folgenden Wahrnehmungen beeinflusst habe" (1878, VR II, 237). Kurz, zwar mag es Gesetze der psychischen Sphäre geben, die, wie Helmholtz sagt, "das Object der philosophischen, philologischen, historischen, moralischen und socialen Wissenschaften ausmachen" (1862/3, VR I, 190); aber diese Gesetze müssen jedenfalls von denen der physischen Sphäre unabhängig sein. Folgerichtig formuliert Helmholtz eine Arbeitsteilung zwischen Philosophie und Naturwissenschaft wie folgt: "Die erstere [...] sucht aus unserem Wissen und Vorstellen auszuscheiden, was aus den Einwirkungen der Körperwelt herrührt, um rein herzustellen, was der eigenen Thätigkeit des Geistes angehört. Die Naturwissenschaft im Gegentheil sucht abzuscheiden, was Definition, Bezeichnung, Vorstellungsform, Hypothese ist, um rein übrig zu behalten, was der Welt der Wirklichkeit angehört, deren Gesetze sie sucht" (1878, VR II, 218).

Ferner rechnet Helmholtz mit Grenzen der Begreiflichkeit überall da, wo es Veränderungen gibt, "die sich dem Gesetz einer nothwendigen Causalität entziehen, die also in das Gebiet einer Spontaneität, Freiheit, fallen" (1847, WA I, 13). Sollten wir nicht annehmen müssen, daß spontan gesteuertes Wahrnehmen, geplantes Experimentieren und die gesamte Geschichte der Naturforschung Gesetzen der psychischen Welt mit Notwendigkeit folgen, wovon bei Helmholtz nicht die Rede ist, wäre der Dualismus von Natur und Geist als eine notwendige Voraussetzung der Begreiflichkeit der Natur erwiesen. Beides, notwendiger Gesetzeszusammenhang des Begriffenen und Freiheit des Begreifenden, müßten zusammenkommen, wenn Natur erkannt werden soll. In der Tat ist schwer vorzustellen, wie wir uns die eingangs genannten Zwecke allen Begreifens - geistiges Erfassen des Naturzusammenhangs und Herrschaft des Menschen über die Natur - bei Verletzung dieser Bedingung sinnvoll denken könnten.

Daß nun allerdings in dieser Gedankenformation die Natürlichkeit des Menschen zum Problem wird, braucht man seit Descartes nicht mehr eigens zu betonen. Freilich ist dies für den metaphysikkritisch eingestellten Naturwissenschaftler des 19. Jahrhunderts ein besonders heikler Punkt. Gehört der Mensch mit zur Natur - und Forscher wie Helmholtz zweifelten daran natürlich nicht -, dann bewirken die Voraussetzungen des Projekts, die Natur völlig zu begreifen, daß just das Gegenteil droht, in den Worten du Bois-Reymonds ausgedrückt: "dass das Naturerkennen, welches vorher als unser Causalitätsbedürfniss vorläufig befriedigend bezeichnet wurde, in Wahrheit dies nicht thut, und kein Erkennen ist" (1872, 18). (Zwar ist diese Bemerkung zunächst auf die dem mechanistischen Weltbild der Natur selbst innewohnenden Erkenntnisgrenzen bezogen; der Sache nach paßt sie indes ebenso gut auf die Unbegreiflichkeit des mechanisch nicht faßbaren Bewußtseins.)

Die Trennung von Geist und objektiver Natur ermöglicht oder erleichtert es Helmholtz nun, die Bedingungen des Begreifens der Natur so weiter auszuarbeiten, daß er über einige ziemlich metaphysisch klingende Zwischenschritte zum bekannten mechanistischen Naturbild gelangt. Da der einschlägige Kerntext, die Schrift *Über die Erhaltung der Kraft* von 1847, wohlbekannt ist, kann ich mich kurz fassen. Von der Deutung der Naturgesetze als Kräfte und Ursachen war schon die Rede. Damit ist begründet, warum für Helmholtz die Aufgabe des Begreifens schließlich die Form annimmt, "dass wir die *Kräfte* aufzusuchen haben, welche die *Ursachen* der Erscheinungen sind" (1869, VR I, 377). Hierbei deutet er, wieder der Etymologie folgend, die Ursache keineswegs, wie es unser und schon der damalige Sprachgebrauch will, als ein zeitlich und räumlich umgrenztes *Ereignis*, sondern als eine zeitüberbrückende *Sache*, "d.h. das hinter dem Wechsel ursprünglich Bleibende und Bestehende", die *res* oder Realität, die auf uns einwirkt und deshalb "das Wirkliche" oder "das Wirksame" heißt (1878, VR II, 241; vgl. 1892, VR II, 354). Gesetzmässigkeit der wahrgenommenen Erscheinungen für sich genommen ist nicht genug; sie soll als die Gesetzmäßigkeit der Wirkungen einer *unverändert bestehenden Wirklichkeit* verstanden werden können. Eben das trennt Helmholtz, so deutlich und im Laufe seines Lebens immer stärker er empiristisch gesonnen gewesen sein mag, als einen Realisten von den phänomenalistischen Nachfahren, etwa Ernst Mach, und verbindet ihn mit der Tradition rationalistischer Naturphilosophie vom 17. Jahrhundert bis hin zu Kant.[2] Erst die zugleich rationalistische und (in einem unter der Einwirkung Kants nicht leicht

2 Der Gegensatz und Übergang zwischen realistischer und phänomenalistischer Auffassung der Physik wird bei Cassirer 1950, Part I, Chap. V, besprochen; vgl. dazu kritisch Anhang 1.

zu bestimmenden Sinne) realistische Ausgestaltung geben Helmholtz' Programm des völligen Begreifens der Natur seine Konkretion und seine Pointe; sie sind es auch, die seine Grenzen deutlich werden lassen. Hiervon soll im folgenden zweiten Teil meines Beitrags die Rede sein.

II

Es ist allenthalben bekannt und braucht deshalb hier nicht erneut entwickelt zu werden, daß für Helmholtz das Ideal der völligen Begreiflichkeit der Natur im mechanistischen Weltmodell seine konkrete Gestalt annahm. Aus der Fülle der historisch wie philosophisch beachtenswerten Aspekte dieses Modells greife ich hier einen einzigen heraus, den ich zur Diskussion stellen möchte: die Frage nach dem Wesen der Materie, sofern sie - um mit den Worten Kants zu reden - als "das Bewegliche im Raume" (1786, I, Erklärung 1) bestimmt werden kann.[3] In dieser Sache wollte Helmholtz etwas begründen, was von Galilei über Descartes bis zu Kant als intuitive Selbstverständlichkeit gegolten hatte: daß das völlig Begreifliche nur im Raum und als räumlich Ausgedehntes vorstellbar sei. Anders als bisher, wo es nur um den abstrakten Begriff der Begreiflichkeit ging, kann ich nunmehr die allmähliche Entwicklung der Helmholtzschen Naturphilosophie nicht länger außer acht lassen. Sie führte von der Übernahme traditioneller Newtonscher Positionen schließlich zu einem bewußt hypothetischen oder empiristischen Realismus.[4]

Um das diese Entwicklung vorantreibende Problem zu profilieren, erinnere ich kurz an ein selten beachtetes Argument Descartes', das Helmholtz' Schwierigkeiten vorwegnimmt, indem es einen wesentlichen Punkt seiner Erkenntnistheorie antizipiert. Es geht dabei um die Beziehung unserer Ideen oder Vorstellungen zur Realität, von der sie Ideen sind. Bekanntlich folgte auch Descartes der verbreiteten Annahme der Naturphilosophie des 17. Jahrhunderts, daß einige dieser Ideen, die primären, Eigenschaften der realen Dinge außerhalb des Bewußtseins abbilden, andere, die sekundären, jedoch lediglich Signale von bestimmten Dispositionen der primären Eigenschaften sind, ohne daß eine irgendwie geartete Ähnlichkeit besteht. Obwohl diese Anschauung

3 Helmholtz' Orientierung an Kant wird untersucht von Heimann 1974; vgl. dazu kritisch Anhang 2.

4 Man vergleiche hierzu auch die Beiträge von Schiemann und Heidelberger, ferner zu den aus der Elektrizitätslehre erwachsenen Schwierigkeiten mit der Begreiflichkeit der Natur den ausführlichen Beitrag von Darrigol in diesem Band.

durchweg mit einer mechanistischen Theorie der Sinneswahrnehmung Hand in Hand geht, werden die aus dieser entspringenden Konsequenzen gewöhnlich übergangen. Nicht so von Descartes. Er macht sich und uns klar, daß im Falle primärer, insbesondere *räumlicher* Eigenschaften des Wahrgenommenen nicht weniger als im Falle sekundärer Qualitäten gilt, daß die Übertragung der Außenreize zum Gehirn und damit zur Schwelle des Bewußtseins jedes Bewahren von Ähnlichkeit illusorisch macht. Er drückt dies so aus, daß *alle* unsere Ideen samt ihrer spezifischen qualitativen Vielfalt "angeboren" seien, *mit Ausnahme allein der Zeit und Gelegenheit ihres Auftretens im Bewußtseins* (1647, AT VIII, 292-294). Dies ist in kaum anderen Worten eine Kurzfassung der Zeichentheorie der Wahrnehmung, wie wir sie von Helmholtz kennen: "Empfindungen sind [...] Wirkungen, welche durch äussere Ursachen in unseren Organen hervorgebracht werden, und wie eine solche Wirkung sich äussert, hängt natürlich ganz wesentlich von der Art des Apparates ab, auf den gewirkt wird". Demnach beschränkt sich die Beziehung von Empfindungen und Objekten darauf, "dass das gleiche Object, unter gleichen Umständen zur Einwirkung kommend, das gleiche Zeichen hervorruft" (1878, VR II, 222). Zeit und Gelegenheit sind gemeinsam, mehr nicht. Eben diese Abhängigkeit von der vorgegebenen Art des physisch-psychischen Apparats belegt Descartes nun mit dem Wort "angeboren".

Um Verwirrung zu vermeiden, muß diese Bedeutung von "angeboren" klar von jener anderen unterschieden werden, die Helmholtz' Polemik gegen den "Nativismus" trägt, also gegen die Auffassung, daß ohne Lernen aus Erfahrung sogleich fertige Vorstellungen von Objekten in uns entstehen (z.B. 1878, VR II, 236). Sie muß mehr noch von einer zweiten Bedeutung von "angeboren" bei Descartes unterschieden werden, die nur die von Gott unserem Verstande eingepflanzten Ideen umfaßt: die des eigenen Ich, die Gottes und die der körperlichen Substanz mit all ihren mathematischen Merkmalen. Erst an dieser Frage der angeborenen Ideen im engeren Cartesischen Sinne scheiden sich die Geister, nämlich die empiristischen wie Locke und - etwas zögerlich - Helmholtz von den rationalistischen wie Descartes und Kant.

Die Zögerlichkeit nun hat durchaus triftige und tiefliegende Gründe, unter denen, soweit die Naturphilosophie betroffen ist, der räumlichen Auffassung der Wirklichkeit der Vorrang gebührt. 1847 sah sich Helmholtz, wie er selbst anmerkt, "durch *Kant's* erkenntnistheoretische Ansichten stärker beeinflusst, als ich jetzt [1881] noch für richtig anerkennen möchte" (1881, WA I, 68). Wie z.B. Heimann gezeigt hat (1974), folgte Helmholtz, wenn nicht dem Text, so doch dem Argumentationsgang der *Metaphysischen Anfangsgründe der*

Naturwissenschaft von 1786, zumindest in groben Zügen. Und ganz wie bei Kant ist dabei über die kausalgesetzliche Zeitfolge hinaus die Räumlichkeit der wahrgenommenen materiellen Natur einfach *vorausgesetzt*: "[...] wir unterscheiden an ihr [der Materie] die räumliche Vertheilung und die Quantität (Masse), welche als ewig und unveränderlich *gesetzt* wird" (1847, WA I, 14; Hervorhebung hinzugefügt). Ferner verknüpfen Kräfte, die objektivierten Formen der Gesetze, nicht *irgendwelche* uns gegebenen Erscheinungen, sondern sind von vornherein angesetzt als solche, die die *Materie* in Sorten differenzieren; ja Helmholtz sieht die unlösbare Verknüpfung von Materie und Kraft geradezu im Begriff des Gesetzes vollzogen, der, wie wir wissen, *die* entscheidende Bedingung des Naturbegreifens ist: "Die [...] Unmöglichkeit beide [Materie und Kraft] isolirt zu denken, ergibt sich [...] einfach daraus, dass das Gesetz einer Wirkung Bedingungen voraussetzt, unter denen es zur Wirksamkeit kommt. Eine von der Materie losgelöste Kraft wäre die Objectivierung eines Gesetzes, dem Bedingungen seiner Wirksamkeit fehlen" (1881, WA I, 68, als Kommentar zu 1847, WA I, 14).

Ferner wird die Befreiung der Gesetzlichkeit von den Schlacken des Zufalls, wovon schon die Rede war, ebenfalls räumlich konkretisiert: "[...] wenn wir die einfachsten und allgemeinsten Wirkungsgesetze [...] kennenlernen und [...] befreien wollen von den Zufälligkeiten der Form, der Grösse und Lage der zusammenwirkenden Körper, so müssen wir zurückgehen auf die Wirkungsgesetze der kleinsten Volumtheile, [...] der Volumelemente. Diese aber sind nicht, wie die Atome, disparat und verschiedenartig, sondern continuirlich und gleichartig" (1871, VR II, 45/46). Kräfte also spielen zwischen "materiellen Punkten", wie es 1847 heißt, und können schon deshalb nichts anderes sein als Zentralkräfte, die allein vom Abstand eines jeweils herausgegriffenen Massenpunktpaares abhängen. Ihre Wirkung kann in nichts anderem bestehen als in räumlicher Veränderung. Sie müssen also allesamt "Bewegungskräfte" sein. Wie tief Helmholtz sich hier auf eine philosophische Betrachtung *a priori* eingelassen hat, zeigt unter anderem, daß er zur Begründung der reinen Distanzabhängigkeit der Kräfte den Satz vom zureichenden Grunde anruft (1847, WA I, 15-16). (Wie der Übergang zu Zentralkräften zwischen Massen*punkten* mit der Differenzierung der Kraftarten nach chemischen Elementen vereinbar zu machen sei, bleibt dabei unklar.)

Das Fazit ist klar und kurz: "Es bestimmt sich also endlich die Aufgabe der physikalischen Naturwissenschaften dahin, die Naturerscheinungen zurückzuführen auf unveränderliche, anziehende und abstossende Kräfte, deren Intensität von der Entfernung abhängt. Die Lösbarkeit dieser Aufgabe ist zugleich die Bedingung der vollständigen Begreiflichkeit der Natur" (1847, WA I, 16).

Das in diesen Worten skizzierte Naturbild sieht einfach und klar aus. Problematisch ist jedoch, daß es mit der sich damals erst entwickelnden Erkenntnistheorie von Helmholtz nicht ohne weiteres in Einklang zu bringen ist. Warum sollte die Objektivierung der als *Zeit*folgen beobachteten Gesetze notwendig eine *räumliche* sein? Was wissen wir aus den bloß als Zeichen zu nehmenden Empfindungen über den Raum? Weshalb sollte er mehr denn bloße Erscheinungsform des sinnlichen Materials, nämlich, wie es 1847 herauskam, subjekttranszendente Form der begriffenen Welt, "die nothwendige Begriffsform der Naturauffassung" (1847, WA I, 17), sein? Bei der Antwort auf diese Frage mochten Rationalisten wie Galilei und Descartes ihre Intuition in den Kontext einer Theorie der von Gott dem Menschen eingepflanzten Verstandesideen stellen; der Empiriker Helmholtz braucht eine transzendentalphilosophische Rechtfertigung, also eine solche, die auf notwendige Bedingungen nun nicht mehr bloß des Gesetzesbegriffs selbst - das war klassische Metaphysik -, sondern unserer *Erkenntnis* von Gesetzen Bezug nimmt. Man findet eine solche Begründung in den Schriften von Helmholtz, wenn auch nicht unzweideutig ausgeführt, so doch deutlich angelegt.

Es war schon davon die Rede, daß nur mit willkürlicher Körperbewegung verbundenes Wahrnehmen, ein Wahrnehmen, das zugleich schon die einfachste Form des Experimentierens ist, uns mehr geben kann als das Muster regelmäßiger Abfolgen von Eindrücken, nämlich ein Objekt. An dieser Auffassung fällt auf (was man übrigens schon bei Descartes' Begründung der Angeborenheit aller Ideen hätte anmerken können), daß bereits eine auf Räumliches bezogene Sprache in Anspruch genommen, daß sie als sinnvoll und sachbezogen und in diesem Sinne bedeutungsvoll vorausgesetzt werden muß. Offenbar noch nicht 1847, wohl aber 1878, als er den Vortrag "Die Thatsachen der Wahrnehmung" hielt und ausdrücklich die Konstruktion unserer Raumanschauung aus Empfindungen erörterte, war Helmholtz zu der Überzeugung gelangt, daß "ein System [...] des extremsten subjectiven Idealismus" als Denkmöglichkeit nicht widerlegt werden könne, daß wir mithin auf eine "realistische *Hypothese*" angewiesen bleiben, wenn wir unserer gewöhnlichen Selbstbeobachtung folgen wollen, derzufolge Objekte unabhängig von unserem Wollen und Handeln den Raum erfüllen (1878, VR II, 283-239).

In eben dieser Arbeit findet sich nun auch die kantkritische Reflexion darüber, daß sehr wohl "die allgemeine Form der Raumanschauung transcendental gegeben" sein könne, *ohne* daß auch die Axiome der Geometrie dieses

Raumes es sind. Er sagt: "Ich möchte [...] hervorheben, dass diese Frage [ob die Axiome transzendental oder empirisch sind] ganz zu trennen ist von der erst besprochenen, ob der Raum überhaupt eine transcendentale Anschauungsform sei oder nicht" (1878, VR II, 229). Im späteren Abdruck von 1884 wird an dieser Textstelle auf die Zusätze verwiesen, die zum Teil wörtlich ebenfalls schon 1878 publizierte Erörterungen über den Status geometrischer Axiome wiedergeben (VR II, 391 und WA II, 640-660, insbesondere 642). Da andere Beiträge in diesem Band dem Thema Geometrie gewidmet sind, gehe ich auf Einzelheiten nicht ein. In meinem Gedankengang kommt es mir allein darauf an, daß Helmholtz seine eigene Frage nach der transzendentalen Natur räumlichen Vorstellens überhaupt schlechterdings nur positiv beantworten kann, wenn er nicht seine gesamte Theorie der Objektivität preisgeben will. So ganz die rechte Klarheit hat er freilich in dieser Sache nicht erreicht, wenn er sich - statt auf eine transzendentale Überlegung - an entscheidender Stelle auf eine "Thatsache des Bewusstseins" beruft, nämlich die, "dass wir wahrzunehmen glauben Objecte, die sich an bestimmten Orten im Raume befinden". Folgerichtig fügt er dann hinzu, daß wir über die für solche Bewußtseinstatsachen verantwortlichen "topogenen Momente" der Realität überhaupt nichts wissen (1878a, WA II, 56-57). Diese Momente setzt er auf eine Stufe mit den sogenannten "hylogenen Momenten", also stofflichen Realitäten, was der Auszeichnung des Räumlichen und der Bewegung zur Sicherung der Objektivität nicht gerecht wird.

Es ist also kaum bestreitbar, daß auf die ursprüngliche - gleichsam reine - Lehre von der "vollständigen Begreiflichkeit der Natur" (1847, WA I, 16) sehr bald gewisse Schatten fallen. Daß diese sich eigentlich schon ab 1860, gar 1850, zu Schlagschatten hätten verdichten können, sei abschließend noch kurz angemerkt; damit kehre ich zugleich zum Innsbrucker Vortrag vor der Versammlung der Naturforscher vom Jahre 1869 zurück. Aus seiner Eröffnung hatte ich den Kernsatz zitiert: "Das *Gesetz* der Erscheinungen finden, heisst sie *begreifen*" (1869, VR I, 375). Wir hatten uns unter Helmholtz' Leitung damit herumgeschlagen, was denn das nun alles impliziert: ein Gesetz der Erscheinungen zu finden. Wir hätten auch einfacher und nicht weniger unter der Leitung von Helmholtz fragen können: Was für Gesetze kann die Wissenschaft uns denn nennen? Bei seiner Durchsicht stößt Helmholtz unter anderem auf diese: (1) den Zweiten Hauptsatz der Thermodynamik und (2) das Gesetz der Evolution der Organismen.

Carnot - so sagt er - und, in berichtigter Form, Clausius haben uns gelehrt, daß der Wechsel der Energie von einer Form zur anderen fortdauernd in einer bestimmten Richtung fortschreitet, derart, daß immer mehr Wärme auf Kosten anderer Energieformen entsteht. Statt nun diesen Sachverhalt auf den die völlige Begreiflichkeit der Natur ausdrückenden Mechanismus der Materiebewegungen zu beziehen (was er freilich spät und mit untauglichen Mitteln nachzutragen versuchte), exemplifiziert Helmholtz den Inhalt dieses Gesetzes anschaulich und einschlägig anhand eines kurzen Abrisses der Kosmogonie vom Urgas zum Wärmetod (1869, VR I, 384). Das Gesetz, so gesehen, drückt dann aber gerade nicht die immer wiederkehrenden Wirkungen unveränderter Kräfte und Materien aus, sondern einen einmaligen globalen Weltlauf, in den diese eingebettet sind.

Sprechender noch ist das zweite Beispiel: Helmholtz identifiziert das Gesetz der Darwinschen Theorie, dessen blindes Walten er gegen die traditionelle Teleologie setzt, als "das Gesetz der Forterbung der individuellen Eigenthümlichkeiten von den Eltern auf die Nachkommen". Auch hier ist der Bezug zum begreiflichen Mechanismus verständlicherweise nicht einmal angedeutet. Helmholtz mag ihn stillschweigend hinzugedacht haben. Aber natürlich ist die Konstanz der Vererbung gerade nicht das einzige und nicht das wichtigste Gesetz der Darwinschen Theorie, wenn es denn überhaupt ein solches Gesetz geben sollte. Helmholtz weitet denn auch seine Beschreibung sogleich aus und kommt, wie unvermeidlich, auf Varietäten der Vererbung, auf Selektion und auf den neuen theoretischen Grundbegriff, die Wahrscheinlichkeit, zu sprechen (1869, VR I, 388). Was aber wäre dann das Gesetz der Evolution? Helmholtz stellt die Frage nicht; nur der Kontext suggeriert, daß es ihm auch hier um die allenthalben wachsende Einsicht in die gesetzliche Ordnung der Natur und um deren Begreiflichkeit geht.

Daß diese Begreiflichkeit unter weit komplizierteren Bedingungen steht, als das mechanistische Weltbild glauben machen konnte, und daß von "vollständiger Begreiflichkeit" (1847, WA, I, 16) vernünftigerweise die Rede nicht sein kann, das konnte auch ein so gewaltiger Denker wie Helmholtz damals, mitten im 19. Jahrhundert, offenbar nicht erkennen. Wie ganz anders stehen wir im 20. Jahrhundert da: mit überwältigend mehr Wissen und gerade deshalb erschüttertem Glauben an die Begreiflichkeit der Natur.

Anhang 1: Zu Cassirer 1950

Da - bei allen Unterschieden des Formats und der Ansprüche - eine gewisse Verwandtschaft besteht zwischen meinem jetzigen Versuch und demjenigen, den Ernst Cassirer im Vierten Teil seines Werkes *Das Erkenntnisproblem* unternommen hat (1950, Part I, Chap. V), füge ich eine Bemerkung darüber an, worauf ich im Unterschied und komplementär zu Cassirer den Akzent lege. Cassirer spricht davon, daß bis zur Mitte des 19. Jahrhunderts kein Zweifel am Newtonschen Wissenschaftsideal bestanden habe, dann jedoch eine "plötzliche Veränderung" eingetreten, nämlich der Realismus der Naturwissenschaft abgelöst worden sei durch einen Phänomenalismus, wie er vornehmlich mit den Namen Mach, Ostwald, Poincaré und Duhem, aber auch Kirchhoff und Hertz verbunden sei. Zum einen kann die komplexe Transformation erkenntnistheoretischer Anschauungen, die sich zusammen mit dem raschen Fortschreiten der Naturforschung vollzog, mittels nur *eines* Gegensatzes, desjenigen zwischen Realismus und Phänomenalismus, bestenfalls in erster grober Näherung gekennzeichnet werden. Es war ja gerade die Pointe der Philosophie Kants gewesen, über diesen Gegensatz hinauszugelangen, wie die doppelte Selbstbeschreibung seiner Position durch die Epitheta 'empirischer Realismus' und 'transzendentaler Idealismus', also seine Abkehr von der vorkritischen Metaphysik, anzeigt. Gerade an Kant jedoch wollte sich Helmholtz orientieren. Es ist also notwendig, in zweiter, ebenfalls noch grober Näherung, die Auffassung des jüngeren Helmholtz, wie oben in Teil I formuliert, zusätzlich durch ihre Nähe zur rationalistischen - im Unterschied zur empiristischen, etwa in der Tradition Humes stehenden - Naturphilosophie zu charakterisieren.

Zum zweiten konzentriert Cassirer sein Augenmerk einseitig auf wichtige *neue* Entdeckungen in der Physik: die der Irreversibilität und der elektrodynamischen Feldtheorie; in ihnen sieht er die treibenden Kräfte der philosophischen Neubestimmung der Physik (1950, 84-85). In diesem Zusammenhang fällt die für unser Thema einschlägige Bemerkung, Helmholtz' Programm des Naturbegreifens wäre sowohl vom empirischen wie vom erkenntnistheoretischen Standpunkt gleichermaßen befriedigend gewesen, wenn es bloß hätte realisiert werden können. Freilich hätte man sich dabei mit der im Begriff der Fernkraft liegenden Absurdität abfinden müssen (1950, 86). Diese Einschätzung des Helmholtzschen Programms durch Cassirer ist allzu optimistisch. Wie schon du Bois-Reymond (1872) hervorgehoben hat, bleiben in ihm das Wesen der Materie und der Ursprung der Bewegung ungelöste Rätsel (vgl. auch 1880, 76). Dem darf man hinzufügen, daß, selbst wenn man instantane Fernwirkungen unproblematisch fände, doch noch Aufklärung verlangen würde, wie die kontingent erscheinende Vielheit qualitativ ganz unterschiedlicher Kräfte zu erklären sei; Helmholtz hatte ja in seiner Arbeit über den Energiesatz angenommen, daß die chemischen Elemente irreduzible Gegebenheiten

- er sagt "unvertilgbare Qualitäten" - seien, welche die an sich qualitätslose Materie differenzieren (1847, WA I, 14-15). Ersichtlich greift Cassirer einen wichtigen Aspekt der Entwicklung der Naturphilosophie im 19. Jahrhundert auf, das Überschreiten der Grenzen der Mechanik. Er scheint jedoch die internen Schwierigkeiten des Mechanismus zu unterschätzen, die schon für sich genommen, auch ohne den Problemdruck der Elektrodynamik und der Thermodynamik von außen, die physikalische und die philosophische Entwicklung weitertreiben konnten. Solche Schwierigkeiten bieten vornehmlich: die Natur der Träger der Bewegung, ihre geometrische Beschaffenheit, Identität, Stabilität und qualitative Vielfalt; die diskrete Vielfalt elementarer Qualitäten; die Rolle kontingenter Rand- und Anfangsbedingungen der Bewegung; schließlich die Existenz komplexer Gestalten. Ohne Auskünfte zu diesen Themen kann ein befriedigendes Begreifen der materiellen Natur schwerlich auskommen.

Anhang 2: Zu Heimann 1974

In seiner Studie über die Orientierung von Helmholtz an Kant hat Heimann herausgestellt, daß aus der These, Gesetzmäßigkeit gemäß dem allgemeinen Gesetz der Kausalität sei die Bedingung der völligen Begreiflichkeit der Natur, keineswegs schon das mechanische Naturbild unter der Wirkung konstanter Kräfte bewegter Materie ableitbar, also der Mechanismus als die "nothwendige Begriffsform der Naturauffassung" erweisbar ist, wie Helmholtz dachte (1847, WA I, 17). Heimann hat sich dabei vor allem auf die von Helmholtz ausgelöste Debatte darüber konzentriert, unter welchen Voraussetzungen die Kräfte nur noch Zentralkräfte sein können. Daß hier Beweislücken - bei Kant so gut wie bei Helmholtz (Heimann konstruiert in *dieser* Hinsicht einen Gegensatz zwischen beiden, der nicht besteht, 228f.) - zu beklagen sind, zeigt natürlich die Elektrodynamik, was Helmholtz durchaus selbst gesehen hat (Zusatz 2 von 1881 zu 1847, WA I, 68-70).

Der Nachweis erstaunlich weitgehender Analogien zwischen den naturphilosophischen Überlegungen von Kant und Helmholtz hat viel Überzeugungskraft, naturgemäß am wenigsten, was den Vierten Teil der *Metaphysischen Anfangsgründe der Naturwissenschaft* Kants, die Phänomenologie, betrifft, in der es um Materie als Gegenstand der Erfahrung geht. Undeutlich ist mir geblieben, wie Heimann die Selbstdistanzierung Helmholtz' von Kant aus dem Jahr 1881 (WA, I, 68) genau verstanden hat. Der Text legt nahe, daß Heimann

Kant eine deutliche Trennung zwischen den transzendentalphilosophischen Argumenten der *Kritik der reinen Vernunft* und der Metaphysik der *Anfangsgründe* zuschreibt (220/1) sowie das vergleichsweise bescheidene Argumentationsziel, die bloße *Möglichkeit* der Newtonschen Physik zu erweisen (229). Demgegenüber hätte dann Helmholtz eine direkte Verknüpfung zwischen transzendentalem Kausalgesetz und der mechanistischen Metaphysik vertreten sowie deren Materiebegriff als notwendig zu erweisen versucht (229). Hiergegen meine ich zum einen, daß Kant mehr beweisen wollte, als Heimann ihm zubilligt; zum anderen erscheint es mir weit plausibler, die Distanzierung Helmholtz' von Kant auf seinen zwischen 1847 und 1881 ausgereiften Empirismus, insbesondere im Hinblick auf die Geometrie, aber auch im Hinblick auf die These, daß alle Kräfte Zentralkräfte seien, zu beziehen.

Was den letzten Punkt betrifft, faßt Helmholtz ins Auge, daß es vielleicht nicht gelingen werde, in der Elektrodynamik allein mit Zentralkräften auszukommen.[5] Ginge daher die Eindeutigkeit der Lösungen der physikalischen Grundgleichungen verloren, so "wäre damit die Aussicht auf vollständige Lösungen der naturwissenschaftlichen Aufgaben aufgegeben". Bliebe die Eindeutigkeit erhalten, würde nach Helmholtz immer noch eine kritische Lage für denjenigen entstehen, der nach der vollständigen Begreiflichkeit der Natur sucht: "Aber ich kann [im eben genannten Fall] das Princip der Begreiflichkeit nicht mehr als zureichend für die Folgerung anerkennen, dass die durch das Zusammenwirken zweier oder mehrerer Bewegungsursachen entstehende Wirkung nothwendig durch (geometrische) Summierung aus denen der einzelnen gefunden werden müsse" (1881, WA, I, 69-70). Sagt Helmholtz hier, daß auch unter komplizierteren Bedingungen, unter denen die Vektoraddition der Kräfte und Geschwindigkeiten nicht gilt, noch die volle Begreiflichkeit der Natur vorliegen könnte, freilich ohne daß diese noch als hinreichendes Prinzip für die Ableitung der "nothwendigen Begriffsform der Naturauffassung" (1847, WA, I, 17) dienen könnte? Oder sagt er vielmehr, daß unter solchen Bedingungen die Begreiflichkeit Grenzen finde, daß sie nie mehr *vollständig* werden könne, mithin die an diesen Begriff zu stellenden Forderungen abzuschwächen seien? Im Haupttext vertrete ich die These, daß der Sache nach für Helmholtz die zweite Lesart näher liegt; unklar erscheint mir, ob er sich dies jemals eingestanden hat.

5 Vgl. hierzu den Beitrag Bevilacquas in diesem Band.

Literatur

Cassirer, Ernst: *The Problem of Knowledge: Philosophy, Science and History since Hegel.* New Haven: Yale UP, 1950. Einschlägig insbesondere I.v. The Goal and Method of Theoretical Physics, 81-117.

Descartes, René: Notae in programma quoddam. *Oeuvres*, ed. Adam/Tannery, Vol. VIII, 1847, 335-370.

du Bois-Reymond, Emil: *Über die Grenzen des Naturerkennens* (1872). Zitiert nach der Ausgabe Leipzig: Veit & Co. 1882.

Heimann, Peter M.: Helmholtz und Kant: The Metaphysical Foundations of *Über die Erhaltung der Kraft. Studies in History and Philosophy of Science 5*, 1974, 205-238.

Helmholtz, Hermann von: *Über die Erhaltung der Kraft* (1847). *WA* I, 12-68. Zusätze von 1881: *WA* I, 68-75.

Helmholtz, Hermann von: Über das Verhältnis der Naturwissenschaften zur Gesammtheit der Wissenschaft (1862). *VR* I, 157-185.

Helmholtz, Hermann von: Über die Erhaltung der Kraft (Vorlesung in Karlsruhe, 1862-63). *VR* I, 187-229.

Helmholtz, Hermann von: Über das Ziel und die Fortschritte der Naturwissenschaft (1869). *VR* I, 367-398.

Helmholtz, Hermann von: Zum Gedächtnis an Gustav Magnus (1871). *VR* II, 33-51.

Helmholtz, Hermann von: Die Thatsachen der Wahrnehmung (1878). *VR* II, 213-247.

Helmholtz, Hermann von: Über den Ursprung und Sinn der geometrischen Axiome; Antwort gegen Herrn Professor Land. Zuerst auf Englisch in *Mind 10* (1878), 212-224; *WA* II, 640-660. (1878a).

Helmholtz, Hermann von: Zusätze von 1881 zu *Über die Erhaltung der Kraft* (1847). *WA* I, 68-75.

Helmholtz, Hermann von: Beilagen von 1884 zu: Die Thatsachen der Wahrnehmung (1878). *VR* II, 387-406. Teil III identisch mit 1878a, 394-406.

Helmholtz, Hermann von: Goethe's Vorahnungen kommender naturwissenschaftlicher Ideen (1892). *VR* II, 335-361.

Helmholtz, Hermann von: *Vorträge und Reden* [VR], 2 Bände. Zitiert nach der 5. Auflage, Braunschweig: Vieweg.

Helmholtz, Hermann von: *Wissenschaftliche Abhandlungen* [WA], 3 Bände. Leipzig: Barth, I (1882), II (1883), III (1895).

Kant, Immanuel: Metaphysische Anfangsgründe der Naturwissenschaft (1786). *Werke*, Akademie-Ausgabe, Bd. 4, Berlin: de Gruyter, 465-565.

Kant, Immanuel: Kritik der reinen Vernunft (1781, 1787). *Werke*, Akademie-Ausgabe, Bd. III, IV. Berlin: de Gruyter.

Helmholtz's Electrodynamics and the Comprehensibility of Nature

Olivier Darrigol

Helmholtz's science was closely tied to the question: What is it to comprehend nature? In 1847 his formulation of the conservation of force and the relevant kind of mechanical reduction strongly depended on a tentative philosophical reply to this question. Later, in the light of his physiology of sensations, Helmholtz modified his epistemological views and found that the comprehensibility of nature involved much less constraint on physical theory than he originally thought.*

The purpose of the present paper is to determine whether counterparts of this evolution can be found in Helmholtz's physics. I have confined the enquiry to electrodynamics, because it was Helmholtz's favorite branch of physics, and because mechanical reducibility was an essential motivation of his researches in this field: "The theory of electrodynamics", he wrote in 1872, "is not only interesting for the comprehension of this rich and important branch of physics, but also, in an even more significant and general manner, for the fundamental principles of mechanics" (Helmholtz 1872: 636).

---

* For a philosophical analysis of Helmholtz's notion of the comprehensibility of nature, see Lorenz Krüger's contribution to this volume.

1. The Comprehensibility of Nature

The conservation of force (1847)

In the philosophical introduction of his memoir on the conservation of force, Helmholtz began with distinguishing between empirical laws, which are nothing but generic concepts embracing visible effects, and the unknown causes of these effects, which ought to exist according to the principle of causality. Whereas experimental physics establishes the empirical laws, theoretical physics unravels chains of hidden causes. At this point, Helmholtz defined the comprehensibility of nature as the possibility of finding the ultimate, *invariable* causes of natural processes (Helmholtz 1847: 13).[1]

Then Helmholtz proceeded, by seemingly deductive means, to a more constraining conception of the comprehensibility of nature, that is to say: all natural processes should be reducible to the action of central forces within pairs of material points. His reasoning went as follows (Helmholtz 1847: 14-17).

Science analyzes the external world in terms of two inseparable abstractions: matter and force. By definition, matter can only change by motion, and force is the cause of motion. The full comprehensibility of nature implies the reduction to forces that depend on spatial relations only, that do not involve time. For spatial relations to exist, there must be at least two bodies present. Hence, the simplest case of forces in action is that of a pair of material points.[2] Further, Helmholtz takes for granted that "the force which two whole masses exert on each other must be resolved into the forces which their parts exert on each other" (Helmholtz 1847: 15). This is a particular case of what I shall call the *principle of decomposition*, according to which all actions in nature must be resolved into actions involving only two elements of volume.

According to the full comprehensibility of nature, the ultimate forces acting between two material points can only depend on the spatial relation between the points, which is nothing but distance. Therefore, these forces must be central forces, that is to say, attractive or repulsive forces, whose intensity only depends on the distance. Taking into account the principle of decomposition,

1 On the philosophical introduction of this memoir (pp. 13-17) and a possible relation with Kant's *Metaphysische Anfangsgründe*, see Heimann 1974a. On other sources of the conservation of force, see Kuhn 1959, Elkana 1974, Heimann 1974b, Lenoir 1982, Bevilacqua 1993, Brain and Wise (in this volume).

2 This does not imply an atomistic conception.

we now reach the second meaning of the comprehensibility of nature, which is the reducibility of all natural actions to pairs of central forces. As Helmholtz put it:

> "The theoretical science of nature shall have fulfilled its mission when the reduction of phenomena to elementary forces is accomplished and when this reduction is known to be the only possible one. Then the reduction would be proved to be the necessary comprehending form of our conception of nature, and it would be regarded as objectively true" (Helmholtz 1847: 17).

After this philosophical introduction, Helmholtz proceeded to show that the central character of the elementary forces implied the impossibility of perpetual motion, as well as the conservation of force. Here is an outline of his reasonings (Helmholtz 1847: 17-27).

If all elementary forces are central, they derive from a potential. Consequently, the theorem of living forces applies, and the living force of the system takes the same value whenever the system returns to the same spatial configuration. Since living force is also ability to perform work, no perpetual motion can exist. Moreover, a conserved quantity can be obtained by taking the sum of the living force and what Helmholtz calls "the sum of tensional force", which is our modern potential energy. This is the conservation of force, in the quantitative sense.

Helmholtz also gave a proof that the impossibility of perpetual motion implied the central character of elementary forces. For a pair of material points, he reasoned, the impossibility of perpetual motion implies that the total living force be a function of their distance only. Now, as a consequence of Newton's second law, the work of forces is equal to the variation of the living force. This seems to imply that the forces acting on the two points are gradients of a function that depends only on the distance of these points. Then the forces should be central (Helmholtz 1847: 19-21).

In sum, in 1847 Helmholtz's notion of a full comprehensibility of nature implied the reduction to central forces between pairs of material points, as well as the uniqueness of this reduction. Enormous as it was, the requirement received a double justification: first a transcendental one, as the necessary implementation of causality; second an inductive one, as a consequence of the empirical impossibility of perpetual motion.

Am Anfang war die That

In the following years Helmholtz substantially reduced his claims, partly because he became aware of gaps in his reasonings, but also because his studies on perception led him to a new, improved conception of cause and force. I will first describe this conception, as found in Helmholtz's epistemological writings from the late 1860s to his death.[3]

Our most primitive experience, says Helmholtz, is made of relations between percepts and motor impulses. Originally, the percepts make no sense. They acquire meaning only after we have detected some regularities in their variations during voluntary movements: "Am Anfang war die That".[4] However, a perceived regularity does not in itself constitute an understanding of nature; it could be contradicted by further experiment and then appear as purely accidental. The regularity must be completed by the trust that it will extend to a wide class of not yet performed experiments. The principle of causality according to Helmholtz is this trust in the existence of universal regularities, and it is the basis of all inductive inference, conscious or unconscious.[5] Gone is the vague conception of cause as antecedent, which Helmholtz uncritically adopted in his earlier writings. Causality is lawfulness (*Gesetzmässigkeit*), and cause is "that which remains invariable behind observed changes", as suggested by a literal rendering of the German *Ursache* (Helmholtz 1878: 241; 1892b: 354).

The principle of causality cannot be proved experimentally, since the proof would require an induction based on the principle itself. It is a *hypothesis* and

3 The relevant epistemological texts are Helmholtz 1868, 1869, 1870c, 1871, 1878, 1892b, 1894a. In "Über das Sehen der Menschen" (1855), Helmholtz's conception of causality was still the same as in 1847; the introductory section of the third vol. (1867) of his *Handbuch der Physiologischen Optik* marks a transition between his first and final views. On Helmholtz's epistemology and its relation to Kant's, see the entry "Epistemology" in the glossary of Chevalley 1991. See also Heimann, 1974a; the editorial comments by P. Hertz, M. Schlick, R.S. Cohen, and Y. Elkana in Helmholtz 1977; Krüger (in this volume); Schiemann (in this volume).

4 Helmholtz quoting from Goethe's *Faust*, for instance in Helmholtz 1892b: 359. Faust's saying is the result of his struggle to translate properly Saint John's "In the beginning was the Word".

5 Conscious induction is not, according to Helmholtz, a passive activity of the mind. Instead, Helmholtz 1894a insists on the necessity of "a certain power of invention" to formulate hypotheses, before their consequences are experimentally checked.

could be contradicted in some fields of experience. We could even live in a world "where all atoms would be different and where nothing would be at rest". But such a world would not be comprehensible: "There would be no regularity at all, and our thinking activity would be inexistent" (Helmholtz 1878: 243-244). Hence causality is a hypothesis of a very special kind: it is a necessary component of our understanding, and in this sense, it is a transcendental law.[6]

Experience teaches us that changes in our sensations can occur independently of our will, sometimes against our will. Whenever such changes are ruled by a well-established law, we imagine an objective power to be acting. This power is what Helmholtz calls force. It is an abstraction, or a grammatical trick through which a verb is turned into a noun. To completely state the law we also need to know what is acted upon. This second abstraction is what Helmholtz calls matter.[7]

As Helmholtz notes, there is a danger that forces could be mistaken as *real* entities, detached from matter; this had happened with the introduction of vital forces. Moreover, the concepts of force and matter are never indispensable to the statement of a physical law. For example, the law of gravitation can be stated in a purely phenomenological manner, namely: "Any two small bodies undergo an acceleration toward each other, the intensity of this acceleration being inversely proportional to their squared distance" (Helmholtz 1892b: 352-353). Yet Helmholtz did not follow his friends Kirchhoff and Hertz who wished to completely eliminate the concept of force from physics. He believed that force and matter, once rid of their metaphysical connotations, were useful

6 Helmholtz's sharpest statement of the hypothetical nature of causality is in a MS reproduced in Koenigsberger 1903: vol. 1, 247. Cf. Schlick's note in Helmholtz 1977: 181. Whether or not Helmholtz's causality is transcendental in Kant's sense is a controversial matter: cf. Schlick, ibid., note 23 on 168, and note 60 on 181.

7 See Helmholtz 1878: 241: "Insofern wir [...] das Gesetz als ein unsere Wahrnehmung und den Ablauf der Naturprozesse Zwingendes, als eine unserem Willen gleichwerthige Macht anerkennen, nennen wir es 'Kraft'"; see also Helmholtz 1892b: 352-354; for the substitution of a noun for a verb, see Helmholtz 1894a: par. 5. The most concise statement of Helmholtz's views on causality, force, and matter is in Helmholtz 1881a: 68: "Das Princip der Causalität ist [...] nichts Anderes als die Voraussetzung der Gesetzlichkeit aller Naturerscheinungen. Das Gesetz als objective Macht anerkannt, nennen wir Kraft. Ursache ist seiner ursprünglichen Wortbedeutung nach das hinter dem Wechsel der Erscheinungen unveränderlich Bleibende oder Seiende, nämlich der Stoff und das Gesetz seines Wirkens, die Kraft".

abstractions, if only because they permitted more concise statements of the laws of physics. In the introduction of his last lectures on mechanics, he still declared: "We can hold to the principle that the law of force is permanent. Under this assumption, the whole theoretical physics can be based on the concept of force" (Helmholtz 1898: 24).

Helmholtz nevertheless abandoned the ideal of a unique reduction to ultimate forces, which he had boldly proclaimed in 1847. In his "Thatsachen in der Wahrnehmung" he wrote:

> "Any reduction of the phenomena to underlying substances and forces pretends to have found something unchanging and final. We are never entitled to such a claim. This is allowed neither by the imperfections of our knowledge, nor by the nature of the inductive inferences on which all our perception of the actual rests" (Helmholtz 1878: 243).

In other words, we can never be sure to have exhausted the proof of a given reduction to force and substance. For example, heat and light, which were originally thought of as substances, are now regarded as forms of motion (ibid.: 240).

From necessary truths to ideals

I now come to the more specific reduction to elementary central forces, which is found at the heart of the memoir on the conservation of force. Through his own reflexions or the criticisms of colleagues like Clausius and Lipschitz,[8] Helmholtz found gaps in his derivations of this requirement. First, in his use of the concept of motive force he had implicitly relied on Newton's laws of motion, without stating their inductive origin, and thereby leaving the impression of a purely a priori derivation of the requirement of central forces.[9] A similar comment could be made about Euclidean space, which at that time Helmholtz surely believed to be transcendentally given. Also, Helmholtz became aware of the fact that the decomposition principle was not *a priori* true. As he explained in 1881, the decomposition is necessary for the masses *on* which forces act, because full comprehensibility implies that the motion of every point of these

8 See Clausius 1853, 1854; Helmholtz 1854b, 1857. On the controversy between Clausius and Helmholtz, see Heimann, 1874a; Bevilacqua (in this volume). On Lipschitz's criticism, see Helmholtz 1881a: 68.

9 See Helmholtz 1881a: 69; and 1898: 25, where the empirical origin of "Newton's axioms" is clearly stated.

masses should be determined; but it is not for the masses *from* which the forces emanate (Helmholtz 1881a: 68-69). Finally, the proof that the impossibility of perpetual motion implies (together with Newton's second law of motion) the centrality of elementary forces turned out to be mathematically wrong. A simple counterexample is given by magnetic forces, which depend on velocity and are nevertheless conservative. In order to save his reasoning, Helmholtz injected Newton's third law at the proper places.[10]

In sum, the reducibility to central forces depended on a number of principles: causality, Euclidean geometry, Newton's laws of motion, and the decomposition principle. The first principle was hypothetical, and the others were inductive inferences. But Helmholtz never abandoned these principles. To the end of his life he remained faithful to the general endeavor of mechanical reduction, which he stated as follows in 1869: "If motion is the ultimate change, on which all other changes in the world rest, then all elementary forces are motive forces, and the final aim of physical science is to find the motions which determine all other changes and to find the corresponding motive forces, in other words: to be reduced to mechanics" (Helmholtz 1869: 379).[11]

Helmholtz remained equally faithful to the decomposition principle. In 1869 he emphasized the historical role of this principle:

> "One had to eliminate the accidents of form and of spatial distribution which the interacting masses could present by seeking to discover from the phenomena observed on large visible masses the laws of the actions of vanishingly small mass-particles; that is to say, expressed objectively, one had to resolve the forces of composite masses into the forces of their smallest elements" (Helmholtz 1869: 376).

Now, Helmholtz could not logically renounce the reducibility to central forces, since he maintained all the principles that led to it. Indeed, we will find that the central character of elementary forces still played a role in his criticism of electrodynamic theories. Moreover, he insisted to the end of his life that energy, unlike mass and other substances, was not spatially divisible.[12] This is a

10 See Helmholtz 1881a: 70 (reply to Lipschitz), and the notes added to Helmholtz 1847, in *WA*1, on 19, 20, 21.

11 The statement occurred in the context of the existence of chemical elements, which Helmholtz took to imply that every change of matter could be reduced to the motion of the elements. On mechanical reducibility, see also Helmholtz's studies on least action, discussed in the last section of this paper.

12 See Helmholtz 1869: 383: "Das Weltall erscheint [...] ausgestattet mit einem Vorrathe an Energie, der durch allen bunten Wechsel der Naturprozesse nicht vermehrt, aber

sure sign that he did not abandon the idea of primitive central forces, which makes the potential energy a holistic property of a pair of mass points.

Phenomenological principles

So far, I have focussed on the idealist aspects of Helmholtz's mature epistemology. But there is, of course, a strong empiricist component, which he himself emphasized in various circumstances: when criticizing the excesses of *Naturphilosophie*, when praising Goethe's rejection of counter-intuitive abstractions, or when flattering pure experimentalists. Quite frequently, Helmholtz tried to play down the difference between theoretical and experimental physics. For instance, in his memorial lecture for Gustav Magnus we read: "Mathematical physics has acquired an entirely different character in the hands of Gauss, of Franz Neumann and his pupils, among the Germans [...]. It is now understood that mathematical physics is a purely experimental science; that it has no other principles to follow than those of experimental physics" (Helmholtz 1871: 45). The reinterpretation of causality in terms of *Gesetzmässigkeit* favored the convergence. Gone was the hierarchy of empirical and theoretical tasks assumed in the memoir of the conservation of force. The theorist no longer kept for himself the privilege of *comprehending* phenomena.[13]

Helmholtz the empiricist could well postpone the reduction to mechanics and central forces. For example, when commenting on the energy principle he usually focussed on mechanical work equivalents - as is done in modern thermodynamics, and did not elaborate on further mechanical reduction.[14] At the

auch nicht vermindert werden kann; der da fortbesteht in stets wechselnder Erscheinungsweise, aber, wie die Materie, von Ewigkeit zu Ewigkeit in unverändlicher Grösse; wirkend im Raume, *aber nicht, wie die Materie, theilbar mit dem Raume*" [emphasis mine]; and a similar statement in "Über dauernde Bewegungsformen und scheinbare Substanzen," MS for an undelivered lecture for the *Naturforscherversammelung* of 1894, in Koenigsberger 1903: vol. 3, 125. N. Wise emphasizes this singularity of Helmholtz's approach in the case of the extended potential theory: "Ironically, Helmholtz, the most notable author of energy concepts on the Continent, could not meet this demand in his own theory, primarily because it did not distribute energy independently through space" (Wise 1981: 301).

13 See also Helmholtz 1892b: 352-354; and 1873a.

14 See for example Helmholtz 1854a, 1863.

most he mentioned the perspective opened by a generalized proof of energy conservation, as in the following extract:

> "It turns out [...] that all natural forces can be measured by the same mechanical standard, and that they are all equivalent to pure motive forces with respect to the production of work. Thus one very significant step has been taken toward the comprehensive theoretical problem of tracing all natural phenomena back to motion" (Helmholtz, 1869: 383).

In the meantime, the energy principle unified experimental and theoretical endeavors. Helmholtz found the same virtue in the decomposition principle, according to which all actions of nature should be resolved into actions involving two elements of volume:

> "In our immediate experience we find bodies variously formed and constituted; only with such can we make our observations and experiments. Their actions are made up of the actions which each of their parts contributes to the sum of the whole; and hence, if we wish to know the simplest and most general law of the action of masses and substances found in nature upon one another, and if we wish to divest these laws of the accidents of form, magnitude, and position of the bodies concerned, we must go back to the laws of action of the smallest particles, or, as mathematicians designate them, the elementary volumes [...]. It is thus admitted that mathematical physics only investigates the laws of action of the elements of a body independently of the accidents of form, in a purely empirical manner, and is therefore just as much under the control of experience as what is called experimental physics. In principle they are not at all different, and the former only continues the function of the latter, in order to arrive at still simpler and still more general laws of phenomena" (Helmholtz 1871: 45-46).

To summarize, in his final epistemology Helmholtz restricted the comprehensibility of nature to the principle of causality, the latter being defined as the requirement of lawfulness. He rejected his earlier view that the energy principle and the decomposition principle resulted from the full comprehensibility of nature, but nevertheless maintained these principles as universal guides in both theoretical and empirical enquiries. He still viewed the reduction to mechanics and the reducibility to central forces as a legitimate horizon, but grew more and more discreet on this matter. He was now aware that no reduction to central forces was either final or unique, and that much physics could be done without it.

2. Electrodynamics

1847

After establishing, in his memoir of 1847, the general foundation for the conservation of force, Helmholtz offered a survey of the whole field of physics, with a double goal: "first to prove the conservation of force in all cases where the laws of the phenomena have been already investigated to our satisfaction; then, with the help of [this principle] and relying on multiple analogy with better known cases, to further determine the laws of the incompletely explored ones, and thus provide the experimenter with a leading thread" (Helmholtz 1847: 17). To serve these aims Helmholtz determined, for each kind of power, the quantity which had a definite mechanical equivalent, and also identified, whenever possible, the underlying motion and central forces. For gravitation, heat, electrostatics, and magnetostatics, the latter reduction was readily available; for chemical effects, Helmholtz could imagine that tensional forces acted between atoms. Hence, for a wide range of phenomena full comprehension could be claimed.

However, no reduction to central forces was available for electrodynamic phenomena. In the preceding year Franz Neumann and Wilhelm Weber had published two mathematical theories of induction, neither of which fitted Helmholtz's criterion for full comprehensibility. Neumann's theory, the archetype of German mathematical phenomenology, sought to relate by the most direct possible mathematical means the various quantities measured in electrodynamic experiments, namely: electric currents, electromotive forces, and mechanical forces. For this purpose Neumann introduced the abstract notion of electrodynamic potential, which is given, for two linear currents I_1 and I_2 by

$$P = -I_1 I_2 \iint \frac{dr_1 \cdot dr_2}{|r_1 - r_2|}.$$

For closed currents, the spatial variations of this potential yield the electrodynamic forces acting on each circuit (in conformity with Ampère's law), while the time derivative of the potential calculated for a unit current in one of the circuits yields the electromotive force of induction in this circuit.[15]

15 Cf. F. Neumann 1847, 1848. On 19th century German theories of electricity, see

In a quite different style, Weber relied on Fechner's picture of a current as a double flow of positive and negative electricity, and proposed a fundamental law for the force acting between two moving particles of electricity. This force is directed along the line joining the two particles and has the intensity

$$f = \frac{ee'}{r^2}\left(1 + \frac{r\ddot{r}}{c^2} - \frac{\dot{r}^2}{2c^2}\right),$$

where e and e' are the charges of the two particles, r their mutual distance, and c a constant with the dimension of a velocity. Proper summations lead to Ampère's electrodynamic forces and to Neumann's induction laws for closed currents. Moreover, the static case retrieves Coulomb's law (Weber 1846).[16]

Helmholtz remained unimpressed by Weber's unification of all electric and magnetic phenomena, no doubt because Weber's forces were not central forces and, thus, did not comply with his criterion for full comprehensibility. Also, he almost certainly believed that Weber's law implied a violation of the conservation of force, by virtue of his general proof that conservation of force implied elementary central forces. In 1847, decency and his genuine admiration for Weber must have prevented him from stating as much.[17]

In the lack of a proper reduction to central forces, Helmholtz contented himself with showing that in simple cases the conservation of energy, together with the existence of electrodynamic and electromagnetic forces deriving from Neumann's potential, implied the induction phenomenon as well as Neumann's laws for it. For example, in the case of a closed circuit near a moving magnet, Helmholtz demanded that the energy provided by the electric source (a battery) should be equal to the Joule heat produced in the circuit plus the work of the electromagnetic forces acting on the magnet. This implied the existence of an electromotive force, the form of which agreed with Neumann's expression (Helmholtz 1847: 61-62).[18]

Whittaker 1951; Woodruff 1962; Kaiser 1981; Buchwald 1985: chap. 21; Darrigol 1993. On Neumann's theory, see Jungnickel and McCormmach 1986: vol. 1, 148-152; and Olesko 1991: 175-178.

16 See Whittaker 1951: 201-206; Darrigol 1993: 226-230 and app. 1. On the development and broader significance of Weber's program, see Wise 1981.

17 But Maxwell, Thomson and Tait soon did it for him. Cf., e.g., Archibald 1989.

18 In Neumann's anterior theory, an undetermined proportionality constant (ε) appeared in the expression of the electromotive force. Helmholtz's reasoning determined the constant.

Helmholtz later admitted that this kind of reasoning did not really provide a *derivation* of induction phenomena, because an energy balance can never be proved to be complete. However, it remained clear that the known induction phenomena *complied with* energy conservation.[19]

The potential theory

Helmholtz did not return to the foundations of electrodynamics until the late 1860s. At that time he had completed his *Physiological optics*, and was tired of his efforts to revise Kant's transcendental philosophy. In March 1869, he confided to his friend Ludwig: "I found that too much philosophizing demoralizes me somehow, and makes my thoughts lax and vague; I shall discipline them again for a while through experiments and mathematics, and then, afterwards, I might well return to the theory of perception".[20]

Helmholtz returned to his first love, physics. More exactly, he decided to enquire into a physical problem encountered in earlier physiological experiments of his. In 1850, while measuring the propagation of the nervous influx, he had noted that electrical shocks had little effect on deeper-lying nerves of the body. Some twenty years later he wanted to prove that the effect was due to self-induction in the conducting mass. For this purpose he performed experiments in which a naked frog-nerve served as a detector for currents travelling through a conducting solution and emerging from electrodes connected to an induction machine.[21] The theoretical analysis of the experiments proved difficult, because the involved currents were open-ended, while all previous electrodynamic experiments were limited to closed, or nearly closed currents. For open currents the various known theories of electrodynamics disagreed, and Helmholtz had to compare them.

The resulting theoretical memoir, published in 1870, covered a hundred pages or so and gave the basis for a new approach to electrodynamics that played a significant role in orienting German researches in the field. Helmholtz expressed his intentions in the following terms: "In the face of conflicting theories, it seemed to me preferable to remain as close as possible to the

---

19 Helmholtz 1873b: 678-679, with the comment: "Als Beweismittel für Thatsächliche Wahrheiten kenne ich nur den Weg der Induktion".

20 Helmholtz to Ludwig, 28 March 1869, quoted in Koenigsberger 1903: vol. 2, 162.

21 Helmholtz 1870a. Cf. Koenigsberger 1903: vol. 2, 164.

ground of facts and to leave undetermined the parts of the theory which could not yet be decided by experiments" (Helmholtz 1870b: 546). Unlike Maxwell or Weber, Helmholtz wished to avoid specific assumptions on the nature of electric charge and current. He looked for the most general theory that complied with the energy principle and with the predictions of other theories for closed currents.[22]

Neumann's notion of potential, Helmholtz realized, offered a suitable basis for this program. Indeed, if mechanical and electromotive forces derived from the same potential, they automatically satisfied the energy principle. Moreover, as Thomson and Maxwell had already noted, the negative of the potential played the role of the kinetic energy corresponding to the electric motion.[23] Accordingly, Helmholtz started with Neumann's potential, which he generalized in the three following manners. First, he introduced the general notion of the potential of two current elements, assuming that, for open currents, the mechanical and electromotive forces were given by the same simple mathematical operations (space and time derivatives) as those used by Neumann in the case of closed circuits. Second, he generalized Neumann's integrand with the constraint that integration over closed circuits should retrieve Neumann's formula (and also that the potential should decrease as the inverse of the distance); the resulting expression for the potential of two current elements depended only on one parameter, denoted by k. Finally, Helmholtz generalized his expressions to three-dimensional currents. I shall henceforth call the product of these three generalizations "the potential theory".

The first generalization may seem to be a straightforward application of the decomposition principle. But it is not, because the principle just tells us that the decomposition should be done, but not *how* it should be done. As Helmholtz noted, it is not a priori obvious that the forces between two current elements should derive from a potential, since electric current is a kinetic phenomenon, for which no potential energy in the ordinary sense exists. For closed currents the property is a consequence of the empirical equivalence (proved by Ampère) with double magnetic sheets. No such equivalence exists for open currents. However, Helmholtz reasoned, to a pair of current elements corresponds a well-defined (though unknown) motion, and therefore a well-

22 On Helmholtz's electrodynamics, see Rosenfeld 1956; Woodruff 1968; Hirosige 1969; Buchheim 1971; Wise 1981; Buchwald 1985; Darrigol 1993, Kaiser 1993.

23 The reference to Maxwell and Thomson is in Helmholtz 1870b: 581.

defined kinetic energy. Consequently, the Neumann potential, which is the negative of this energy, should also exist in this case.[24]

As Helmholtz would later realize, the reasoning is flawed. That one meaning of the potential extends to current elements does not prove that the other meaning does. Nevertheless, the idea of generating all electrodynamic interactions from the energy function of pairs of current elements remained, as Buchwald noted, the cardinal feature of Helmholtz's theory (Buchwald 1990). Some further comments on the origins and legitimation of this idea are therefore in order.

Helmholtz consistently attributed the generalization of Neumann's potential to current elements to Franz Neumann himself.[25] In reality Neumann believed that electrodynamic and electromotive forces derived from a potential *only* in the case of closed currents. Hence, Helmholtz's attribution may only be taken as a sign that he judged the generalization to current elements to be extremely natural. He was not the only one. Carl Neumann, the son of Franz, wondered "why this apparently so obvious path ha[d] not been earlier trodden, although it certainly seem[ed] worth a try" (C. Neumann 1871: 452).

The mathematical simplicity of Helmholtz's procedure was indeed appealing: when a property holds for finite volumes, one generally tries to extend it to volume elements. But Helmholtz had deeper motivations: the potential theory represented a middle way between the excessive demand of a reduction to central forces and a purely phenomenological approach. It combined, in the

24 Helmholtz 1870b: 562: "Dass für geschlossene Ströme [die elektrodynamischen] Kräfte auf ein Potential zurückgeführt werden können, ist durch die Ergebnisse der Versuche erwiesen. Für ungeschlossene Ströme könnte dies zweifelhaft erscheinen. Eben deshalb ist es wichtig, dass [das Potential zweier Ströme *i* und *j*] noch die zweite, von den Bewegungen der Stromleiter unabhängige Bedeutung hat. Sie ist derjenige Theil des vorhandenen Arbeitsäquivalentes, der von dem gleichzeitigen Vorhandensein der beiden Ströme *i* und *j* herrührt".

25 This comes out in the title of Helmholtz 1873c: "Vergleich des Ampère'schen und des Neumann'schen Gesetzes für die elektrodynamischen Kräfte", and in his repeated praises of Franz Neumann's supposed innovation: "Die Auffindung des Potentialgesetzes durch Gauss [in a posthumously published MS] und Neumann senior kann in der That als eine der glücklichsten und glänzendsten Errungenschaften der mathematischen Physik angesehen werden" (Helmholtz 1874a: 712), and "Ich habe die Entdeckung [des Potentialgesetzes] durch Herrn F.E. Neumann, den Vater, stets für einen der glücklichsten und fruchtbarsten Gedanken gehalten, welchen die neuere mathematische Physik aufzuweisen hat" (Helmholtz 1874b: 773).

simplest manner, two honorable residues of the old-style comprehensibility of nature: the decomposition principle and the energy principle.[26]

I now return to the contents of the 1870 memoir. Helmholtz limited his considerations to the electric motion in bodies at rest, and first asserted, without proof, that the choices $k = 1$, $k = 0$, and $k = -1$ retrieved, respectively, the predictions of Neumann's theory, Maxwell's theory and Weber's theory (the claim is strictly valid only in the third case). Then he showed that a negative value of k implied the energetic instability of electric equilibrium. The argument excluded Weber's theory. For a further determination of the value of k, Helmholtz discussed the possibility of various experiments, with a negative conclusion: the expected effects were too small to be measured. Finally, Helmholtz treated the case of polarizable media by simply inserting polarization currents in his potential law. When vacuum itself is extremely polarizable, he could show that his theory implied, after proper charge-renormalization, the same propagation phenomena as Maxwell's theory, which confirmed the possibility of an electromagnetic nature of light.

Program shifts

A passive reader of Helmholtz's memoir could gain the impression that all previous theories of electrodynamics were particular cases of the potential theory, and that the only important issue at stake was the determination of k. The rejection of the value $k = -1$, corresponding to Weber's theory, certainly in-

26 Helmholtz referred explicitly to the decomposition principle in this context: "Nun lag es in der Natur der Sache, dass man bei den Wirkungen, die ein so zusammengesetztes und variables Gebilde, wie ein elektrischer Stromkreis, hervorbringt, nicht stehen bleiben konnte. Man musste versuchen, die Wirkungen des Ganzen in die Wirkungen seiner einzelnen Elemente aufzulösen" (Helmholtz 1875: 776). He also repeatedly asserted that the potential theory provided the simplest possible decomposition, e.g., in Helmholtz 1874a: 712, 760.
Buchwald 1993 puts much emphasis on Helmholtz's anti-romanticism as a proper context for the potential theory. I agree that anti-romanticism contributed to Helmholtz's sympathy for Franz Neumann's and Kirchhoff's mathematical phenomenology, and to his hostility toward Weberians. But, as Buchwald himself admits, the idea of bipartite interactions had another source: the crumbled metaphysics of the *Erhaltung der Kraft*, adapted for the needs of the phenomenologist.

duced strong defensive reactions from the elderly Weber and his disciples.[27] But Helmholtz's critics also noticed that the potential theory, whatever the values of *k*, sharply contradicted previous theories, as soon as one considered electrodynamic forces and moving bodies. In addition to Ampère's forces, the potential theory implied forces acting on the extremities of current. For the induction in moving conductors, it yielded predictions incompatible with Faraday's law or with earlier German theories. As a consequence, Helmholtz shifted his program from the determination of *k* to the discrimination between the potential theory and other theories, although he still believed the potential theory to be the most probable one. He now recognized that, despite the existence of an energy function for two elements of current, the forces between the elements did not necessarily derive from a potential, and he even developed the consequences of a strict validity of Ampère's force law, for which no potential exists (Helmholtz 1873b: 679-683; 1873c; 1874a).

Around 1875 Helmholtz and a couple of visitors to his laboratory performed experiments that did not fit the expectations of the potential theory (Helmholtz 1875). Helmholtz was himself responsible for the most decisive of these, in which a conducting segment rotated under a magnet without changing its potential, but nevertheless crossed the lines of force of the magnet. According to Faraday's rules an induction had to occur in this case, while the potential theory gave no effect at all. By a clever device Helmholtz managed to detect the inductive effect without closing the circuit, and thus endangered his cherished potential law.

Yet Helmholtz did not abandon the potential law. He immediately noted that his theory could absorb the conflicting evidence if only vacuum was endowed with a sufficiently high polarizability (Helmholtz 1875: 788). More generally, he realized that in this limit he could retrieve all electrodynamic predictions of Maxwell's theory. As a consequence, Helmholtz shifted his program for the second time. The competing theories were now, on the one hand, the potential theory with vacuum polarization - which I shall call "the extended potential theory" - and, on the other hand, the theory derived from the assumption of strict validity of Ampère's force law. Again, Helmholtz had no faith in the second alternative, which seemed much too complicated.[28] He neverthe-

27 On the ensuing polemics, see Archibald 1989; Wise 1981; Kaiser 1981: 100-106; Hoppe 1884; Wiedemann, 1885: vol. 4, 1049ff; Cahan (in this volume).

28 See Helmholtz 1874a: 760. Helmholtz's preference also appears in his (never realized)

less felt the need of discriminating experimentally between the two theories. If vacuum was polarizable, he reasoned, a current should exist between the plates of a condenser during its charge or discharge. If this current varies very quickly, induced currents should occur in a nearby secondary circuit. Helmholtz proposed the electrodynamic effects of polarization currents as a prize subject for the Berlin Academy in 1879. Hertz was the one who, with some delay, picked up the subject, with the consequences that everyone knows.[29]

Elementary forces

Helmholtz's fidelity to the potential law betrays the persistence of some fragments of the full comprehensibility of nature. We may wonder whether the more constraining reducibility to central forces also played a role in his electrodynamic works of the 1870s. The answer is a moderate yes. As we saw Helmholtz rejected Weber's theory for leading to absurd instabilities. To this reproach he added another: Weber's elementary forces were not central. As predicted in the memoir on the conservation of force, this defect led to the possibility of a perpetual motion. More exactly, Helmholtz conceded that Weber's forces did not allow cyclic perpetual motion, but showed that a finite motion of the electric particles could lead to the production of an infinite amount of work, which in itself permitted perpetual motion. Although the argument failed to convince Weberian physicists, it certainly spoke against the elementary and fundamental character of Weber's law.[30]

---

project of a complete electrodynamics of moving conductors and insulators, with a "complete connection with Maxwell's theory", announced in Helmholtz 1876: 788. See also his Faraday lecture (Helmholtz 1881b: 59) in which he bets on the truth of Faraday's assumption of dielectric polarization currents.

29 For the Academy problem see the introduction of Hertz 1892. On Hertz's experiments, see d'Agostino 1975; Cazenobe 1982; Buchwald 1990, 1994; Doncel 1991.

30 Helmholtz 1870b: 552-554; 1873b: 653-679. For secondary literature, see note 27. Helmholtz later detected another defect of Weber's law: it does not comply with the following decomposition principle for accelerations: when several causes act together, the acceleration of a given mass-point is the resultant of the accelerations which the various causes would have separately implied (Helmholtz 1881a: 68-69). This peculiarity had already been mentioned in Weber 1846: 168, but only to suggest that a third body, or a medium, was involved.

In this context Helmholtz evoked Maxwell's ether model of 1862, as proof that Weber's adventure beyond accepted norms of mechanical reduction was unnecessary:

> "Electromagnetic and electrodynamic phenomena do not differ at all from the usual processes of inorganic nature in such a way that their explanation would require a new sort of fundamental forces, namely, forces which do not depend only on the position of interacting masses but also on their motion. This results from Maxwell's theory, which explains the above phenomena not in terms of action at a distance, but in terms of a motion continuously propagating in ether. In this theory the postulated interaction between two contiguous volume elements of ether [...] can be simulated by simple mechanical contrivances; it can therefore be analyzed in terms of standard mechanical forces" (Helmholtz 1873b: 674).

The contact forces operating in Maxwell's model could in turn be reduced to central forces acting only at very small distances, so that the full reducibility of electrodynamics to central forces was indirectly proven.[31]

In sum, in the construction of his own electrodynamics Helmholtz relied on the simplest possible combination of the energy principle and the decomposition principle, which were residues of his first discussion of the comprehensibility of nature. In his criticism of other theories, he further evoked the reducibility to central forces. That is to say, he remained convinced that the reduction should in principle remain possible, but he no longer believed that an explicit reduction was necessary.

Why not a field theory?

Helmholtz remained faithful to the extended potential theory until Hertz's experiments. Since in the limit of infinite vacuum polarizability this theory led exactly to the same observable predictions as Maxwell's theory, one may wonder why Helmholtz waited so long to adopt the field conception. A first reason was his phenomenological inclination, which prevented him from basing electrodynamics on Maxwell's early mechanical model of ether:

> "English physicists, like Lord Kelvin with his theory of vortex atoms and Maxwell with his assumption of a system of cells with rotating content [...] have obviously drawn more satisfaction from such explanations than from the plain representation of facts and their laws, as it is given by the system of the differential

31 See also Helmholtz 1872: 639. On Maxwell's model, see Siegel 1991.

> equations of physics. For my part, I must admit that so far I have confined myself to the latter kind of representation, and thus felt on the safest ground; yet I would not object in principle to the path chosen by such outstanding physicists" (Helmholtz 1894b: 377).

After all Maxwell himself was not satisfied with his early model of the electromagnetic field, and turned, in 1865, to a more abstract dynamical approach in which the mechanical nature of field processes was only implicit.[32] Helmholtz could have been converted to this theory, if only he had understood Faraday's and Maxwell's bewildering conceptions of charge and current. But he did not, as he confessed in a lecture of 1881 in Faraday's honor.[33]

Most important, Helmholtz had no prejudice against action at a distance in favor of contact action. Only on one occasion, in his foreword to Hertz's *Mechanics*, did he manifest an interest for universal reduction to contact actions (Helmholtz 1994b: 374)[34]; and even here he might only have reflected Hertz's opinion on the subject. Clearly, for Helmholtz the reducibility to central forces remained most basic, and contact action was only a limiting case of action at a distance. As already mentioned, in his latest written notes he still declared that fundamentally energy was not spatially divisible. With this in mind, we can better understand why Helmholtz waited so long to adopt Maxwell's theory, and why in his extended potential theory he tolerated a finite value of vacuum polarizability.

---

32 Cf. Siegel 1991 and Buchwald 1985.

33 Helmholtz 1881b: 60: "I confess I should really be at a loss to explain without the use of mathematical formulae, what [Maxwell] considers as a quantity of electricity, and why such a quantity is constant, like that of a substance."

34 In the Faraday lecture Helmholtz did not really endorse the ideal of a complete reduction to contact forces. He considered (Helmholtz 1881b: 59) that Faraday's "deciding assumption" was the existence of dielectric currents that made all currents closed, and treated the dominant character of contact actions as a subsidiary consequence: "An *inference* from this supposition is that the forces working at a distance do not exist, *or are at least unimportant*, when compared with the tensions and pressures of the dielectric medium" (my emphasis). Clearly, the statement referred not to Faraday's real views on the matter, but to Helmholtz's extended potential theory.

Least action

Heinrich Hertz originally shared his mentor's theory and attitude. In a late stage of his experiments on electric waves he believed that he had measured a finite value for the polarizability of vacuum. But in the end he reached the opposite conclusion and decided that Maxwell's theory was now superior, if only because of its greater analytical simplicity. Hardly more successful than Helmholtz in understanding the Maxwellian concepts of charge and current, Hertz reformulated Maxwell's theory without imagery, simply postulating the field equations and deriving the concepts of charge and current by symbolic operations. In this process, Hertz focussed on observable field quantities and lost the dynamical underpinning of Maxwell's theory.[35]

Helmholtz promptly acknowledged the superiority of the Maxwell-Hertz theory and adopted it in his own researches as well as in his lectures. Yet he did not approve of Hertz's abandonment of Maxwell's dynamical foundation. Since the mid-1880s he had convinced himself that all physics should be subsumed under the principle of least action,[36] and he found in Maxwell's dynamical theory of the electromagnetic field an essential step in this direction. There Maxwell had shown that the energy of a system of currents could be regarded as a Lagrangian, for which the currents played the roles of generalized velocities. The corresponding Lagrange equations yielded all that needed to be known, namely, the expressions of electromotive and mechanical forces. Maxwell thus drew all the profits of assuming the mechanical nature of field processes, without appealing to any specific mechanism (Maxwell 1865).[37]

Helmholtz could easily adapt this fascinating aspect of Maxwell's theory to his potential theory, since the electrodynamic potential was, for closed currents, identical to Maxwell's electrokinetic energy. The resulting insight was that Neumann's rule for deriving the electromotive force from the potential was nothing but a Lagrange equation (Helmholtz 1886: 208). As Helmholtz himself explained, an essential motivation of his general researches on the action principle was precisely the understanding of the form of action encountered in

35 See Heimann 1971; Buchwald 1985: 192-193; Darrigol 1993.

36 See Klein 1972.

37 On this aspect of Maxwell's theory, see Buchwald 1985: chap. 5, and further references there.

electrodynamic theory.[38] No wonder then that his first contribution to the Maxwell-Hertz theory was, in 1892, a full derivation of the equations of this theory from the action principle (Helmholtz 1892a). Little before his death he was still working on improving the derivation (Helmholtz 1894c).[39]

What precise function did Helmholtz attribute to the action principle? In 1886, he wrote:

> "The general validity of the principle of least action seems to me so widely secured that it can have a high value as a heuristic principle and as a lead in our striving to formulate the laws of new classes of phenomena. Besides, the principle has the advantage of condensing in just one formula all the relations that are relevant to the investigated class of phenomena; which gives us a complete overview of everything essential" (Helmholtz 1886: 210).

The action principle was, like the energy principle, an organizing principle, but a much more powerful one. Where the energy principle only provided constraints, the action principle provided the full equations of evolution of the system. Moreover, it synthesized all invariant properties of the system in one function, the action. This could be a serious advantage in complicated cases, as Helmholtz found in his dispersion theory of 1892.[40]

This is not all. The action principle being, from the start, a principle of mechanics, its application to phenomena beyond the realm of visible motions implied a mechanical reduction. But in what sense? The answer may be inferred from the more precise contents of Helmholtz's memoirs on least action. The starting point is an ordinary mechanical system, which is defined as a set of material points interacting through a potential and eventually submitted to

38 Helmholtz 1886: 209: "Es waren gerade Untersuchungen über die Form des kinetischen Potentials, welches Maxwell's Theorie der Elektrodynamik fordert, die mich auf die vorliegenden vorbereitenden Untersuchungen geführt haben".

39 In 1892 Lorentz also provided a dynamical foundation (based on d'Alembert's principle of virtual work) of the Maxwell-Hertz theory, but without giving the derivation of mechanical forces. Before that, in his Sorbonne lectures of 1888, Poincaré had put great emphasis on the virtues of Lagrange's methods in Maxwell's electrodynamics: see Darrigol 1993.

40 Helmholtz 1893: 508: "Ich habe es vorgezogen, statt von den Maxwell'schen Gleichungen auszugehen, die neu hinzukommenden Einflüsse in die von mir entwickelte Form des Princips der kleinsten Wirkung aufzunehmen, weil man dadurch von dem Übersehen einzelner nothwendig vorhandener Gegenwirkungen in dem hier schon ziemlich verwickelten Spiel der Kräfte geschützt wird, und dadurch die Anzahl der unabhängigen Hypothesen von zweifelhafter Richtigkeit wesentlich vermindert wird".

integrable constraints. At this level we are very close to a full reduction to central forces. The action principle of course applies, and the Lagrangian is the sum of a quadratic function of the velocities only and of a function of the coordinates only. Then Helmholtz introduces various kinds of elimination of coordinates, with the effect that the Lagrangian can take a more general form with respect to the remaining coordinates and their derivatives. For example, using the constraints to eliminate redundant variables, the coefficients of the quadratic function of the velocities become functions of the coordinates. Less trivially, the elimination of cyclic variables introduces terms in the Lagrangian that are linear in the velocities (Helmholtz 1886: 213-217).[41]

With such considerations Helmholtz aimed to show that restrictions on the form of the Lagrangian (except that it should be a function of coordinates and velocities) did not need to survive in the application of the action principle to systems in which hidden motions occur. We might speculate that, conversely, every system whose evolution can be derived by varying an action function can be regarded as a mechanical system with hidden motions, and is in principle reducible to a set of material points acting in pairs by central forces. Helmholtz did not assert so much: to him the elimination of coordinates in a fully comprehended mechanical system was more like Wittgenstein's ladder: it helped reach the level of a general action principle, but was better thrown away after the climb.[42] Yet Helmholtz's studies on least action could hardly be understood without referring to his older ideal of a full comprehensibility of nature.[43]

41 Such eliminations already occurred in Helmholtz's works on the analogy between monocyclic mechanical systems and thermodynamic systems: see Klein 1972.

42 This view may be inferred from some remarks in Helmholtz 1886: 207, 215.

43 An interesting parallel can be drawn between Helmholtz's studies on least action and Hertz's mechanics. Both thinkers tried to justify the universal validity of the general principles of mechanics (especially the energy principle and the action principle) in terms of more primitive mechanical concepts, and thereby introduced concealed motions. But they differed in the choice of the primitive concepts. Whereas Helmholtz kept thinking in terms of the inseparable abstractions of force and matter, Hertz went back to the oldest origins of mechanics and thought in terms of connecting mechanisms. For Hertz the concept of force belonged to a metaphysics of action at a distance, which the recent discoveries in the field of electricity had made obsolete. See Hertz to Helmholtz, Dec. 1892, quoted in Koenigsberger 1903: vol. 3, 48; and Helmholtz 1894b: 375.

Conclusion

Helmholtz's epistemological positions and the general principles under which he subsumed electrodynamics evolved in parallel. At the beginning the full comprehensibility of nature involved an explicit, unique reduction to central forces. Electrodynamics did not meet this criterion, but Helmholtz's hoped that it would do so in the future, just as gravitation, heat, electro- and magnetostatics already did.

By 1870 Helmholtz admitted that the comprehensibility of nature only implied the causality principle, reinterpreted as the assumption of lawfulness. From his earlier notion of comprehensibility, he rescued two efficient organizing principles, the energy principle and the decomposition principle, which lent themselves very well to a more phenomenological approach to physics. As we saw, the potential theory resulted from the simplest possible combination of the energy principle and the decomposition principle. It was in a sense a softer version of the earlier demand of comprehensibility. Helmholtz also kept in mind the reduction to central mechanical forces, but only meaning that it should exist, not that it should be done. His reactions to Weber's and Maxwell's theories in the 1870s illustrate this conception.

Finally, Helmholtz renounced the potential theory and replaced the relevant organizing principles with the almighty action principle. The latter mechanical principle completely determined the evolution of the system without introducing unobservable mechanisms. In assuming its full generality, Helmholtz reached the best possible harmonization of his mechanistic ideal with his more phenomenological inclination.

Abbreviations:

WA: Helmholtz, Hermann von: *Wissenschaftliche Abhandlungen.* 3 vols.. Leipzig, 1882, 1883, 1895.

VR: Helmholtz, Hermann von: *Vorträge und Reden.* 2 vols.. Braunschweig, 1903.

Literature:

d'Agostino, Salvo: Hertz's researches on electromagnetic waves. *Historical studies in the physical sciences*, 6, 1975, 261-323.

Archibald, Thomas: Energy and the mathematization of electrodynamics. *Archives internationales d'histoire des sciences*, 39, 1989, 276-308.

Bevilacqua, Fabio: Helmholtz's *Über die Erhaltung der Kraft*: The emergence of a theoretical physicist. In: D. Cahan (ed.): *Hermann von Helmholtz and the foundations of nineteenth century science*. Berkeley, 1993, 291-333.

Buchheim, Gisela: Hermann von Helmholtz und die klassische Elektrodynamik. *NTM*, 8, 1971, 26-36.

Buchwald, Jed Z.: *From Maxwell to microphysics. Aspects of electromagnetic theory in the last quarter of the nineteenth century*. Chicago, 1985.

Buchwald, Jed Z.: The background to Heinrich Hertz's experiments in electrodynamics. In: T.H. Levere and W.R. Shea (eds.): *Nature, experiments, and the sciences*. Amsterdam, 1990, 275-306.

Buchwald, Jed Z.: Electrodynamics in context: object states, laboratory practice and anti-idealism. In: D. Cahan (ed.): *Hermann von Helmholtz and the foundations of nineteenth century science*. Berkeley, 1993.

Buchwald, Jed Z.: *The creation of scientific effects: Heinrich Hertz and electric waves*. Chicago, 1994.

Cazenobe, Jean: Les incertitudes d'une découverte: L'onde de Hertz de 1888 à 1900. *Archives internationales d'histoire des sciences*, 32, 1982, 236-265.

Chevalley, Catherine: Glossary for Niels Bohr, *Physique atomique et connaissance humaine*. Paris: Gallimard, 1991, 422-442.

Clausius, Rudolph: Über einige Stellen der Schrift von Helmholtz 'Über die Erhaltung der Kraft'. *Annalen der Physik und der Chemie*, 89, 1853, 574.

Clausius, Rudolph: Über einige Stellen der Schrift von Helmholtz 'Über die Erhaltung der Kraft', zweite Notiz. *Annalen der Physik und der Chemie*, 91, 1854, 604.

Darrigol, Olivier: The electrodynamic revolution in Germany as documented by early German expositions of 'Maxwell's' theory.' *Archive for the history of exact sciences*, 45, 1993, 189-280.

Doncel, Manuel: On the process of Hertz's conversion to Hertzian waves. *Archive for the history of exact sciences*, 43, 1991, 1-27.

Elkana, Yehuda: *The discovery of the conservation of energy*. London, 1974.

Heimann, Peter M.: Maxwell, Hertz, and the nature of electricity. *Isis*, 62, 1971, 149-157.

Heimann, Peter M.: Helmholtz and Kant: The metaphysical foundations of *Über die Erhaltung der Kraft*. *Studies in the history and philosophy of science*, 5, 1974, 205-238. (1974a).

Heimann, Peter M.: Conversion of forces and the conservation of energy. *Centaurus*, 18, 1974, 147-161. (1974b).

Helmholtz, Hermann von: *Über die Erhaltung der Kraft, eine Physikalische Abhandlung*. Berlin, 1847. Also in: *WA* 1, 12-68.

Helmholtz, Hermann von: Über die Wechselwirkung der Naturkräfte und die darauf bezüglichen neuesten Ermittelungen der Physik. In: *VR*1, 48-84. (1854a).
Helmholtz, Hermann von: Erwiderungen auf die Bemerkungen von Hrn. Clausius. In: *WA*1, 76-93. (1854b).
Helmholtz, Hermann von: Über das Sehen der Menschen. In: *VR*1, 85-118. (1855).
Helmholtz, Hermann von: Theorie der Wärme. *Fortschritte der Physik im Jahre 1854*. Berlin, 1857, 369.
Helmholtz, Hermann von: Über die Erhaltung der Kraft (Karlsruhe lectures, 1862-1863). In: *VR*1, 187-230. (1863).
Helmholtz, Hermann von: *Handbuch der Physiologischen Optik*. Vol. 3. Heidelberg, 1867.
Helmholtz, Hermann von: Über die Thatsachen die der Geometrie zu Grunde liegen. In: *WA*2, 618-636. (1868).
Helmholtz, Hermann von: Über das Ziel und die Fortschritte der Naturwissenschaft. In: *VR*1, 367-400. (1869).
Helmholtz, Hermann von: Über die physiologische Wirkung kurzdauernder elektrischer Schläge im Innern von ausgedehnten Leitern. In: *WA*1, 526- 530. (1870a).
Helmholtz, Hermann von: Über die Theorie der Elektrodynamik. Erste Anhandlung: Über die Bewegungsgleichungen der Elektricität für ruhende Körper. In: *WA*1, 545-628. (1870b).
Helmholtz, Hermann von: Über den Ursprung und die Bedeutung der geometrischen Axiome. In: *VR*2, 1-32. (1870c).
Helmholtz, Hermann von: Zum Gedächtnis an Gustav Magnus. In: *VR*2, 33-52. (1871).
Helmholtz, Hermann von: Über die Theorie der Elektrodynamik. Vorläufiger Bericht. In: *WA*1, 636-646. (1872).
Helmholtz, Hermann von: Induction und deduction. Foreword to the second part of the first volume of the German translation (1873) of Thomson and Tait's *Treatise on natural philosophy*. In: *VR*1, 413-421. (1873a).
Helmholtz, Hermann von: Über die Theorie der Elektrodynamik. Zweite Abhandlung. Kritisches. In: *WA*1, 646-687. (1873b).
Helmholtz, Hermann von: Vergleich des Ampère'schen und des Neumann'schen Gesetzes für die elektrodynamischen Kräfte. In: *WA*1, 688-701. (1873c).
Helmholtz, Hermann von: Über die Theorie der Elektrodynamik. Dritte Abhandlung. Die elektrodynamischen Kräfte in bewegten Leitern. In: *WA*1, 702-762. (1874a).
Helmholtz, Hermann von: Kritisches zur Elektrodynamik. In: *WA*1, 763-773. (1874b).
Helmholtz, Hermann von: Versuche über die im ungeschlossenen Kreise durch Bewegung inducirten elektromotorischen Kräfte. In: *WA*1, 774-790. (1875).
Helmholtz, Hermann von: Bericht betreffend Versuche über die elektrodynamische Wirkung elektrischer Convection, ausgeführt von Hrn. Henry A. Rowland. In: *WA*1, 791-797. (1876).
Helmholtz, Hermann von: Die Thatsachen in der Wahrnehmung. In: *VR*2, 213-248. (1878).
Helmholtz, Hermann von: *Zusätze* to the memoir on the conservation of force. In: *WA*1, 68-75. (1881a).
Helmholtz, Hermann von: On the modern development of Faraday's conception of electricity. In: *WA*3, 52-87. (1881b).

Helmholtz, Hermann von: Über die physikalische Bedeutung des Princips der kleinsten Wirkung. In: *WA*3, 203-248. (1886).

Helmholtz, Hermann von: Das Princip der kleinsten Wirkung in der Elektrodynamik. In: *WA*3, 476-504. (1892a).

Helmholtz, Hermann von: Goethe's Vorahnungen kommender naturwissenschaftlicher Ideen. In: *VR*2, 335-362. (1892b).

Helmholtz, Hermann von: Elektromagnetische Theorie der Farbenzerstreuung. In: *WA*3, 505-525. (1893).

Helmholtz, Hermann von: *Einführung zu Vorlesungen über Theoretische Physik*. Leipzig, 1894. (1894a).

Helmholtz, Hermann von: Introduction to Hertz, Heinrich: *Prinzipien der Mechanik*. Leipzig, 1894. Also in: *VR*2, 365-378. (1894b).

Helmholtz, Hermann von: Nachtrag zu dem Aufsatze: Über das Princip der kleinsten Wirkung in der Elektrodynamik (unfinished MS, 1894). In: *WA*3, 597-604. (1894c).

Helmholtz, Hermann von: *Vorlesungen über theoretische Physik. Vol. I.2: Die Dynamik discreter Massenpunkte*. Leipzig, 1898.

Helmholtz, Hermann von: *Epistemological writings*. Dordrecht, 1977.

Hertz, Heinrich: *Untersuchungen über die Ausbreitung der elektrischen Kraft*. Leipzig, 1892.

Hirosige, Tetu: Origins of Lorentz's theory of electrons and the concept of the electromagnetic field. *Historical studies in the physical sciences*, 1, 1969, 151-209.

Hoppe, E.: *Geschichte der Elektricität*. Leipzig, 1884.

Jungnickel, Christa, and McCormmach, Russel: *Intellectual mastery of nature: Theoretical physics from Ohm to Einstein*. 2 vols.. Chicago, 1986.

Kaiser, Walter: *Theorien der Elektrodynamik im 19. Jahrhundert*. Hildesheim, 1981.

Kaiser, Walter: Helmholtz's instrumental role in the formation of classical electrodynamics. In: D. Cahan (ed.): *Hermann von Helmholtz and the foundations of nineteenth century science*. Berkeley, 1993, 374-402.

Klein, Martin: Mechanical explanation at the end of the nineteenth century. *Centaurus*, 17, 1972, 58-82.

Koenigsberger, Leo: *Hermann von Helmholtz*. 3 vols.. Braunschweig, 1902-1903.

Kuhn, Thomas: Energy conservation as an example of simultaneous discovery. In: Clagett, M. (ed.), *Critical problems in the history of science*. Madison, 1959, 321-356.

Lenoir, Timothy: *The strategy of life: Teleology and mechanics in 19th century biology*. Dordrecht, 1982.

Maxwell, James Clerk: A dynamical theory of the electromagnetic field. *Transactions of the Royal Society of London*, 155, 1865, 459.

Neumann, Carl: Über die von Helmholtz in die Theorie der elektrischen Vorgänge eingeführte Prämissen, mit besonderen Rücksicht auf das Princip der Energie. *Berichte über die Verhandlungen der Königlichen Sächsischen Gesellschaft der Wissenschaften zu Leipzig*, 1871, 450-478.

Neumann, Franz: Allgemeine Gesetze der inducirten elektrischen Ströme (read in 1845). *Sitzungsberichte der deutschen Akademie der Wissenschaften zu Berlin*, 1847.

Neumann, Franz: Über ein allgemeines Princip der mathematischen Theorie inducirter elektrischer Ströme. *Sitzungsberichte der deutschen Akademie der Wissenschaften zu Berlin*, 1848.

Olesko, Kathryn: *Physics as a calling: Discipline and practice in the Königsberg seminar for physics*. Ithaca, 1991.

Rosenfeld, Leon: The velocity of light and the evolution of electrodynamics. *Nuovo Cimento*, 4, 1956, suppl., 1630-1669.

Siegel, Daniel: *Innovation in Maxwell's electromagnetic theory: Molecular vortices, displacement current, and light*. Cambridge, 1991.

Weber, Wilhelm: *Elektrodynamische Maassbestimmungen. I. Über ein allgemeines Grundgesetz der elektrischen Wirkung*. Leipzig, 1846.

Whittaker, Edmund: *A History of the theories of Aether and electricity*. Vol.1. London, 1951.

Wiedemann, Gustav: *Die Lehre der Elektricität*, 4 vols.. Braunschweig, 1885.

Wise, Norton: German concepts of force, energy, and electromagnetic ether. In: Cantor G. and Hodge M. (eds.): *Conceptions of ether*. Cambridge, 1981, 269-307.

Woodruff, Arthur: Action at a distance in nineteenth century electrodynamics. *Isis*, 53, 1962, 439-459.

Woodruff, Arthur: The contributions of Hermann von Helmholtz to electrodynamics. *Isis*, 59, 1968, 300-311.

IV
Helmholtz über Geometrie

Apriorische Funktion und aposteriorische Herkunft: Hermann von Helmholtz' Untersuchungen zum Erfahrungsstatus der Geometrie

Renate Wahsner

Daß Geometrie auf Erfahrung beruht, ist unserem Jahrhundert eine inzwischen abgestanden erscheinende Einsicht. Das war aber nicht so zu Helmholtz' Zeiten.[1] Die diesbezüglichen Arbeiten von Mathematikern und Physikern wie Gauß, Riemann, Klein und Helmholtz belegen nicht nur die Genese dieser Einsicht, sondern auch ihre Problematik. Denn so leicht, wie man heutzutage

1 Charakteristisch hierfür sind die das Motiv für seine geometrischen Untersuchungen offenbarenden Worte, mit denen Helmholtz seinen Vortrag "Über den Ursprung und die Bedeutung der geometrischen Axiome" einleitet: "Die Thatsache, daß eine Wissenschaft von der Art bestehen und in der Weise aufgebaut werden kann, wie es bei der Geometrie der Fall ist, hat von jeher die Aufmerksamkeit aller derer, welche für die prinzipiellen Fragen der Erkenntnistheorie Interesse fühlten, im höchsten Grade in Anspruch nehmen müssen. Unter allen Zweigen menschlicher Wissenschaft gibt es keine zweite, die gleich ihr fertig, wie eine erzgerüstete Minerva aus dem Haupte des Zeus, hervorgesprungen erscheint, keine vor deren vernichtender Aegis Widerspruch und Zweifel so wenig ihre Augen aufzuschlagen wagten. Dabei fällt ihr in keiner Weise die mühsame und langwierige Aufgabe zu, Erfahrungsthatsachen sammeln zu müssen, wie es die Naturwissenschaften im engeren Sinne zu tun haben [...]. So ist denn auch die Thatsache, daß Geometrie besteht und solches leistet, [...] immer benutzt worden, um an einem imponierenden Beispiele zu erweisen, daß ein Erkennen von Sätzen realen Inhalts ohne entsprechende aus der Erfahrung hergenommene Grundlage möglich sei [...]. Ich beabsichtige [...] , Ihnen Bericht zu erstatten über eine Reihe sich aneinanderschließender neuerer mathematischer Arbeiten, welche die geometrischen Axiome, ihre Beziehungen zur Erfahrung und die logische Möglichkeit, sie durch andere zu ersetzen, betreffen" (Helmholtz 1876: 21 f.).

meist glaubt, ist Kant nicht zu widerlegen. Und darum ging es eigentlich: ob der von Kant behauptete Raum-Apriorismus haltbar ist oder nicht. Eingebettet war dies in die Frage nach dem Charakter der Axiome und dem Begriff der Anschauung.

Helmholtz zeichnet sich dadurch aus, daß er sich voller Respekt und ernsthaft mit Kant auseinandersetzt. Er faßt das Ergebnis seiner Untersuchungen in die These: Der Raum kann transzendental sein, ohne daß es die Axiome sind.[2] Inwiefern Helmholtz Kant wesentlich mißverstanden hat, wird in einer umfangreichen, ab etwa 1870 entstehenden und auf alle Fälle bis in die zwanziger Jahre unseres Jahrhunderts hineinreichenden Literatur bewiesen (vgl. z.B. Cohen 1871; Tobias 1875; Krause 1878; Heyfelder 1897; Goldschmidt 1898; Riehl 1904; Warburg et al. 1922). Dennoch ermöglicht die Helmholtzsche Diskussion des Kantischen Apriorismus eine Schlußfolgerung, die auch Kant korrigiert.

Das Ziel seiner durch die vorangegangenen sinnesphysiologischen Untersuchungen veranlaßten Arbeiten über die Ursprünge der allgemeinen Raumanschauung bestimmte Helmholtz dahingehend, herauszufinden, "welche Sätze der Geometrie Wahrheiten von thatsächlicher Bedeutung aussprechen, welche dagegen nur Definitionen oder Folgen von Definitionen und deren besonderen gewählten Ausdrucksweise sind" (Helmholtz 1883: 610).

Diese Untersuchung soll zunächst offenlassen, woher unsere Kenntnis der "Sätze von thatsächlicher Bedeutung" stammt, hingegen aber mit prüfen, ob die aufgeführten Voraussetzungen vollständig sind. Hierbei tut sich nun - wie Helmholtz bemerkt - eine Schwierigkeit auf, die sich daraus ergibt, daß geometrische Körper Ideale sind, denen sich die körperlichen Gebilde der wirklichen Welt stets nur in einem gewissen Maß annähern, und die Überprüfung, ob dieses Maß hinreicht, gerade mittels derselben geometrischen Sätze stattfinden muß, die überprüft werden sollen (Helmholtz 1883: 610). Helmholtz glaubt, dieses grundsätzliche Hindernis durch die Diskussion der genannten Frage im Rahmen der analytischen Geometrie überwinden zu können. Hinter diesem Glauben steht die Überzeugung, daß bei der konstruktiven Geometrie die Anschauung eine wesentliche Rolle spielt, in diese aber immer *unbewußt*

2 So lautet der Titel der Beilage, die Helmholtz der Überarbeitung seines Vortrages "Ueber die Thatsachen in der Wahrnehmung" anfügte (vgl. auch Helmholtz 1903: 229 f.; Helmholtz 1895: 356 f.). Es sei hier schon vorab bemerkt, daß Helmholtz' Begriff *transzendental* von dem Kants wesentlich unterschieden ist (vgl. dazu auch die in dem der Fußnote folgenden Satz zitierte Literatur).

Erfahrungen mit eingehen (Helmholtz 1903: 340 f.). Die analytische Geometrie hingegen gestatte es, genau zu überschauen, was vorausgesetzt und was abgeleitet wird, und gäbe so die Möglichkeit, unbewußte, durch die Anschauung vermitttelte Voraussetzungen auszuschließen (Helmholtz 1883: 610; 618-660; Helmholtz 1876: 21-54).

Ausgehend vom Begriff der Kongruenz fixiert Helmholtz nun die analytischen Eigenschaften des Raumes, die seines Erachtens vorausgesetzt werden müssen, um die (analytisch gefaßte) Euklidische Geometrie herleiten zu können. "Mein Ausgangspunkt war", schreibt er 1868 in seinem Aufsatz "Ueber die Thatsachen, die der Geometrie zum Grunde liegen",

> "dass alle Raummessung auf Beobachtung der Congruenz beruht [...] . Von Congruenz kann man aber überhaupt nicht reden, wenn nicht feste Körper oder Punktsysteme in unveränderlicher Form zu einander bewegt werden können, und wenn Congruenz zweier Raumgrössen nicht ein unabhängig von allen Bewegungen bestehendes Factum ist. Die Möglichkeit der Raummessung durch Constatirung von Congruenz habe ich also von Anfang an vorausgesetzt und mir die Aufgabe gestellt, die allgemeinste analytische Form einer mehrfach ausgedehnten Mannigfaltigkeit zu suchen, in der die dabei verlangte Art der Bewegung möglich ist" (Helmholtz 1883: 621).[3]

So verfahrend will Helmholtz insbesondere zeigen, daß die Hypothese, von der Bernhard Riemann 1854 in seinem Habilitationsvortrag "Über die Hypothesen, welche der Geometrie zugrunde liegen" ausgegangen war, als Folgerung hergeleitet werden kann. Riemann war - grob gesprochen - mit der allgemeinsten analytischen Form einer mehrfach ausgedehnten Mannigfaltigkeit gestartet und durch eine spezifische einschränkende Bedingung zur Euklidischen Geometrie, zum sogenannten wirklichen Raum, gelangt. Indem Helmholtz nun von Riemanns einschränkender Bedingung ausging, von der Bedingung, daß endliche Raumgebilde ohne Formveränderung überallhin beweglich und in jede Richtung drehbar sein müssen (um Kongruenz konstatieren zu können), konnte er in der Tat Riemanns Ausgangshypothese ableiten. Er hielt sein Verfahren für sinnvoller, da er meinte, seine eigenen Ausgangssätze nicht als Hypothesen unterstellen zu müssen, da sie als Erfahrungen bestätigte Wahrheiten seien.

Gemäß seiner Forderung, daß das System der Raummessung diejenigen Bedingungen voraussetzen muß, unter denen allein Kongruenz konstatiert

---

3 Daß die Geometrie, die die Bedingungen erfüllt, die erforderlich sind, um die Kongruenz feststellen zu können, eine besondere Raumstruktur ist, hatte Gauß im Jahre 1827 in der Arbeit *Disquisitiones circa superficies curvas* erkannt.

werden kann, gibt Helmholtz vier Voraussetzungen seiner Untersuchung an, die diesen Ausgangspunkt garantieren sollen. Sie betreffen die Kontinuität und die Dimensionen, die Existenz beweglicher und in sich fester Körper, die freie Beweglichkeit und die Unabhängigkeit in sich fester Körper von der Drehung (Monodromie). Unter den zusätzlichen Postulaten, daß der Raum drei Dimensionen haben und unendlich ausgedehnt sein soll, resultiert hieraus - wie Helmholtz zunächst meinte - die Euklidische Geometrie. Später fügte er hinzu, daß auf diese Weise zwischen ebenem und pseudosphärischem Raum noch nicht unterschieden werden kann (Helmholtz 1883: 613-617; 619-625, 637-639). Aber bis auf diese Unentschiedenheit hatte sich die "wirkliche" Raumstruktur ergeben.

Helmholtz war es so erstmals gelungen, die Geometrie aus dem Begriff der Kongruenz, den er als einzige Grundrelation unterstellt, abzuleiten, und er hatte somit einen Grund für ihre Gültigkeit oder "Brauchbarkeit" offenbart. Der etwas in Vergessenheit geratene Zusammenhang von Geometrie und Messung, Messung und Beschaffenheit der Natur wurde in das Bewußtsein zurückgeholt.

Helmholtz selbst sieht hierin den Vorzug seines Weges gegenüber dem Riemanns. In seiner Arbeit von 1866 schreibt er:

> "Hervorzuheben ist, dass hierbei deutlicher heraustritt, wie ein bestimmter Charakter der Festigkeit und ein besonderer Grad von Beweglichkeit der Naturkörper vorausgesetzt wird, damit ein solches Messungssystem wie das in der Geometrie gegebene überhaupt eine thatsächliche Bedeutung haben könne".

Und er schlußfolgert:

> "Die Unabhängigkeit der Congruenz fester Punktsysteme von Ort, Lage und relativer Drehung derselben ist die Thatsache, auf welche die Geometrie gegründet ist. [...] Wie jede physikalische Messung muß auch die des Raumes sich auf ein unveränderliches Gesetz der Gleichförmigkeit in den Naturerscheinungen stützen" (Helmholtz 1883: 616).

Inwiefern ist in diesem Fazit eine Antwort auf die Ausgangsfrage zu sehen? Es erwies sich, daß die zu Tatsachen erklärten Vorraussetzungen hinreichen, um aus ihnen die Geometrie abzuleiten. (Ob die Kenntnis dieser Tatsachen apriorischer oder aposteriorischer Herkunft ist, sollte ja zunächst nicht erörtert werden.) Was hat man damit nun gewonnen? Aufklärung über den hier zugrunde liegenden Gedankengang bringt vermutlich eine Passage der berühmten Rektoratsrede aus dem Jahre 1878 "Die Thatsachen in der Wahrnehmung", in der Helmholtz in der Frage "Was ist Wahrheit in unserem Anschauen und Denken? In welchem Sinne entsprechen unsere Vorstellungen der Wirklichkeit?" das Grundproblem der Erkenntnistheorie sieht und dazu ausführt:

"Auf dieses Problem stoßen Philosophie und Naturwissenschaft von zwei entgegengesetzten Seiten; es ist eine gemeinsame Aufgabe beider. Die erstere, welche die geistige Seite betrachtet, sucht aus unserem Wissen und Vorstellen auszuscheiden, was aus den Einwirkungen der Körperwelt herrührt, um rein hinzustellen, was der eigenen Tätigkeit des Geistes angehört. Die Naturwissenschaft im Gegenteil sucht abzuscheiden, was Definition, Bezeichnung, Vorstellungsform, Hypothese ist, um rein übrig zu behalten, was der Welt der Wirklichkeit angehört, deren Gesetze sie sucht. Beide suchen dieselbe Scheidung zu vollziehen, wenn auch jede für einen anderen Teil des Geschiedenen interessiert ist" (Helmholtz 1903: 218).

Man sieht hiernach, wie für Helmholtz die Lösung der selbstgestellten obengenannten Aufgabe, Vorausgesetztes und Abgeleitetes in der Geometrie zu fixieren, einhergeht mit dem Auffinden des Wirklichen resp. des Tatsächlichen. Doch es liegt hierin ein entscheidender Fehler. Denn einen Wirklichkeitsbezug hat ein theoretisches System nur mit Einschluß seiner Voraussetzungen. (In obigem Zitat müßte der Schlußsatz lauten: Beide nehmen einen anderen Teil des Unterschiedenen als unbezweifelte Voraussetzung.) Dies zu verkennen, heißt genau das verfehlen, was Kant im Sinne hatte, als er Raum und Zeit als reine Anschauungsformen a priori festschrieb. Denn nach Kant ist es die "Erfahrung, durch die uns allein Gegenstände gegeben werden" (Kant 1974: 252). Erfahrung ist jedoch "eine Erkenntnisart, die Verstand erfordert, dessen Regel ich in mir, noch ehe mir Gegenstände gegeben werden, mithin a priori voraussetzen muß" (Kant 1974: 26).

Soviel man nun auch an diesem Standpunkt Kants aussetzen kann, so ist es doch richtig, daß eine empirisch bestätigte naturwissenschaftliche Theorie nur mit Einschluß ihrer Voraussetzungen Aussagen über die Wirklichkeit macht. Gegenteilig zu denken, die Voraussetzungen einer Theorie zu vergessen oder abzuscheiden und nur ihre Lösungen als Wirklichkeitsaussagen zu nehmen, heißt, diese Theorie ontologisch zu verfälschen. Physikalisch fand diese Einsicht später ihren Niederschlag in der sogenannten Poincaréschen Summe. Hiernach kann nur die Summe von Raum-Zeit-Struktur und physikalischem Gesetz bzw. von Geometrie und Dynamik der Erfahrung unterworfen werden, keiner der Summanden für sich allein (Poincaré 1906: 51 f.; Einstein 1979). Doch diese Erkenntnis ist im wesentlichen eine unseres Jahrhunderts, die aber auch in diesem Jahrhundert nicht so gefestigt ist, daß Einstein nicht noch Ende der vierziger Jahre mit Reichenbach in einen Disput darüber geriet (Einstein 1979a: 502-506).

Bevor sich der Poincaré-Einsteinsche Standpunkt durchgesetzt hatte (und nachdem man die Erkenntnisse der griechischen Antike vergessen hatte), ging man irrtümlich davon aus, daß die Voraussetzungen oder Axiome für sich genommen gerechtfertigt sein müßten (sei es durch logischen Zwang oder durch Erfahrung), nicht aber verstand man sie als das Vorausgesetzte bezüglich einer bestimmten Theorie, auf dessen Grundlage die dynamischen Erklärungen dieser Theorie (wenn es sich z.B. um Physik handelt) möglich sind. Derartige Voraussetzungen können durchaus nur gesetzt sein. Ihre "Richtigkeit" erweist sich an der Gültigkeit (experimentellen Bestätigung) der Gesamttheorie.

Die Helmholtzsche Sicht ergibt sich, weil die Frage nach dem Wirklichkeitsgehalt nicht mit Bezug auf eine bestimmte Theorie gestellt wird, sondern "an sich". Kants Raum-Zeit-Apriorismus, den Helmholtz mit seinen geometrischen und sinnesphysiologischen Untersuchungen widerlegt zu haben meint, resultierte jedoch gerade aus der positiven Antwort auf die Frage nach den Bedingungen der Möglichkeit von Wissenschaft, der Möglichkeit einer wissenschaftlichen Theorie. (Man muß aber auch zugeben, daß Kant die Apriorität des Raumes für etwas an sich Gegebenes ansah, nicht für etwas auf eine *bestimmte* Theorie Bezogenes, als etwas, wodurch Wissenschaft schlechthin ermöglicht werden sollte. Insofern verfehlt das Helmholtzsche Abtrennungsstreben nur den rationalen Kern des Kantschen Apriorismus, nicht dessen Erscheinungsform.)

Wenn Helmholtz' oben zitiertes Fazit besagt: die Euklidische Geometrie hat deshalb Bedeutung, einen Wirklichkeitsbezug, weil sie die Bedingungen erfüllt, ohne die die Kongruenz geometrischer Gebilde nicht feststellbar ist, ohne die nicht gemessen werden kann; die Euklidische Geometrie hat einen Wirklichkeitsbezug, weil sie Bedingungen unterstellt, die durch die Naturkörper realisiert sein müssen, wären sie es nicht, so könnte man nicht messen, dann werden im Grunde genommen Meßanforderungen mit Tatsachen gleichgesetzt, Postulate als Fakten charakterisiert. Diese Gleichsetzung basiert vermutlich auf dem Gedankengang: Man kann messen, also sind die Bedingungen der Messung realisiert. Was heißt es aber, messen zu können? Es heißt, die Kongruenz muß feststellbar sein. Damit sie es ist, müssen starre Körper, freie Transformation und freie Drehung gewährleistet sein. Es gibt also feste Körper und die genannten Bewegungsverhältnisse. Die Wirklichkeit ist folglich so beschaffen, wie die Messung es erfordert. Letzteres ist allerdings ein Schluß, keine empirische Überprüfung. Wollte man den Schluß empirisch überprüfen, müßte man die Geometrie als ganze unterstellen (und zwar in ihrer axiomatisierten Form), was zu der von Helmholtz selbst genannten Schwierigkeit führt, die Beschaffenheit der Natur mittels Sätzen überprüfen zu müssen, die gerade getestet werden sollen. Und diese Schwierigkeit ist prinzipieller Natur; sie ist nicht durch eine Auflösung in einzelne Voraussetzungen bzw. durch eine logische Ableitung zu umgehen.

Helmholtz glaubt jedoch, es sei dies möglich, indem er überlegt: Wenn ich aufzeige, daß die Euklidische Geometrie an ganz bestimmte Eigenschaften der Naturkörper und an ganz bestimmte Bewegungsverhältnisse gekoppelt ist, dann beweise ich ihren empirischen Ursprung. Doch es bleibt hierbei nicht nur die Frage offen, wie man beweist, daß diese Verhältnisse in der Natur vorhanden sind, ohne die Geometrie schon zu unterstellen, sondern es wird auch übersehen, daß aus demselben Nachweis z.B. auch auf eine prästabilierte Harmonie geschlossen werden könnte.

Helmholtz' Leistung besteht jedoch zweifellos darin, Voraussetzungen aufzuzeigen, die in den theoretischen Konzepten, die wir benutzen, enthalten sind. Dabei wird jedoch das Notwendige mit dem Wirklichen (bzw. das Hinreichende mit dem Notwendigen *und* Hinreichenden) identifiziert. Es gilt daher das Reich des Raumes und der Zeit, der Größe und der Zahl als das einzige Gebiet, wo unsere Vorstellungen mit der Wirklichkeit übereinstimmen (Helmholtz 1871: 98). Die Welt der primären Qualitäten wird als Welt der Dinge an sich genommen - zwar nicht dem Bekenntnis nach, aber in der Konsequenz.[4] Eine so konzipierte Wirklichkeit könnte nun allerdings aus sich selbst heraus nicht bestehen. Die Wirklichkeit ist nicht an sich so beschaffen, wie die Physik sie denkt, sondern man muß und kann sie so präparieren. Zu diskutieren, was notwendig ist, um die Natur so präparieren zu können, daß man messen kann, das spart Helmholtz aus. Mithin spart er nicht nur die Metaphysik, sondern auch den Prozeß der Bildung physikalischer Begriffe aus. Die den Meßnotwendigkeiten entsprechenden theoretischen und gegenständlichen Konstrukte werden von ihm konsequenterweise nicht als Meßmittel angesehen, sondern als Meßobjekte. Die Raumstruktur oder die Geometrie gilt ihm daher nicht als Mittel der Physik, sondern als Objekt der Naturforschung.

Die noch im 18. Jahrhundert selbstverständliche Identifizierung von geometrischem und wirklichem Raum wurde durch die Diskussion verschiedener möglicher Geometrien problematisiert. Die Diskussion führte zu dem Ergebnis: Es gibt einen physikalischen, nicht aber anschaulich erfaßbaren Raum, dessen Struktur und Eigenschaften nur mathematisch-geometrisch beschreibbar sind. Von hier aus mußte man auch nach der Anschauung zurückfragen -

4 In diesem Zusammenhang sei auf die Analogie des Helmholtzschen zum d'Alembertschen Konzept hingewiesen, wonach die Natur nur insoweit erkennbar ist, inwieweit ihre Gegenstände als Meßkörper aufgefaßt werden können (vgl. Wahsner 1988).

und zwar in dem Sinne, ob der bisherige Begriff resp. Vollzug der Anschauung nicht einer Vorentscheidung unterliegt, die auf gewissen Vorurteilen beruht. Das hieß, man mußte nicht nur den gängigen Raumbegriff, sondern auch die gängige Auffassung von Anschauung in Frage stellen. Dieser neue Ansatz wurde zunächst in der sinnesphysiologischen Forschung versucht. Dabei ergab sich allerdings kein neuer Gesichtspunkt, unter dem die Anschauung zu fassen gewesen wäre.

Helmholtz hat sich - fußend auf den Arbeiten von Gauß und Beltrami - zu zeigen bemüht, daß auch die nicht-euklidischen Geometrien anschaulich gemacht werden können. Hierbei unterstellt er implizit die sinnliche Wirksamkeit eines geometrischen Objektes bzw. die eindeutige Beziehung der Eigenschaften eines geometrischen Objektes zum bestimmten Verhalten eines physikalischen Objektes (vgl. z.B. Helmholtz 1876: 28; Helmholtz 1903: 229-232). Indem Helmholtz zeigen wollte, daß die Geometrie durch die Physik begründet wird, war er darum bemüht, die angestrebte (neue) Anschaulichkeit an die physikalische Denkmöglichkeit zu knüpfen (ohne den *Begriff* der Anschauung neu fassen zu wollen).

"Die geometrischen Axiome", schreibt er, "sprechen also gar nicht über die Verhältnisse des Raumes allein, sondern gleichzeitig auch über das mechanische Verhalten unserer festesten Körper bei Bewegungen" (Helmholtz 1876: 49). Würde man sie jedoch wie die strengen Kantianer als a priori durch transzendentale Anschauung gegebene Sätze betrachten, dann könnten sie keine synthetischen Sätze a priori (im Sinne Kants) sein.

> "Denn sie würden dann nur etwas aussagen, was aus dem Begriffe der zur Messung notwendigen festen geometrischen Gebilde analytisch folgen würde, da als feste Gebilde nur solche anerkannt werden könnten, die jenen Axiomen genügen" (Helmholtz 1876: 50).[5]

Helmholtz sieht das Verhältnis von Geometrie und Physik als Verhältnis von Theorie und Wirklichkeit. Doch die Physik ist selbst eine Theorie. Und ihr Wirklichkeitsbezug ist nur über ihr Verhältnis zur Messung, mithin zur Geometrie (und zwar zur axiomatisierten) zu bestimmen. Helmholtz glaubt jedoch,

5 Hier ergibt sich wieder der schon genannte Zirkel: Fest sind Körper dann, wenn sie den in den geometrischen Axiomen angegebenen Bedingungen entsprechen. Die in den geometrischen Axiomen ausgesagten Verhältnisse sind dann verwirklicht, wenn es feste Körper gibt, mittels derer festgestellt werden kann, daß eben diese Verhältnisse vorliegen. Der Zirkel kann nur durchbrochen werden, wenn man akzeptiert, daß es keine absolute Festigkeit gibt, sondern nur eine, die bezogen auf das, was gemessen werden soll, hinreichend ist. Ob das zutrifft, entscheidet die physikalische Theorie.

ein Verfahren zu haben, das es ihm ermöglicht, eine empirische, von der sogenannten transzendentalen unterschiedene, Geometrie zu begründen und damit die transzendentale, vorgeblich apriorische Geometrie überflüssig machen zu können. In der schon erwähnten Rektoratsrede erklärt er: "Eine solche rein und absichtlich durchgeführte Geometrie würde möglich sein und vollständig den Charakter einer Naturwissenschaft haben" (Helmholtz 1903: 395). Schon deren erste Schritte würden seines Erachtens auf Sätze führen, welche den geometrischen Axiomen entsprächen, wenn nur statt der transzendentalen Gleichheit der Raumgrößen ihre physische Gleichwertigkeit gesetzt wird. Dies würde geschehen, sobald wir eine passende Methode gefunden hätten, um zu bestimmen, ob die Entfernung je zweier Punktpaare einander physisch gleichwertig sind (Helmholtz 1903: 395). Physisch gleichwertig nennt Helmholtz Raumgrößen, "in denen unter gleichen Bedingungen und in gleichen Zeitabschnitten die gleichen physischen Vorgänge ablaufen können" (Helmholtz 1903: 395; 1883: 648). *Ob* man das, ohne die "transzendentale Geometrie" vorauszusetzen, herausfinden kann, prüft Helmholtz nicht. Sein *Sobald* resp. *Wenn* ist ihm nicht problematisch. Die Helmholtzsche Definition der physischen Gleichheit basiert nur auf einer gedachten Gleichheit, nach deren Feststellbarkeit nicht gefragt wird.

Eigentlich hätte Helmholtz skeptisch sein müssen, daß eine solche passende Methode gefunden werden kann. Denn in anderem Zusammenhang zeigt er explizit, daß physikalische Größen stets nur eine Gleichheit in einer bestimmten, durch ein Verfahren wohldefinierten Hinsicht erfassen können (Helmholtz 1903a: §§ 10, 11).[6] Zugleich fordert er völlig zu Recht (wobei hier nicht diskutiert werden soll, inwiefern man sich dabei, "auf ein ganz allgemeines Ergebnis unserer Erfahrung" berufen kann) bezüglich der physikalischen Gleichheit, die die Konstanz der geometrischen Entfernung simulieren soll, daß "wenn die Gleichwertigkeit zweier Raumgrößen durch irgendwelche dazu aus-

6 Insbesondere schreibt er dort: "Da stoßen wir zunächst auf den Begriff des Gleich, und es fragt sich, wo ist denn der Begriff des Gleich hergekommen und was bedeutet er? [...] Sollen wir sagen, zwei Größen sind einander gleich, wenn sie für einander gesetzt werden können? Dagegen ist einzuwenden, daß man zwei Größen niemals *in allen Fällen* für einander setzen kann. [...] In allen diesen Fällen sind die als gleich erklärten Größen nur *in gewisser Beziehung* einander gleich. Es gehört, um über die Gleichheit zu entscheiden, die Kenntniß der Methode dazu, nach der die Vergleichung ausgeführt werden soll. [...] Zugleich bemerken wir, daß die Ungleichheit der allgemeinere Fall ist und die Gleichheit einen besonderen Ausnahmefall bildet" (Helmholtz 1903a: 26 f.).

reichende Methode physikalischer Messung erwiesen worden ist, dieselben sich auch allen anderen bekannten physikalischen Vorgängen gegenüber als gleichwertig erweisen" (Helmholtz 1903: 395). Soll die physische Gleichheit die Geometrie simulieren, dann ist dies als unablässige Bedingung zu stellen. Doch weiß Helmholtz eben, daß das im allgemeinen nicht garantiert ist. Der Gedanke selbst ist tiefsinnig. Das sieht man an der Einsteinschen Allgemeinen Relativitätstheorie, die ja von einer Äquivalenz der Geometrie mit der Gravitation ausgeht. Doch diese Äquivalenz kennt man nicht aus der allgemeinen Erfahrung, sie besteht auch nicht schlechthin, sondern nur partiell bzw. lokal, und zudem erübrigt sie nicht die axiomatisierte Euklidische Geometrie. Man kann es natürlich auch so lesen, daß Helmholtz mit seiner Vision die relativistische Gravitationstheorie antizipiert hat.

Sein Ziel war das jedoch nicht. Er wollte durch seine "physische Geometrie" kein neues (nicht-euklidisches) Meßsystem und damit keine von der klassischen Mechanik abweichende physikalische Theorie konstruieren. Er wollte nur die bekannte Geometrie physikalisch nachbauen und war mit Selbstverständlichkeit davon überzeugt, daß sein der Euklidischen Geometrie entstammender Kongruenzbegriff durch keine Raumstruktur negiert wird. Die Selbstverständlichkeit seiner Überzeugung bewirkt, daß er bei all seinen Überlegungen die Euklidische Geometrie mitdenkt, sie quasi unterstellt. Doch letzteres ist nicht im Sinne des angegebenen Ziels.

Eine rein empirische Geometrie ist als physikalische Raumstruktur nicht möglich. Gewiß ist es richtig, daß die geometrischen Sätze nur deshalb gelten, weil bestimmte Naturverhältnisse vorliegen (die auch empirisch, aber nicht ohne Mittel, also nicht nach empiristischer Vorstellung, erfahren werden können), aber dennoch muß man transzendentale Schlüsse ziehen, vom Endlichen auf das Unendliche schließen, vom Unvollkommenen auf das Vollkommene oder vom Sinnlich-Konkreten auf das Ideale, um eine allgemeingültige Theorie und um ein Mittel zu haben, das die angestrebte physikalische Erkenntnis ermöglicht.

Kant brauchte die apriorische Anschauung, um Erfahrungsgegenstände konstruieren zu können, um das Wirkliche nicht aus dem Begriff ableiten zu müssen. Denn er hatte erkannt, daß es nicht gelingt, von der Idealität unvermittelt in die Realität überzugehen. Etwas physikalischer gesprochen, heißt dies: Ein lediglich logisch widerspruchsfreies System der Mechanik ist noch keine Physik. Damit es dazu wird, muß man messen können, muß man also die Euklidische Geometrie als gültig verwenden und kann deshalb nicht mit diesem mechanischen System wiederum eine empirische Geometrie modellieren (durch die man die sogenannte transzendentale dann ersetzen könnte). Und auch nachdem heute die Denkmöglichkeit und sogar physikalische Bedeutsamkeit nicht-euklidischer Geometrien nachgewiesen ist, behält der Kantsche Gedanke seine Berechtigung, der Gedanke: Um die Möglichkeit allgemeingültiger notwendiger Gesetzesaussagen bzw. die Möglichkeit von Wissenschaft zu begründen, müssen Raum und Zeit als reine Anschauungsformen a priori angenommen werden (oder unverfänglicher: müssen Raum und Zeit in irgendeiner Weise apriorisch gesetzt werden), andernfalls kann keine Erfahrung gemacht werden, sondern sind nur Wahrnehmungen möglich (Kant 1974: insbes. 26). Diese Berechtigung drückt sich in der (gerade durch die Relativitätstheorie und Quantenmechanik ermöglichten) Erkenntnis aus, daß jede physikalische Theorie einen meßtheoretischen Anteil enthält, den sie nicht "physikalisieren", nicht aus der Dynamik ableiten kann. Eine Erklärung findet dieser Sachverhalt in der Einsicht, daß die Mittel der Erkenntnis erst einmal unterstellt werden müssen, bevor man mittels ihrer das Objekt der Erkenntnis untersuchen kann. Dabei sind die Mittel, soll die Untersuchung erfolgreich sein, selbstredend durch den beabsichtigten Zweck, mithin indirekt auch durch das Objekt (mit)bedingt. Daraus folgt jedoch nicht, daß die Zwecksetzung bzw. das Objekt das Mittel erzeugt (v. Borzeszkowski/Wahsner 1980; v. Borzeszkowski/Wahsner 1989; Wahsner/v. Borzeszkowski 1992: insbes. 239-285; Wahsner 1992).[7] Anders gesprochen: Die Poincaré-Einsteinsche Summe bedeutet nicht die Möglichkeit, den einen Summanden auf den anderen reduzieren bzw. den einen aus dem anderen ableiten zu können. Der "Summenstandpunkt" widerspricht nicht der Notwendigkeit, meßtheoretische Mittel voraussetzen zu müssen.[8]

7 Ein Vergleich: Heißt der Zweck "Hände waschen", so ist das Mittel "Messer" ausgeschlossen, das Mittel "Seife" jedoch bedingt - ohne daß die Seife dadurch erzeugt werden könnte.

8 Bei Kant findet sich die hierin enthaltene Problematik in der Feststellung, daß Sinnlichkeit und Verstand zwei völlig verschiedene Erkenntnisvermögen sind, Erkenntnis aber nur aus ihrer Synthese entspringt. Auf die hiermit implizierte Frage, wie diese Synthese möglich sei, antwortet Kant, "daß es ein Drittes geben müsse, was einerseits mit der Kategorie, andererseits mit der Erscheinung in Gleichartigkeit stehen muß, und die Anwendung der ersten auf die letzte möglich macht". "Diese vermittelnde Vorstellung", fährt er fort, "muß rein (ohne alles Empirische) und doch einerseits intellektuell, andererseits sinnlich sein. Eine solche ist das transzendentale Schema" (Kant 1974: 187 f.). Dieses gibt den Kategorien ihre Bedeutung und der Anschauung ihre kategoriale Struktur, wodurch zum einen die Anschauung erst als An-

Es bedeutet dies freilich nicht, daß die meßtheoretischen Mittel in jeder Hinsicht apriorisch sind. Das glaubte nun Kant allerdings. Und *hiergegen* ist Helmholtz eindeutig im Recht: Die Euklidische Geometrie wurde durch menschliche Erfahrung gewonnen,[9] durch die Erfahrung der Land- oder Erdvermessung und deren Extrapolation auf den Himmel resp. auf das Universum. Diese in Jahrtausenden aposteriorisch gewonnenen Erkenntnisse wurden dann um 300 v.u.Z. von Euklid axiomatisiert - und erschienen dann als logisch zwingende. (Eine maßgebliche Vorbedingung dieser Axiomatisierung war aber nicht nur das angesammelte Erfahrungswissen, sondern auch die von den Eleaten initiierte Bestimmung der Kriterien von Theorie, vor allem die Erkenntnis, daß ein jedes theoretische System Prinzipien voraussetzen muß, die es nicht beweist, und die Einsicht, warum dies so ist und sein kann (Szabó 1971).)[10]

Doch ebenso wie die Euklidische Geometrie, die, wie Helmholtz es nennt, spezifische Raumbestimmung, hat sich natürlich - so muß man gegen Helmholtz einwenden - der Begriff des Raumes überhaupt herausgebildet, wenn auch gewiß nicht in der Weise, daß sich zunächst der Begriff des Raumes überhaupt und dann erst die Begriffe seiner verschiedenartig möglichen Strukturen entwickelten. Wahr ist jedoch, daß man die Herausbildung des Raumbegriffs als solchen durch eine Analyse der Physik oder der axiomatisierten Geometrie nicht erkunden kann. Denn beide wurden lange nach der Bildung eines Raumbegriffes überhaupt begründet. Sie setzen ihn in der Tat voraus.

Dies alles im Blick habend, kann man Helmholtz nicht bestätigen, den empirischen Ursprung der geometrischen Axiome bewiesen zu haben. Er hat die Idee ihres empirischen Ursprungs diskutiert und damit propagiert. Er hat damit vielleicht die *Möglichkeit* des empirischen Ursprungs der geometrischen Axiome bewiesen.

---

schauung fungieren kann, zum anderen die Begriffe erst als Begriffe fungieren können (vorher sind sie das nur virtuell oder im Prinzip).

9 Es muß natürlich der Mensch als Gattung und nicht als Individuum gedacht sein.

10 Helmholtz' Überbetonung der Aposteriorität, die einen radikalen Abbau der Apriorität beabsichtigt, ihn aber nicht konsistent durchführen kann, findet eine getreue Parallele in dem behaupteten Verhältnis von Induktion und Deduktion: Die Argumentation für die überragende Rolle der Induktion unterstellt de facto die Möglichkeit der Deduktion, fragt aber nicht nach den Voraussetzungen dieser Möglichkeit (vgl. z.B. Helmholtz 1903: 413-421). Geschähe dies, so müßte sowohl die Fundierung (selbstredend nicht die Reduktion) der Induktion durch die Deduktion als auch die der Aposteriorität durch die Apriorität bewußt werden.

Um die Geometrie - wie es Helmholtz vorschwebte - als erste und vollendetste Naturwissenschaft darstellen zu können, bedarf es theoretisch-historischer Forschungen. Nimmt man die Geometrie nicht in ihren historischen Erfahrungsformen, sondern in ihrer axiomatisierten Form (was Helmholtz ja tut), dann ist sie ein Meßmittel und insofern keine Naturwissenschaft. Newton als Zeugen hiergegen anzurufen, verkennt den Sinn der Newtonschen Worte (Helmholtz 1883: 659).[11] Beide Sichten auf die Geometrie, die Objekt- und die Mittelsicht, miteinander zu vermengen, bringt keine Lösung. Man muß zwischen Herkunft und Funktion, zwischen der aposteriorischen Genesis und der apriorischen Funktion der Euklidischen Geometrie unterscheiden. Helmholtz setzt jedoch Herkunft, Funktion und theoretischen Status der Geometrie gleich. Seine Ursprungsanalysen geraten daher zur logischen Zerlegung der vorhandenen Theorie und ihrer sukzessiven Rekonstruktion aus den so gewonnenen Elementen.

Der tiefere Grund hierfür liegt darin, daß Helmholtz das *Problem* des Überganges vom Empirischen zum Logischen nicht sieht. Mit anderen Worten: Er abstrahiert davon, daß auch die Erfahrung zu erklären ist. Er hält sie für ein "Gebiet", das per se allen hinreichend bekannt ist, dessen Status nicht erforscht werden muß; Helmholtz hält Erfahrung für etwas an sich Gegebenes. "Wie ist Erfahrung möglich?" - genau das war aber Kants Frage (die ihn notgedrungen zu seinem Raum-Zeit-Apriorismus geführt hatte). Und sie muß auch noch heute gestellt werden, verbunden mit der, inwiefern man in dem Kantischen Konzept einen Sinn finden kann unter der Voraussetzung, daß die Euklidische Geometrie nicht als die einzige Geometrie, auch nicht als die einzig physikalisch bedeutsame, angesehen werden kann. Die Differenzierung zwischen ihrer Funktion und ihrer Genesis ermöglicht es, die Kantische Antwort konstruktiv aufzuheben. Daß man aber den Gedanken dieser Differenzierung fassen kann, ist maßgeblich durch die Helmholtzschen Arbeiten befördert worden.[12]

11 Helmholtz nimmt hier Bezug auf Newtons Erklärung: "Die Geometrie hat demnach ihre Basis in der praktischen Mechanik, und sie ist derjenige Teil der allgemeinen Mechanik, welcher die Kunst, genau zu messen, aufstellt und beweist" (Newton 1872: 1). Zu dieser Passage vgl. auch Wahsner 1994.

12 Etwas ausführlicher hierzu siehe Wahsner 1992a.

Literatur

Borzeszkowski, Horst-Heino v. und Wahsner, Renate: Zur Beziehung von experimenteller Methode und Raumbegriff. *Deutsche Zeitschrift für Philosophie*, 28, 1980, 685-696.

Borzeszkowski, Horst-Heino v. und Wahsner, Renate: *Physikalischer Dualismus und dialektischer Widerspruch: Studien zum physikalischen Bewegungsbegriff.* Darmstadt: Wissenschaftliche Buchgesellschaft, 1989.

Cohen, Hermann: *Kants Theorie der Erfahrung.* Berlin: Cassirer, 1871.

Einstein, Albert: Geometrie und Erfahrung. *Sitzungsberichte der Preuß. Akademie der Wissenschaften,* Januar 1921; auch in: *Wiederabdruck der Akademie-Vorträge Albert Einsteins* Nr. 64951/1 bis Nr. 6495/46. Berlin: Akademie-Verlag, 1979.

Einstein, Albert: Bemerkungen zu den in diesem Bande vereinigten Arbeiten. In: Paul Arthur Schilpp (Hg): *Albert Einstein als Philosoph und Naturforscher.* Braunschweig/ Wiesbaden: Vieweg, 1979(a), 493-511.

Goldschmidt, Ludwig: *Kant und Helmholtz.* Hamburg/ Leipzig: Verlag von Leopold Voss, 1898.

Helmholtz, Hermann von: Die neueren Fortschritte in der Theorie des Sehens. In: Ders.: *Populäre wissenschaftliche Vorträge,* Zweites Heft. Braunschweig: Vieweg, 1871.

Helmholtz, Hermann von: Ueber den Ursprung und die Bedeutung der geometrischen Axiome. In: Ders.: *Populäre wissenschaftliche Vorträge*, Drittes Heft. Braunschweig: Vieweg, 1876, 21-54.

Helmholtz, Hermann von: Ueber die thatsächlichen Grundlagen der Geometrie. In: Ders.: *Wissenschaftliche Abhandlungen,* Zweiter Band. Leipzig: Joh. A. Barth, 1883, 610-617.

Helmholtz, Hermann von: Ueber die Thatsachen, die der Geometrie zum Grunde liegen. In: Ders.: *Wissenschaftliche Abhandlungen*, Zweiter Band. Leipzig: Joh. A. Barth, 1883, 618-639.

Helmholtz, Hermann von: Ueber den Ursprung und Sinn der geometrischen Sätze; Antwort gegen Herrn Professor Land. In: Ders.: *Wissenschaftliche Abhandlungen*, Zweiter Band. Leipzig: Joh. A. Barth, 1883, 640-660.

Helmholtz, Hermann von: Zählen und Messen, erkenntnistheoretisch betrachtet. In: Ders.: *Wissenschaftliche Abhandlungen*, Dritter Band. Leipzig: Joh. A. Barth, 1895, 356-391.

Helmholtz, Hermann von: Die Thatsachen in der Wahrnehmung. In: Ders.: *Vorträge und Reden*, Zweiter Band. Braunschweig: Vieweg, 1903, 213-247.

Helmholtz, Hermann von: Goethes Vorahnungen kommender naturwissenschaftlicher Ideen. In: Ders.: *Vorträge und Reden.* Braunschweig: Vieweg, 1903, 335-361.

Helmholtz, Hermann von: Die Anwendbarkeit der Axiome auf die physische Welt (Beilage zu dem Vortrag "Die Thatsachen in der Wahrnehmung"). In: Ders.: *Vorträge und Reden*, Zweiter Band. Braunschweig: Vieweg, 1903, 394-411.

Helmholtz, Hermann von: Induction und Deduction. In: Ders.: *Vorträge und Reden*, Zweiter Band. Braunschweig: Vieweg, 1903, 413-421.

Helmholtz, Hermann von: *Einleitung zu den Vorlesungen über theoretische Physik.* Leipzig: Joh. A. Barth, 1903(a).

Heyfelder, Victor: *Über den Begriff der Erfahrung bei Helmholtz.* Berlin: R. Gaertners Verlagsbuchhandlung, 1897.

Kant, Immanuel: *Kritik der reinen Vernunft.* In: Wilhelm Weischedel (Hg.): Immanuel Kant, Werke in 12 Bänden, Bände III und IV. Frankfurt a. M.: Suhrkamp, 1974.

Krause, Albrecht: *Kant und Helmholtz über den Ursprung und die Bedeutung der Raumanschauung und der geometrischen Axiome.* Hamburg: Verlag von C. Boysen, 1878.

Newton, Isaac: *Mathematische Prinzipien der Naturlehre.* Hg. von Ph. J. Wolfers. Berlin: Oppenheim, 1872.

Poincaré, Henri: *Wissenschaft und Hypothese.* Leipzig: Teubner, 1906.

Riehl, Aloys: Helmholtz in seinem Verhältnis zu Kant. *Kant-Studien*, 9, 1904, 261-285.

Szabó, Árpad: Die Philosophie der Eleaten und der Aufbau von Euklids *Elementen. FILOSOFIA*, 1, Athen 1971, 195-228 (Research Center for Greek Philosophy at the Academy of Athens).

Tobias, Wilhelm: *Grenzen der Philosophie, constatirt gegen Riemann und Helmholtz, vertheidigt gegen v. Hartmann und Lasker.* Berlin: G. W. F. Müller, 1875.

Wahsner, Renate: Eigenschaft und Verhalten - Zur Beziehung von Mathematik und Physik. In: Renate Wahsner (Hg.): *Gravitation und Kosmos: Beiträge zu Problemen der Allgemeinen Relativitätstheorie.* Berlin: Akademie-Verlag, 1988, 132-140.

Wahsner, Renate und Borzeszkowski, Horst-Heino v.: *Die Wirklichkeit der Physik: Studien zu Idealität und Realität in einer messenden Wissenschaft.* Frankfurt a. M./ Berlin/ Bern/ New York/ Paris/ Wien: Peter Lang Verlag, 1992.

Wahsner, Renate: Erkenntnistheoretischer Apriorismus und neuzeitliche Physik. *Deutsche Zeitschrift für Philosophie*, 40, 1992, 24-35.

Wahsner, Renate: *Prämissen physikalischer Erfahrung: Zur Helmholtzschen Kritik des Raum-Apriorismus und zur Newton-Marxschen Kritik des antiken Atomismus.* Berlin: Verlag für Wissenschaft und Bildung 1992 (1992a).

Wahsner, Renate: Die Newtonsche Vernunft und ihre Hegelsche Kritik. Vortrag, gehalten auf dem Internationalen Hegel-Kongreß 1993 "Vernunftbegriffe der Moderne", Stuttgart 10.-13. 6. 1993. Preprint des FSP Wissenschaftstheorie und Wissenschaftsgeschichte Nr. 02/94. Berlin, 1994.

Warburg, Emil, Max Rubner und Moritz Schlick: *Helmholtz als Physiker, Physiologe und Philosoph.* Drei Vorträge, gehalten zur Feier seines 100. Geburtstages im Auftrage der Physikalischen, Physiologischen und der Philosophischen Gesellschaft zu Berlin. Karlsruhe i. Br.: C. F. Müllersche Hofbuchhandlung, 1922.

Das Helmholtz-Liesche Raumproblem und seine ersten Lösungen

Volkmar Schüller

Kants Ansichten über die Stellung der Mathematik und Geometrie im System der Wissenschaften waren in den erkenntnistheoretischen Auseinandersetzungen des vorigen Jahrhunderts von zentraler Bedeutung, denn seit Kant hatten die Mathematik und auch die Naturwissenschaften eine Entwicklung genommen, die eine Auseinandersetzung der Mathematiker und Naturwissenschaftler mit den Kantschen Ansichten geradezu provozieren mußte. Es sei nur an einige wenige Äußerungen Kants erinnert, an denen sich jene Auseinandersetzungen entzündeten. In der Einleitung zur zweiten Auflage von 1787 seiner *Kritik der reinen Vernunft* schreibt er:

> "In allen theoretischen Wissenschaften der Vernunft sind synthetische Urteile a priori als Prinzipien enthalten. [...] Zuvörderst muß bemerkt werden: daß eigentliche mathematische Sätze jederzeit Urteile a priori und nicht empirisch sind, weil sie Notwendigkeit bei sich führen, welche aus der Erfahrung nicht abgenommen werden kann. [...] Ebensowenig ist irgendein Grundsatz der reinen Geometrie analytisch. [...] Einige wenige Grundsätze, welche die Geometer voraussetzen, sind zwar analytisch und beruhen auf dem Satze des Widerspruchs; sie dienen aber auch nur wie identische Sätze zur Kette der Methode und nicht als Prinzipien. [...] Naturwissenschaft (Physica) enthält synthetische Urteile a priori als Prinzipien in sich" (Kant, KrV: B 14ff).

Gleichgültig, ob man diesen Ansichten folgen wollte oder nicht: Es war im vorigen Jahrhundert für jeden die Beantwortung der berühmten Kantischen Frage "Wie sind synthetische Urteile a priori möglich?" (Kant, KrV: B 19) das entscheidende erkenntnistheoretische Problem, welches es gemäß des Erkenntnisstandes der Mathematik, Geometrie und Naturwissenschaft zu lösen galt. Kein geringerer als Hermann von Helmholtz stellte gerade diese Kantische Frage immer wieder in den Mittelpunkt seiner Reflexionen über die Ergebnisse seiner eigenen wissenschaftlichen Arbeit, und wir wollen, als ein Beispiel hierfür, hier anhand seiner Arbeiten zur Geometrie verfolgen, in welchem mathematikhistorischen Zusammenhang Hermann von Helmholtz' Gedanken über das Raumproblem stehen, und zeigen, wie bereits vor Helmholtz ein inzwischen vergessener Philosoph des vorigen Jahrhunderts, dessen Überlegungen von einer von der Helmholtzschen Fragestellung ganz verschiedenen ausgingen und ein ganz anderes Ziel als Helmholtz zu erreichen suchten, zu einem geometrischen Ergebnis gelangt ist, welches weitgehend mit dem Helmholtzschen Ergebnis übereinstimmt.

Im Verständnis des vorigen Jahrhunderts setzt die Geometrie sowohl den Begriff des Raumes als auch die elementarsten Grundbegriffe voraus und gibt von den letzteren nur Nominaldefinitionen, wohingegen die wesentlichen geometrischen Aussagen in Form von Axiomen auftreten. Dabei bleibt es vollkommen ungewiß, ob und inwieweit die in den Axiomen angegebenen Verbindungen der geometrischen Grundbegriffe notwendig und a priori möglich sind. Darum bewegte Anfang des vorigen Jahrhunderts viele Mathematiker und Erkenntnistheoretiker die Frage, wie es möglich sei, daß sich die Sätze der Geometrie - obwohl diese nicht gezwungen ist, Erfahrungstatsachen zu sammeln, wie es von den Naturwissenschaften im engeren Sinne des Wortes verlangt wird, sondern ihre Sätze ausschließlich durch Deduktion aus den ihr zugrundeliegenden Axiomen gewinnt - so außerordentlich erfolgreich auf die uns umgebende Wirklichkeit anwenden lassen. In dem Streit um die Frage, ob ein Erkennen von Sätzen realen Inhalts möglich sei, ohne dabei auf eine empirisch gewonnene Grundlage zurückgreifen zu müssen, kam der Diskussion über die Grundlagen der Geometrie eine besondere Bedeutung zu. Gerade die geometrischen Axiome schienen die überzeugendsten Beispiele zu sein, um Kants berühmte Frage "Wie sind synthetische Sätze a priori möglich?" positiv zu beantworten und zu sichern, daß überhaupt synthetische Sätze a priori möglich sind.

Wer sich mit der Geometrie zu beschäftigen beginnt, stellt meist mit Überraschung fest, daß zu Anfang einige Sätze als Axiome aufgestellt werden, von denen die Geometrie selbst erklärt, daß sie diese nicht beweisen könne und man deren Richtigkeit anerkennen müsse. Woher kommen aber nun solche Sätze, die zwar unbeweisbar, aber doch zweifellos richtig sind? Wieviel an den geometrischen Sätzen kommt objektive Gültigkeit zu, wieviel ist nur Definition bzw. eine Folge von Definitionen?

Die seit der Antike der Geometrie vorgeschriebene Methode, die geometrischen Probleme durch eine Konstruktion zu lösen, birgt immer die Gefahr in sich, daß sich in die Schlußverfahren allzu leicht Ansichten der alltäglichen Erfahrung als angebliche Denknotwendigkeiten einschleichen, da man sich bei den geometrischen Konstruktionen selbstverständlich von der Anschauung leiten läßt. Der erfahrene Geometer wird sich darum bei jedem geometrischen Beweisschritt fragen, ob die für irgendeinen Beweis zu ziehende Linie mit einer geforderten Eigenschaft auch wirklich möglich ist. Das Wesentliche an einer geometrischen Konstruktion ist, daß durch sie die Existenz des konstruierten Gebildes feststeht. Ferner zeigt die Kontruktion auch, ob die Punkte, Linien, Kreise oder dergleichen, wie sie in der Konstruktionsaufgabe zu konstruieren verlangt werden, entweder unter allen Bedingungen möglich sind oder nur unter bestimmten Einschränkungen. Die Grundlage aller Beweise gemäß dieser synthetischen Methode ist der Nachweis der Kongruenz (Deckungsgleichheit) der betreffenden Linien, Winkel, Flächen, Körper u.ä. Diese Kongruenz setzt aber voraus, daß die betreffenden geometrischen Gebilde zueinander hinbewegt werden können, ohne daß sich dabei ihre Form und Größe verändert. Daß dies in der Tat stets möglich und ausführbar sei, erscheint dem Nichtmathematiker in der Regel als selbstverständlich, doch wenn man diese Annahme der freien Beweglichkeit (in sich) fester geometrischer Gebilde nach jeder Stelle des Raumes hin als Grundlage für die geometrischen Beweise benutzen will, so muß man sich selbstverständlich davon überzeugt haben, daß diese Annahme keine logisch unerwiesene Voraussetzung enthält. Es wird sich aber bald zeigen, daß in jener Annahme doch eine solche enthalten ist, was weitreichende Konsequenzen nach sich ziehen wird.

Der Gefahr, bei der Benutzung der synthetischen Methode und der mit ihr verbundenen Anschauung unbewußt in die geometrischen Beweise logisch unerwiesene Voraussetzungen hereinzutragen und dadurch zu falschen Ergebnissen zu kommen, kann man entgehen, wenn man sich der analytischen Geometrie bedient, die keine konstruierende, sondern eine rechnende Geometrie ist. Da die Ausführung der Rechnungen eine rein logische Operation ist, können sich in die Ergebnisse auch keine unerwiesenen Voraussetzungen einschleichen, und die sich ergebenden mathematischen Beziehungen zwischen den der Rechnung unterworfenen Größen können nur das enthalten, was in den Gleichungen, die den Ansatz der Rechnung bilden, enthalten ist.

In seinem berühmten Habilitationsvortrag "Ueber die Hypothesen, welche der Geometrie zu Grunde liegen", den Bernhard Riemann am 10. Juni 1854 in Göttingen gehalten hat und der erst 1868 veröffentlicht wurde, eröffnete Riemann der Geometrie vollkommen neue Möglichkeiten zur Beschreibung und Untersuchung ihrer Probleme. Eine interessante Darstellung, inwieweit Riemann zu seinem Mannigfaltigkeitsbegriff durch philosophische oder mathematische Erwägungen veranlaßt worden ist, findet man bei Scholz (1982). Mit der Einführung des Begriffes der n-fach ausgedehnten Größe bzw. der n-fach ausgedehnten Mannigfaltigkeit schuf Riemann ein Begriffssystem, welches sich auch außerhalb der Geometrie in der Mathematik außerordentlich erfolgreich bewähren sollte, aber in der Geometrie zuerst seine Bewährungsprobe bestand. Eine n-fach ausgedehnte Mannigfaltigkeit ist eine Menge, deren einzelne Elemente (die Orte) durch n veränderliche Größen x_1, x_2, ... x_n (die Koordinaten der Orte) festgelegt sind, dies bedeutet, daß sich die Ortsbestimmung in diesem Raum auf die Bestimmung von n Größen zurückführen läßt. Sind diese n Größen noch nicht selbst die Koordinaten, so ist gemeint, daß sich jene n Koordinaten aus den n bestimmten Größen eindeutig berechnen lassen. Ferner verlangt Riemann, daß sich die Elemente einer Mannigfaltigkeit stetig verändern können und zwischen den einzelnen Elementen ein Maßverhältnis bestehen soll, welches in seiner konkreten mathematischen Gestalt sehr verschieden sein kann, je nachdem von welcher mathematischen Beschaffenheit die Elemente der n-fach ausgedehnten Mannigfaltigkeit sind. Der Raum, den die Geometrie betrachtet, ist für Riemann nur ein Spezialfall einer n-fach ausgedehnten Mannigfaltigkeit, nämlich einer dreifach ausgedehnten Mannigfaltigkeit mit einer geeigneten Metrik. Eine notwendige Folge davon ist, daß sich die Sätze der Geometrie nicht aus allgemeinen Größenbegriffen und allgemeinen, zwischen den Größen bestehenden Maßverhältnissen (Metriken) ableiten lassen, sondern daß diejenigen Eigenschaften, durch welche sich der Raum der Geometrie von anderen denkbaren dreifach ausgedehnten Mannigfaltigkeiten unterscheidet, eine Folge jener speziellen Metrik ist, die zwischen den Punkten dieses Raumes besteht. Welche spezielle Metrik aber besteht zwischen den Punkten dieses Raumes? Für Riemann

> "[...] entsteht die Aufgabe, die einfachsten Thatsachen aufzusuchen, aus denen sich die Massverhältnisse des Raumes bestimmen lassen - eine Aufgabe, die der Natur der Sache nach nicht völlig bestimmt ist; denn es lassen sich mehrere Systeme einfacher Thatsachen angeben, welche zur Bestimmung der Massverhältnisse des Raumes hinreichen; am wichtigsten ist für den gegenwärtigen Zweck das von Euklid zu Grunde gelegte. Diese Thatsachen sind wie alle Thatsachen nicht notwendig, sondern nur von empirischer Gewißheit, sie sind Hypothesen".

Folgerichtig muß sich Riemann der Frage zuwenden, welcher Maßverhältnisse (Metriken) eine solche Mannigfaltigkeit fähig ist und welche Bedingungen hinreichend sind, um diese Maßverhältnisse zu bestimmen. Um diese Aufgabe zu lösen, verfolgt Riemann folgenden Gedanken:

> "Maßbestimmungen erfordern eine Unabhängigkeit der Größen vom Ort, die in mehr als einer Weise stattfinden kann; die zunächst sich darbietende Annahme, welche ich hier verfolgen will, ist wohl die, dass die Länge der Linie unabhängig von der Lage sei, also jede Linie durch jede messbar sei. Wird die Ortsbestimmung auf Grössenbestimmungen zurückgeführt, also die Lage eines Punktes in der gegebenen *n*fach ausgedehnten Mannigfaltigkeit durch *n* veränderliche Grössen x_1, x_2, x_3, und so fort bis x_n ausgedrückt, so wird die Bestimmung einer Linie darauf hinauskommen, dass die Grössen *x* als Function Einer Veränderlichen gegeben werden. Die Aufgabe ist dann, für die Länge der Linien einen mathematischen Ausdruck aufzustellen, zu welchem Zwecke die Grössen *x* als in Einheiten ausdrückbar betrachtet werden müssen".

Unter einigen naheliegenden zusätzlichen Annahmen kommt Riemann zu dem Ergebnis: "[...] folglich ist *ds* = der Quadratwurzel aus einer immer positiven ganzen homogenen Function zweiten Grades der Grössen *dx*, in welcher die Coeffizienten stetige Functionen der Grössen *x* sind [...]; der Raum ist also unter diesem einfachsten Falle enthalten". Riemann zeigt nun noch: "Die Massverhältnisse dieser Mannigfaltigkeiten, wo das Linienelement durch die Quadratwurzel aus einem Differentialausdruck zweiten Grades dargestellt wird, lassen sich so auf eine von der Wahl der veränderlichen Grössen völlig unabhängige Weise ausdrücken". Dies bedeutet, daß eine solche Metrik unabhängig von den benutzten Koordinaten ist und wirklich nur ein Maßverhältnis zwischen den Elementen der *n*-fach ausgedehnten Mannigfaltigkeit darstellt, eben die ganz unabhängig von den benutzten Koordinaten bestehenden "geometrischen" Eigenschaften der Elemente der Mannigfaltigkeit beschreibt. Diese wichtigen von Riemann für eine *n*-fach ausgedehnte Mannigfaltigkeit hergeleiteten Ergebnisse wendet er schließlich auf den Raum an, indem er sie für den Fall $n = 3$ spezifiziert. Sein Ergebnis lautet:

> "Nach Untersuchungen über die Bestimmung der Massverhältnisse einer *n*fach ausgedehnten Grösse lassen sich nun die Bedingungen angeben, welche zur Bestimmung der Massverhältnisse des Raumes hinreichend und nothwendig sind, wenn Unabhängigkeit der Linien von der Lage und der Darstellbarkeit des Linienelements durch die Quadratwurzel aus einem Differentialausdruck zweiten Grades, also Ebenheit in den kleinsten Theilen vorausgesetzt wird.
> Sie lassen sich erstens so ausdrücken, dass das Krümmungsmass in jedem Punkte in drei Flächenrichtungen = 0 ist, und es sind daher die Massverhältnisse des Raumes bestimmt, wenn die Winkelsumme im Dreieck allenthalben gleich zwei Rechten ist.

> Setzt man aber zweitens, wie Euklid, nicht bloss eine von der Lage unabhängige Existenz der Linien, sondern auch der Körper voraus, so folgt, dass das Krümmungsmass allenthalben constant ist, und es ist dann in allen Dreiecken die Winkelsumme bestimmt, wenn sie in Einem bestimmt ist".

Durch seine Untersuchungen über die räumliche Anschauung des Menschen beim Sehen wurde Hermann von Helmholtz angeregt, auch über die Frage nach dem Ursprung und dem Wesen unserer allgemeinen Anschauung von dem Raum nachzudenken. Da man es in der Geometrie immer mit idealen Gebilden zu tun hat, deren körperliche Darstellung immer nur eine Annäherung an die Gegenstände in der Wirklichkeit ist, und wir unsere Entscheidung darüber, ob ein Körper fest, seine Seiten eben und seine Kanten gerade sind, mit Hilfe geometrischer Sätze treffen, deren Richtigkeit erst durch ihre Prüfung in der Wirklichkeit gezeigt werden müßte, ist es sehr wichtig zu wissen, durch welche Geometrie die Wirklichkeit richtig beschrieben wird. Denn durch die Arbeiten von Gauss, Bolyai und Lobatschewsky wußte man inzwischen, daß es auch andere Geometrien als die Euklidische gibt, sogenannte nichteuklidische, die zumindest denkmöglich sind, deren Beziehung zur Wirklichkeit aber noch vollkommen unklar war. In seiner berühmten Abhandlung "Ueber die Thatsachen, die der Geometrie zum Grunde liegen", für deren Titel ganz zweifellos der Titel von Riemanns Habilitationsvortrag zum Vorbild diente - insbesondere wenn man bedenkt, daß für Riemann die der Geometrie zugrundeliegenden Tatsachen Hypothesen sind -, versucht Helmholtz herauszufinden, welche Geometrie im wirklichen Raume gilt. Dabei schlägt Helmholtz einen sehr ähnlichen Weg wie Riemann ein, um dieses Problem analytisch lösen zu können, weil dieses Verfahren nicht auf die Anschauung baut und somit den auf diesem Gebiet so schwer zu vermeidenden Täuschungen entgeht. Ein weiterer Vorteil der analytischen Methode ist ihre Möglichkeit, auch ein von der Euklidischen Geometrie abweichendes Axiomensystem relativ leicht überblicken zu können. Während Riemann den geometrischen Raum durch die naheliegenden Hypothesen charakterisiert, daß der geometrische Raum eine dreidimensionale stetige Mannigfaltigkeit sei und das Längenelement einer Linie dieser Mannigfaltigkeit gleich der Quadratwurzel einer homogenen Funktion zweiten Grades der Koordinatendifferentiale sei, und aufgrund dieser Hypothesen die Euklidische Struktur des Raumes zeigt, versucht Helmholtz, mithilfe von Annahmen, die auf den ersten Blick wesentlich plausibler und einfacher erscheinen und für jedermann Tatsachen im wirklichen Raume sind, das Problem zu lösen. Sein Ausgangspunkt ist, daß die Ausmeßbarkeit des Raumes die Existenz (in sich) fester Körper zur Grundlage habe, welche vermöge ihrer freien Beweglichkeit im Raum zur Deckung gebracht werden können, wobei dies unabhängig von dem Orte und der Richtung der sich deckenden Raumgebilde möglich ist und auch von dem Wege, auf welchem sie zueinander geführt worden sind.[1] Helmholtz setzt also die Möglichkeit der Raummessung durch Kongruenz von Anfang an voraus und stellt sich die Aufgabe, auf analytischem Wege die allgemeinste n-fach ausgedehnte Mannigfaltigkeit ($n = 3$) zu bestimmen, in welcher die verlangten Bewegungen möglich sind.

Die Hypothesen, welche Helmholtz seinen Betrachtungen zugrunde legt, sind:

I. Der Raum von n Dimensionen ist eine n-fach ausgedehnte stetige Mannigfaltigkeit.

II. Es wird die Existenz von beweglichen, aber in sich festen Körpern vorausgesetzt, wie sie notwendig ist, um einen Vergleich der Raumgrößen durch Kongruenz vornehmen zu können, wobei unter einem in sich festen Körper ein Körper verstanden wird, bei welchen zwischen den $2n$ Koordinaten eines jeden Punktepaares, welches zu diesem Körper gehört, eine von der Bewegung des Körpers unabhängige Gleichung besteht, die für alle kongruenten Punktepaare die gleiche ist. Kongruent sind solche Punktepaare, die gleichzeitig oder nacheinander mit demselben Punktepaar des Raumes zusammenfallen können.
   Dies hat zur Folge, daß zwischen den Koordinaten zweier fester Punkte nicht jede beliebige Art von Gleichungen bestehen kann, sondern nur Gleichungen mit ganz bestimmten Eigenschaften, welche analytisch näher bestimmt werden müsssen.

III. Es wird vollkommen freie Beweglichkeit der festen Körper vorausgesetzt, das heißt jeder Punkt eines Körpers kann stetig in jeden anderen Ort des Raumes übergehen, soweit er nicht durch die Gleichungen, die zwischen ihm und den übrigen Punkten des festen Körpers bestehen, zu welchem er gehört, in seiner Bewegungsfreiheit eingeschränkt ist.
   Zusammen mit der Hypothese II folgt daraus die Unabhängigkeit der Kongruenz zweier geometrischer Gebilde von ihrer Lage im Raum.

IV. Eine vollständige Drehung ohne Umkehr führt jeden festen Körper immer wieder in seine Anfangslage zurück (Monodromieprinzip).

---

1 Für das von Helmholtz aufgeworfene Problem, nämlich sowohl die Euklidische Geometrie als auch die nichteuklidische Geometrien - Riemannsche Räume konstanter Krümmung - durch die Forderung möglichst großer Beweglichkeit starrer Körper zu charakterisieren, hat sich in der Mathematik die Bezeichnung Helmholtz-Liesches Raumproblem eingebürgert, an dessen Lösung viele bedeutende Mathematiker, wie z.B. Lie, Poincaré und Weyl, beteiligt waren.

Helmholtz glaubt, daß diese vier Hypothesen voneinander unabhängig seien und zur vollständigen Charakterisierung der Geometrie des Raumes ausreichten. Unter der Voraussetzung von drei Dimensionen ($n = 3$) kommt er zu dem Schluß, daß zwischen den Koordinatendifferentialen ein homogener Ausdruck zweiten Grades existieren muß, welcher bei jeder Bewegung zweier miteinander fest verbundener Punkte von verschwindend kleinem Abstand unverändert bleibt. Dies stellt aber gerade den Ausgangspunkt der Riemannschen Überlegungen bezüglich der Metrik des Raumes dar.

Helmholtz deutet dieses Ergebnis in dem Sinne, daß die Kongruenzbewegungen im Raume Isometrien der Riemannschen Metrik sind. Damit kann er sich auf Riemanns Ergebnisse über Mannigfaltigkeiten mit Riemannscher Metrik stützen und kommt so zu dem Resultat, daß der geometrische Raum eine Mannigfaltigkeit mit konstanter Krümmung ist. Unter der zusätzlichen Annahme, daß der tatsächliche Raum im Riemannschen Sinne unendlich ist, meinte Helmholtz schließlich für den tatsächlichen Raum die Gültigkeit der Euklidischen Geometrie feststellen zu können.[2] Mit diesem Ergebnis glaubte Helmholtz, eine auf Tatsachen gestützte Begründung dafür geliefert zu haben, daß die "richtige" Geometrie des Raumes die Euklidische Geometrie ist. Daß sich dadurch die Kantianer zum Streit herausgefordert sahen, nimmt nicht wunder; Helmholtz entgegnete ihnen vor allem in seinen beiden Schriften "Über den Ursprung und die Bedeutung der geometrischen Axiome" und "Über den Ursprung und Sinn der geometrischen Sätze", mit welchen er nicht nur viel zur Anerkennung und Verbreitung der nichteuklidischen Geometrien beigetragen hat, sondern in denen er auch die physikalischen und philosophischen Konsequenzen gründlich analysiert hat. Für den Wissenschaftshistoriker

2 Gegen die mathematischen Herleitungen, so wie sie von Helmholtz durchgeführt wurden, lassen sich viele Einwände geltend machen. Die erste umfassende Kritik und Korrektur der Helmholtzschen Schlußweisen formulierte wohl Sophus Lie in seiner *Theorie der Transformationsgruppen*, wo er schreibt: "[Wir liefern] eine Kritik der Schlüsse, die Herr v. Helmholtz für den Fall des dreifach ausgedehnten Raumes aus seinen Axiomen zieht, und zeigen, dass er *zu seinem Endergebnis gelangt, indem er stillschweigend eine Reihe Voraussetzungen macht, die sämmtlich unrichtig sind*" (Lie 1893: 437). Die wichtigsten Einwände sind, daß die für das Endliche aufgestellten Axiome nicht ohne weiteres auf das unendlich Kleine angewendet werden dürfen, wie es Helmholtz bei seinen Herleitungen tut, daß das Monodromieprinzip entbehrlich ist und daß ein dreidimensionaler Riemannscher Raum mit konstanter Krümmung, der im Riemannschen Sinne unendlich ist, nicht unbedingt ein Euklidischer Raum sein muß, sondern durchaus auch ein Raum mit negativer Krümmung sein könnte, wovon Beltrami in seinem Brief vom 24.4.1869 Helmholtz in Kenntnis setzte (siehe Koenigsberger 1903: 153ff).

ist es aber nun besonders interessant festzustellen, daß es bereits vor Helmholtz in einem anderen ideengeschichtlichen Zusammenhang einen anderen Lösungsversuch des Helmholtz-Lieschen Raumproblems gegeben hat.

Es ist ein weit verbreitetes Vorurteil, daß es vor der Entdeckung der Möglichkeit nichteuklidischer Geometrien keine Versuche gegeben habe, die Euklidische Geometrie bzw. die Formulierung der Geometrie, wie sie uns von Euklid überliefert worden ist, in ihren Grundlagen zu verbessern. Es sei nur an Friedrich Fries (1773-1843) erinnert, der einen großen Teil seines mathematischen und philosophischen Werkes dieser Aufgabe gewidmet hat. Dabei bestand zu dieser Zeit nicht das Problem, die "richtige" Geometrie des "tatsächlichen" Raumes zu ermitteln, wie es später nach der Entdeckung der nichteuklidischen Geometrien zu lösen war und wofür Helmholtz einen so außerordentlich wichtigen Beitrag geliefert hat, sondern man suchte nach einer logisch konsistenten Formulierung der Geometrie, um auf diese Weise zum Beispiel die gegenseitige Abhängigkeit oder Unabhängigkeit wichtiger geometrischer Axiome besser erkennen zu können.

Als einem wichtigen Beispiel für solche Bemühungen wollen wir uns an dieser Stelle nun einer interessanten mathematischen Arbeit zuwenden, die in der damaligen mathematischen Literatur überhaupt nicht beachtet worden ist, obwohl der in ihr vorgeschlagene Lösungsversuch für eine neue Grundlegung der Geometrie sehr bemerkenswert ist. Allerdings ist für uns hier der vorgeschlagene Lösungsversuch, was die Grundlagen der Geometrie angeht, von geringerer Bedeutung, sondern uns interessiert vielmehr die Tatsache, daß bei diesem neuen Grundlegungsversuch der Euklidischen Geometrie bereits vor Riemann und Helmholtz das Helmholtz-Liesche Problem aufgeworfen worden ist. Wir meinen Friedrich Ueberwegs[3] Schrift "Die Principien der Geometrie, wissenschaftlich dargestellt" aus dem Jahre 1851.

---

3 Mit seinen beiden Hauptwerken *System der Logik* (1. Aufl. Bonn 1857) und *Grundriß der Geschichte der Philosophie* (3 Bände, 1. Aufl. Berlin 1862-1866), die viele Auflagen erlebten, hatte der Philosoph Friedrich Ueberweg (1826-1871) seinerzeit einen beträchtlichen Einfluß auf die philosophische Lehre in Deutschland. Er studierte in Göttingen und Berlin vorwiegend Philologie und Philosophie; nachdem er sich 1852 in Bonn habilitiert hatte, schlug er sich als Privatdozent mühsam durch und wurde 1862 in Könisberg außerordentlicher Professor, wo er 1868 schließlich zum ordentlichen Professor für Philosophie ernannt wurde. Es wird berichtet, daß er in Berlin bei Dirichlet und möglicherweise auch bei Jacobi mathematische Vorlesungen besucht habe. Sein gesamtes philosophisches Schaffen ist polemisch gegen Kant gerichtet, welches er im Vorwort zur 2. Auflage (Berlin 1865) seines *System der Logik* mit folgenden Wor-

Da man in der Mitte des vorigen Jahrhunderts von einer systematischen und vollständigen axiomatischen Darstellung der Geometrie noch weit entfernt war und im wesentlichen noch ganz auf den Spuren Euklids wandelte, was von vielen damaligen Mathematikern und Philosophen als ein großer Mangel beklagt wurde, setzte sich Ueberweg das Ziel, sämtliche geometrischen Sätze auf möglichst wenige fundamentale Sätze der Geometrie zurückzuführen, so daß sich alle geometrischen Sätze aus diesen fundamentalen Sätzen durch strenge Beweise herleiten lassen. Um ein solches vollständiges System fundamentaler Sätze für die Geometrie zu finden, geht Ueberweg unter Berufung auf Kant davon aus, daß die geometrischen Sätze keine rein analytischen Sätze, das heißt lediglich formal-logische Folgerungen aus Begriffsdefinitionen, sein können - dann nämlich bedürfte man in der Geometrie keiner Axiome und alle Lehrsätze der Geometrie würden ihren Beweis in ihrer rein formal-begrifflichen Ableitung aus den zugrundeliegenden Definitionen finden - sondern synthetische Sätze sein müssen, da in der Geometrie die Anschauung des Raumes als ein synthetisches Element hinzukomme. Will man nun die wesentlichen Merkmale des Raumes, deren Begriffe als die geometrisch einfachsten keine genetischen Definitionen zulassen, in geometrische Sätze fassen, so ist dies nur mit Hilfe von geometrischen Sätzen möglich, die selbst nicht beweisbar sind, was zur Folge hat, daß in den Prinzipien der Geometrie immer noch etwas Unbeweisbares enthalten ist, was sich nicht vollständig eleminieren läßt. Darum sieht sich Ueberweg vor zwei Aufgaben gestellt: zum einen durch eine gemeinschaftliche Ableitung einen möglichst engen Zusammenhang zwischen den in der Geometrie benutzten Grundbegriffen und Axiomen

ten umreißt: "Der Kern meines Gegensatzes zu Kant liegt in dem durchgeführten Nachweis, wie die wissenschaftliche Einsicht, welche die bloße Erfahrung in ihrer Unmittelbarkeit noch nicht gewährt, nicht mittels aprioristischer Formen von rein subjektivem Ursprung, die nur auf die im Bewußtsein des Subjektes vorhandenen Erscheinungsobjekte Anwendung finden, gewonnen wird (auch wie Hegel u.a. wollen, a priori und doch mit objektiver Gültigkeit), sondern durch die Kombination der Erfahrungsthatsachen nach logischen Normen, deren Befolgung unserer Erkenntnis eine objektivreale Gültigkeit sichert. Ich suche zu zeigen, wie insbesondere die räumlich-zeitliche und kausale Ordnung, auf deren Erkenntnis die Apodiktizität beruht, nicht erst von dem anschauenden und denkenden Subjekte in einen chaotisch gegebenen Stoff hineingetragen, sondern aus der (natürlichen und geistigen) Realität, in der sie ursprünglich ist, successive durch Erfahrung und Denken in das subjektive Bewußtsein aufgenommen wird".

herzustellen und zum anderen die Gewißheit der benutzten Axiome nachzuweisen.

Eine solche gemeinschaftliche Ableitung könnte man entweder aus der allgemeinen philosophischen Bestimmung des Raumes zu gewinnen versuchen oder aus einer konkreten empirischen Anschauung. Der erste Weg verlangt aber die Beantwortung der beiden philosophischen Grundfragen über den Raum, nämlich erstens die erkenntnistheoretische Frage, woher unsere Anschauung des Raumes stamme, ob aus den Dingen, die außer uns existieren (Lockescher Realismus) oder aus uns selbst allein a priori (Kantischer Idealismus) oder gemeinschaftlich aus den Dingen und uns, und zweitens die metaphysische Frage, was der Raum als solcher sei und welche Stelle er in der Totalität alles Seienden einnehme. Aus gutem Grund will Ueberweg diesen Weg nicht einschlagen, sondern die von ihm gesuchte gemeinschaftliche Ableitung der benutzten Grundbegriffe und Axiome anhand einer konkreten Anschauung vornehmen, die er mit der Absicht analysiert, allgemeine Begriffe und Sätze der Geometrie zu gewinnen, welche er als Grundlage für seinen Aufbau der Geometrie benutzen kann.

Zu diesem Zweck geht Ueberweg von der These "die Anschauung, zu wissenschaftlichem Zwecke veranstaltet, ist ein Experiment" aus, so daß seine Grundlegung der Geometrie eine experimentell-empirische Grundlegung sein wird. Einem solchen Vorgehen steht aber der Einwand entgegen, daß eine solche empirische Ableitung der geometrischen Prinzipien diesen keine wahre, innere Begründung geben kann, wie es eine begriffliche Ableitung aus der allgemeinen philosophischen Bestimmung des Raumes vermocht hätte. Doch die Berechtigung für sein Vorgehen sieht Ueberweg darin, daß der Erkenntnisprozeß in jeder Wissenschaft in zwei Teile zerfällt, nämlich beim ersten Schritt wird von irgendeinem empirisch gegebenen Einzelnen ausgegangen, um daraus die allgemeinen Prinzipien zu gewinnen, aus welchen die Wissenschaft beim zweiten Schritt das Einzelne begreift, so daß jede Wissenschaft in ihre analytische Grundlegung und in die sich anschließende synthetische Ausführung zerfällt, wobei es aber für die analytische Grundlegung einer Wissenschaft durchaus verschiedene Möglichkeiten gibt. Für Ueberweg gehört der Nachweis, wie aus bestimmten empirischen Daten die Prinzipien einer Wissenschaft abgeleitet werden, wesentlich mit zur Wissenschaft. Sein Ziel ist es, "diese analytische Grundlegung [...] der Geometrie in wissenschaftlicher Strenge zu geben, das eben suchen wir durch Lösung unserer Aufgabe zu erreichen", und mit nicht zu übersehenden Bezug auf Kant schreibt er weiter: "Man hat geglaubt, die Geometrie bedürfe solcher empirisch-analytischen

Grundlegung nicht, weil uns ihre Principien a priori gegeben seien. Diese vermeintliche Apriorität lassen wir dahingestellt sein, weil zu rein philosophischen Discussionen hier nicht der Ort ist", denn er versteht seine Arbeit trotz aller mit ihr verknüpften philosophischen Überlegungen als eine rein geometrische Arbeit.

Auch bei dieser Grundlegung der Geometrie wird in ihren Grundlagen etwas mathematisch Unbewiesenes zurückbleiben, was Ueberweg vollkommen zu Recht nicht weiter beunruhigt, wohingegen er die Gewißheit der von ihm gefundenen fundamentalen Sätze der Geometrie durch folgende Überlegung zu sichern meint. Der Grund für die Gewißheit der geometrischen Axiome ist seiner Meinung nach derselbe wie bei den physikalischen Hypothesen, nämlich die fortlaufende approximative Bestätigung ihrer Konsequenzen durch die Erfahrung. Die aus den geometrischen Axiomen abgeleiteten Sätze lassen durch eine wirkliche Konstruktion einen Vergleich mit der Erfahrung zu. Da wir zwar nicht vollkommen genau konstruieren können, ist eine vollkommene Übereinstimmung natürlich nicht möglich, aber so weit wie unsere Erfahrung reicht, ist eine um so genauere Übereinstimmung feststellbar, je genauer wir konstruieren. Falls nun in den Hypothesen auch nur die geringste Abweichung verborgen sein sollte, so wäre doch zu erwarten, daß sie sich wenigstens in irgendeinem der vielen abgeleiteten geometrischen Sätze so deutlich zeigt, daß der Widerspruch zur Erfahrung unübersehbar ist.

In der im Jahre 1860 erschienenen französischen Übersetzung seiner Arbeit, die er in ihrem mathematischen Teil zwar unverändert ließ, in ihrer Einleitung aber erweiterte, legt Ueberweg seinen empiristischen Standpunkt zur Begründung der Geometrie noch ausführlicher dar und setzt sich kritisch mit der damals vorherrschenden Kantischen Theorie von der Apriorität der Raumanschauung auseinander. Bei der Beantwortung der Frage, woher die Gewißheit der geometrischen Fundamentalsätze stamme, welche doch selbst geometrisch nicht beweisbar sind, nimmt Ueberweg einen zu Kants Theorie der Apriorität entgegengesetzten Standpunkt ein. Für unsere Zwecke hier ist es vollkommen belanglos, inwieweit Ueberweg bei seiner Kantinterpretation den Kantischen Ansichten tatsächlich gerecht wird; für uns ist seine Interpretation nur insofern wichtig, als er sich selbst in einen bewußten Gegensatz zu Kant stellen will. Nach Kant liegt die Anschauung des Raumes in uns a priori als die Form des "äußeren Sinnes", also vor aller Erfahrung und unabhängig von aller Erfahrung, obwohl sie uns erst mit der Erfahrung zu Bewußtsein kommt. Diese Form wird, wie die Formen der Anschauung und des Denkens überhaupt, von dem Ich aus sich selbst heraus durch eine ursprüngliche Tat erzeugt und in sie wird von dem Ich jede empirische Wahrnehmung gelegt, so daß alle Erscheinungen diese Formen an sich tragen müssen. Jedoch was von dem Ich unabhängig existiert und unsere Sinne affiziert, existiert nicht in diesen Formen. Die sogenannten "Dinge an sich" sind unräumlich und unzeitlich und unterliegen nicht dem Gesetz der Kausalität. Raum und Zeit sollen also a priori gegeben sein und also als Formen der Anschauung nur von dem Ich ohne Mitwirkung der Dinge erzeugt werden, und trotzdem sollen auch die Dinge selbst räumlich und zeitlich existieren. Die Apriorität der Anschauung des Raumes folgert Kant aus der Apodiktizität der mathematischen Sätze, und das Auffinden der geometrischen Sätze geschieht nicht durch eine bloße Zergliederung der jeweiligen Subjektsvorstellungen, sondern nur durch ein Zurückgehen auf die Anschauung des Raumes, so daß sie keine analytischen, sondern synthetische Urteile sind. Wäre nun die Raumanschauung empirisch, so würden nach Kant die geometrischen Sätze auch nur empirische Gültigkeit besitzen, also nicht apodiktisch gewiß sein; ist hingegen der Raum eine Anschauung a priori, so kommt jenen Sätzen strenge Allgemeinheit und Notwendigkeit zu. Ueberweg will nun die apodiktische Gültigkeit der mathematischen Sätze mit Kant durchaus anerkennen, aber meint im Gegensatz zu Kant, daß diese apodiktische Gültigkeit mit dem empirischen Ursprung der Raumanschauung durchaus vereinbar sei. In Kants Überlegungen vermißt Ueberweg nämlich den Nachweis, wie und wodurch die von Kant vorausgesetzte Apriorität der Raumanschauung die Apodiktizität der geometrischen Fundamentalsätze, und zwar jener bestimmten, welche uns von Euklid überliefert worden sind, sichere. Außerdem vermißt Ueberweg bei Kant einen Beweis dafür, daß die Apodiktizität der mathematischen Sätze mit dem empirischen Ursprung der Raumanschauung unvereinbar sei. Zwar räumt er gegenüber Kant ein, daß die einzelnen Grundsätze der Geometrie für sich allein keineswegs apodiktische, sondern nur assertorische Gültigkeit besitzen, doch hält er Kant entgegen, daß dieser nicht beachte, daß sich die aus den Grundsätzen abgeleiteten Sätze nicht nur auf jene Grundsätze stützen, sondern diesen Grundsätzen auch zur Stütze dienen. Denn die gegenseitige Übereinstimmung der aus hypothetisch angenommenen Prämissen gefolgerten Konsequenzen und ihre Übereinstimmung mit der Erfahrung, so weit wie ein solcher Vergleich mit der Erfahrung möglich ist, dient den Prämissen selbst zur Bestätigung und verleiht ihnen eine stets wachsende Gewißheit, die dann zu einer absoluten wird, sobald sich zeigen läßt, daß sich das tatsächlich Gegebene nur aufgrund dieser Prämissen erklären läßt. Für Ueberweg ist die Raumanschauung empirisch begründet, und die Vorstellungen räumlicher Gebilde sind zum Teil aus der Erfahrung gewonnene Abstraktionen und Idealisierungen, zum Teil aufgrund der Erfahrung frei geschaffene Konstruktionen. Zwar gelten die Fundamentalsätze für sich allein nur mit assertorischer Gewißheit, aber dem mathematischen System als Ganzem kommt apodiktische Gewißheit zu, wodurch auch den Elementen des Systems diese apodiktische Gewißheit verliehen wird.

Für die Geometrie bedeutet dies insbesondere, wie für alle anderen Wissenschaften auch, daß sie aufgrund des empirisch Gegebenen in einem analytischen Teil ihre Realprinzipien bestimmen muß, um dann aus diesen auf synthetischem Wege alle ihre einzelnen Sätze abzuleiten. Wäre der empirische Ursprung der mathematischen Fundamentalerkenntnisse schon allgemein anerkannt, so könnten Euklids Axiome als das bezeichnet werden, was sie nach Ueberwegs Meinung sind, nämlich als der Ausdruck wahrgenommener Verhältnisse mit hypothetischer Steigerung zur absoluten Gültigkeit, welche sich mehr und mehr an der Verkettung der aus ihnen gefolgerten Konsequenzen bewährt. Aber diese Anerkennung soll ja erst noch gegenüber der zu jener Zeit vorherrschenden Kantischen Aprioritätstheorie durchgesetzt werden. Darum will Ueberweg von Euklids Axiomen, denen infolge unserer langen Gewöhnung das Vorurteil der Apriorität fest anhafte, vorläufig absehen und statt ihrer einmal ganz andere Sätze als geometrische Fundamentalsätze aufstellen, und zwar solche, deren empirischer Ursprung offensichtlich und unbestreitbar ist. Sobald man zeigen kann, daß auch aus diesen Sätzen die gesamte Geometrie abgeleitet werden kann, so läßt es sich nach Ueberwegs Meinung auch nicht mehr bestreiten, daß eine Deduktion der Geometrie auf naturwissenschaftlichem Wege ohne Zuflucht zu der Kantischen Aprioritätstheorie möglich ist. Ist dies alles erst einmal geleistet, so Ueberwegs Hoffnung, dann wird sich niemand mehr seiner richtigeren Ansicht über den logischen und erkenntnistheoretischen Charakter der Euklidischen Axiome und der Anerkennung der Geometrie als Naturwissenschaft verschließen können.

Ueberweg wird nicht müde, immer wieder zu betonen, daß es für die Wahl der empirisch zu konstatierenden Tatsachen, die als Grundlage der Betrachtungen im analytischen Teil einer Wissenschaft dienen können, in der Regel verschiedene Möglichkeiten gibt, die uns einen gewissen Spielraum beim systematischen Aufbau einer Wissenschaft geben. Ausgangspunkt bei seiner Begründung der Geometrie ist die sinnliche Gesamtanschauung, bei welcher Raum und Materie voneinander noch keine Trennung erfahren haben. Um nun aber diese Trennung zu erreichen, bedarf es der Abstraktion, als deren Grundlage die Bewegung dienen möge. Ein Körper geht bei seiner Bewegung von einem Ort in einen anderen über, wobei etwas unverändert geblieben ist und sich etwas verändert hat. Was sich verändert hat, nennt man Ort. Auf diese Weise gewinnt man durch Abstraktion das Bewußtsein von dem Ort als etwas Räumlichem im Unterschied von dem Materiellen, welches den Ort einnehmen und verlassen kann. Für Ueberweg ist bei seiner systematischen Formulierung der Geometrie die Möglichkeit der Bewegung starrer Körper im Raum die entscheidende empirische Grundlage, so daß er seine Formulierung der Geometrie auf das folgende Experiment gründen will, dessen vollkommene Gültigkeit er als Hypothese bzw. Axiom annimmt:

Ein materieller fester Körper kann nach dem Zeugnis der Sinne:

(1) wenn er unbefestigt ist, überallhin bewegt werden, wo sich nicht schon ein anderer fester Körper befindet;
(2) wenn er an einer einzelnen Stelle festgehalten wird, nicht mehr unbeschränkt überallhin bewegt werden, aber er ist noch nicht jeder Bewegungsmöglichkeit beraubt;
(3) wenn er außerdem noch an einer zweiten Stelle festgehalten wird, zwar nicht mehr alle nach (2) möglichen Bewegungen ausführen, aber er kann immer noch bewegt werden;
(4) wenn er aber noch an einer dritten Stelle festgehalten wird, die nach (3) noch bewegt werden konnte, überhaupt nicht mehr bewegt werden.

Unter der Annahme, daß dieses Experiment mit absoluter Genauigkeit gilt, muß Ueberweg nun mit mathematischer Strenge analytisch auf die Grundbestimmungen des Raumes zurückschließen, wenn er sein Versprechen, eine empirisch-experimentelle Grundlegung der Geometrie zu geben, einlösen will. Jede Anschauung, jede Annahme bzw. jede Bestimmung, die nicht eine direkte Folge aus dem Experiment ist, muß dabei vermieden werden. Während er die einzelnen Teilschritte seines Experimentes nacheinander betrachtet und auf ihnen aufbauend die geometrischen Grundbegriffe, wie z.B. *Punkt*, *Gerade*, *Parallele*, *Länge*, *Lage*, *Winkel* und dergleichen, definiert, versucht er zu zeigen, daß für diese die von Euklid überlieferten Axiome gelten, insbesondere auch das berühmte Parallelenaxiom. Leider müssen wir es uns hier versagen, Ueberwegs Gedankengänge im einzelnen darzustellen, und wollen nur darauf hinweisen, daß ebenso wie Helmholtz' Beweise auch Ueberwegs Beweise einige mathematische Trugschlüsse enthalten, die wohl als erster Torretti aufgedeckt hat. Für den Mathematikhistoriker entsteht nun sofort die Frage, auf die wir aber keine überzeugende Antwort wissen, warum das von Ueberweg bereits im Jahre 1852 aufgeworfene Problem bei den zeitgenössischen Mathematikern kein Interesse hervorgerufen hat. Die mathematischen Unzulänglichkeiten in Ueberwegs Überlegungen können nicht der Grund gewesen sein, denn dann hätten Helmholtz' Überlegungen auch kein Interesse finden dürfen. Vielleicht ist auch ein Grund darin zu suchen, daß Helmholtz nicht nur das mathematische Problem behandelt hat, sondern auch die physikalischen und philosophischen Konsequenzen daraus gezogen hat, oder vielleicht müssen wir den Grund darin suchen, daß Ueberweg seine Überlegungen nicht in einer mathematischen Zeitschrift veröffentlicht hat. Daß Ueberweg ein wesentliches mathematisches Problem aufgeworfen hat, ist gewiß unbestritten, und die Ähnlichkeit zwischen Ueberwegs Experiment und Helmholtz' Hypothesen ist nicht zu übersehen; während Helmholtz sich der analytischen Methode bedient hat, versuchte Ueberweg das Problem auf synthetischem Wege zu lösen.

Literatur

Helmholtz, Hermann von: Ueber die Thatsachen, die der Geometrie zum Grunde liegen. In: *Nachrichten von der Königlichen Gesellschaft der Wissenschaften und der Georg-August-Universität.* 1868, Nr. 9, 193-221.

Helmholtz, Hermann von: Über den Ursprung und die Bedeutung der geometrischen Axiome. In: *Vorträge und Reden.* Band II. 3. Aufl., Braunschweig 1884, 1-31.

Helmholtz, Hermann von: Über den Ursprung und Sinn der geometrischen Sätze; Antwort gegen Herrn Professor Land. In: *Wissenschaftliche Abhandlungen.* Band II, Leipzig 1883. Nr. LXXIX, 640 - 660.

Kant, Immanuel: *Kritik der reinen Vernunft.* 2. Aufl., Akademie-Ausgabe, Bd. III.

Koenigsberger, Leo von: *Hermann von Helmholtz,* Band II. Braunschweig 1903.

Lie, Sophus: *Theorie der Transformationsgruppen,* Band III. Leipzig 1893.

Riemann, Bernhard: Ueber die Hypothesen, welche der Geometrie zu Grunde liegen. In: *Abhandlungen der Königlichen Gesellschaft der Wissenschaften zu Göttingen* 13 (1868), 132 - 152.

Scholz, Erhard: Herbart's Influence on Bernhard Riemann. In: *Historia Mathematica* 9 (1982), 413 - 440.

Torretti, Roberto: *Philosophy of Geometry from Riemann to Poincaré.* Dordrecht-Boston-London 1978, 260-264.

Ueberweg, Friedrich: Die Principien der Geometrie, wissenschaftlich dargestellt. In: *Neue Jahrbücher für Philologie und Pädagogik,* Supplement-Band XVII, Leipzig 1851, 20-54.

Ueberweg, Friedrich: Exposition scientifique des principes de la géométrie, précédée d'une discussion sur le fondement de la certitude des propositions premières de cette science. In: Joseph Delboeuf, *Prolégomènes philosophiques de la géometrie.* Liége 1860, 269-305.

Geometric Facts and Geometric Theory: Helmholtz and 20th-Century Philosophy of Physical Geometry[1]

Martin Carrier

Helmholtz's geometric investigations are remembered not so much for the one technical result that actually came out of it. Rather, what attracts contemporary attention is his impact on the philosophical interpretation of physical geometry. As I will argue, Helmholtz actually fathered three important schools of geometric thought, one of them intentionally and two unwittingly. What he explicitly tried to buttress was an empiricist *philosophy* of geometry, and what he came close to formulating - although he stopped short of actually doing so - was *conventionalism* and *metrogenic apriorism*. My aim is to provide a sketch of what became of Helmholtz's views on geometric facts and geometric theory in 20th-century philosophy of geometry. I begin by giving a brief outline of Helmholtz's geometric project and subsequently turn to a discussion of the three lines of geometric reasoning that this project spawned. Finally, I address the question how Helmholtz's general approach fares in light of contemporary philosophy of geometry.

1 I am grateful to Peter McLaughlin for valuable suggestions.

1. Helmholtz's Problem of Space

There is one item on the venerable agenda of time-honored philosophical problems that refers to geometry: Why are the theorems of geometry applicable to reality? In other words, how is it possible that a theory whose claims can apparently be checked by reason alone can yet be employed to describe relations among empirically given locations? This problem had assumed an entirely new shape through the development of non-Euclidean geometries from the 1820's onward. This new development had found its expression in Bernhard Riemann's celebrated 1854 lecture "On the hypotheses which lie at the foundation of geometry". In this lecture Riemann formulated what was later to be called Riemannian geometry.

Characteristic of Riemann's approach is, first, a rather abstract conception of space. Riemann began by introducing what is now known as an *n*-dimensional manifold. In such a manifold every element is uniquely singled out by assigning *n* coordinate values to it, each of which is supposed to be capable of smooth variation. Second, geometric relations are defined in this manifold by giving a general expression for lengths. This expression relates differences in the coordinate values to distances; it is based on the assumption that the *Pythagorean theorem* holds within infinitesimal regions of the manifold. In such regions the squared distance ds^2 between two points is given by the sum of their squared coordinate differences dx_i^2 (that is, $ds^2 = \Sigma dx_i^2$). This assumption is distinguished by its simplicity among the alternatives (such as $ds^4 = \Sigma dx_i^4$) (Riemann 1854: 9). Moreover, it is tantamount to the supposition that in these infinitesimal regions an *n*-dimensional generalization of Euclidean geometry holds. However, this does not apply to extended regions. The infinitesimally Euclidean metrics may be conjoined such that they yield non-Euclidean distance relations in finite-sized regions. Deviations from Euclidean flatness are expressed by what came to be called the Riemann curvature tensor. Riemann allowed for the possibility that curvature might vary with the position; moreover, he deduced the tenet that in manifolds of constant curvature geometric figures can be displaced and rotated without an alteration of shape (Riemann 1854: 14). Riemann concluded that the "independence of bodies from their locations" requires and entails the constancy of curvature (Riemann 1854: 19).

Non-Euclidean geometry in general and Riemann's work in particular brought an entirely new question in its train; the question, namely, which one of the abstractly conceivable geometries is actually realized in physical space.

Naturally, this question could only be asked after it had become clear that Euclidean geometry is not the only possible option. In Riemann's view this question had to be decided by recourse to *experience* (Riemann 1854: 2).

Helmholtz was in full agreement with this view, and the declared aim of his geometric research project was to further elaborate the empirical basis of geometry.[2] In Helmholtz's eyes, Riemann had started his endeavor from the wrong end: He had begun by simply postulating the Pythagorean nature of the metric and subsequently derived theorems about displacement and rotation of geometric figures. Helmholtz felt that this line of reasoning had to be reversed in order to bring out more clearly the empirical foundations of geometry. Physical geometry needs to be based on a principle whose truth can be checked by direct observation. This principle was supposed to be furnished by the alleged fact of observation that the congruence of spatial figures is independent of place, direction, and path of transportation.

This *principle of free mobility* was spelled out by Helmholtz through the following four axioms.

(1) Space is an n-dimensional manifold.
(2) Point-pairs can be moved without thereby changing their relative separation.
(3) Every point can be moved continuously so as to coincide with any other point.
As Helmholtz pointed out, it follows from axioms (2) and (3) that if two bodies are congruent at one location, they are congruent at any other location as well.
(4) A body can be rotated such that it eventually resumes its original state.

From these axioms Helmholtz derived the existence of a metric of quadratic, Pythagorean form, and concluded he had captured Riemann's vantage point.[3]

2 This agreement is veiled by the title of one of Helmholtz's papers, namely, "On the facts which lie at the foundation of geometry", which is deliberately antithetical to the title of Riemann's lecture. In fact, however, Riemann used the word "hypothesis" in order to refer to empirical facts. Facts are not known with certainty and are hypothetical in this sense (Riemann 1854: 2).

3 Helmholtz 1866: 110-111; Helmholtz 1868: 37-41; see also Torretti 1978: 158; DiSalle 1993: 509-510. As Lie later showed, axiom (4) is actually superfluous. The quadratic form of the metric can be deduced from the first three axioms alone.

This derivation of the Pythagorean metric constitutes the one technical result Helmholtz managed to accomplish.

Helmholtz was at pains to stress that these axioms are supposed to apply to *infinitesimal* regions in the first place. In other words, he was quite aware of the fact that he only needed infinitesimal free mobility so as to obtain Riemann's general formula for the length element (Helmholtz 1868: 41, 57-58). But he accepted the additional claim that finite-sized bodies are also freely movable and concluded, by drawing on the Riemannian result mentioned, that the curvature of space is constant (Helmholtz 1868: 58).

On the whole, then, Helmholtz took double advantage of the condition of free mobility. He showed, first, that free mobility - even if understood infinitesimally - suffices to yield the Pythagorean nature of the metric, and he stressed, second, that if free mobility is postulated to extend to finite-sized bodies, it follows that the curvature of space is constant. The challenge this approach engendered was to clarify the conceptual status of the principle of free mobility. It could be regarded as a *fact of nature*, a *convention*, or a *condition of the possibility of geometric measurement*. These three options define the agenda for much of early 20th-century philosophy of geometry; all of them were at least foreshadowed in Helmholtz's work.

2. The Empiricist Construal of Physical Geometry

As already mentioned, Helmholtz emphasized that the determination of the actually realized geometry of space is a matter of experience. His reasoning was as follows. Riemann's research had opened up a wide range of geometric systems, and physical space can correspond to only one of them. The actual structure of space is provided by *two facts*, namely, free mobility and vanishing curvature. It is a fact of experience that finite-sized bodies can be moved without change of shape. This observation narrows down the range of empirically possible spaces to those possessing constant curvature. Astronomical measurements confirm, then, that the actual curvature is "indiscernible from zero" (Helmholtz 1868: 60; Helmholtz 1870: 7, 19, 23). It follows that the geometry of physical space is Euclidean geometry, and this result is established by drawing on experience. Helmholtz's approach thus displays a strongly *empiricist* leaning.

Helmholtz quite naturally operated within the framework of Newtonian physics. After the advent of general relativity theory early in this century, the relevant scientific situation has changed drastically. General relativity is characterized by the *geometrization* of gravitation. Gravitation is construed as a manifestation of space-time structure. The metric incorporates the gravitational field. Along with the demise of Newtonian physics, Helmholtz's unchangeable, flat space was overthrown as well. It gave way to the notion of variably curved space - or space-time for that matter. In the framework of geometrized gravity, variable curvature indicates changes in the gravitational field which find its empirical expression in the appearance of tidal forces. It follows that space (or space-time) lacks the homogeneity attributed to it by Helmholtz. In the space-time of general relativity finite-sized bodies are not freely movable in Helmholtz's sense.

Given this difference in the relevant physical situation, it is clear that Helmholtz's general approach can only be sustained if the substantive details are essentially modified. His approach can no longer be followed literally, but at most in spirit. One might, namely, attempt to capture its gist by delineating those facts on which this altered geometric structure rests.

A project of this sort was first explicitly advocated by Hans Reichenbach in 1924 - albeit restricted to the space-time of special relativity theory. Reichenbach specified a large number of supposedly simple facts of observation regarding light propagation and the running of clocks, and he set out to derive the Minkowski-structure of space-time from them (Reichenbach 1924). Although Reichenbach's account is less than felicitous technically - as shown by Hermann Weyl (Weyl 1924) - its chief philosophical merit consisted in transferring Helmholtz's project to relativistic physics and by providing an appropriate label for it. What Helmholtz's project is all about, namely, is to yield a *constructive axiomatization* of physical geometry.

In an axiomatization of a theory a certain subset of all the propositions making up this theory is placed at the top and the remainder is derived from this distinguished subset. It is of high philosophical significance what kind of propositions are accepted as basic. Reichenbach introduced, and at the same time recommended, a type of axiomatization in which the propositions placed at the top express directly observable facts. This type Reichenbach called *constructive axiomatization*; and his reason for favoring it over approaches which proceed by singling out more abstract and theoretically far-reaching principles as fundamental is that in constructive axiomatizations the axioms are more directly amenable to empirical scrutiny. Constructive axioms are thus interpreted as capturing elementary facts (Reichenbach 1924: 2-5). Translated into this terminology, what Helmholtz had intended to furnish was a constructive axiomatization of the space of classical physics.

As Jürgen Ehlers, Felix Pirani and Alfred Schild demonstrated in 1972, this Helmholtzian program of a constructive axiomatization can indeed be extended to the space-time of general relativity. Against the changed physical background Helmholtz's question takes the form: Which facts are suited to show that Riemannian geometry rather than an even more general geometric system represents the actual geometry of space-time? In accordance with the essentially spatiotemporal spirit of general relativity, Ehlers/Pirani/Schild single out the motions of freely falling particles and the paths of light rays as basic.[4] Among the supposedly fundamental facts they invoke are the principle of equivalence, the invariance of the speed of light, or the fact that a freely falling particle always lags behind a light pulse. Moreover, freely falling particles are distinguished as force-free (Ehlers/Pirani/Schild 1972: 63-68; Ehlers 1973: 85-87; Ehlers 1988: 146-158).

Ehlers/Pirani/Schild explicitly place their project in the Helmholtzian tradition. As they explain, they endeavor to achieve for general relativity what Helmholtz had accomplished for classical spaces of constant curvature: The structure of space-time is intended to be derived from phenomena that are strongly suggested by experience (Ehlers/Pirani/Schild 1972: 65). Their project can thus be construed as being aimed at a constructive axiomatization of the space-time of general relativity. The empiricist leaning of this project leaps to the eye. In a sense it constitutes the up-dated and refined successor of Helmholtz's enterprise.

3. The Conventionality of Physical Geometry

As described, Helmholtz considers free mobility as a fact of nature; observation testifies that the standard of rigidity can be transported without alteration. On the other hand, Helmholtz realized that this supposed fact was somewhat shaky in that truly rigid bodies are difficult to recognize. It is possible, namely, to come up with deviant standards of rigidity without running into empirical trouble. The root of this problem is, as Helmholtz argued, that the adequacy of the distinction of certain bodies as rigid and certain lines as straight is jugded by the appropriateness of the results they furnish (Helmholtz 1866: 108; Helmholtz 1868: 32-33). Put more concretely, if the angles of a triangle do not add up to 180^0, one may either assume that one is dealing with non-Euclidean geometric relations or that one has failed to draw a straight line correctly. Analogously for rigid bodies. As Helmholtz explained, the acceptance of a de-

4 Weyl had demonstrated in 1923 that metric relations can be derived on this basis but he had failed to specify any procedure actually suitable to achieve this objective; Weyl 1923a: 228-229; Weyl 1923b: 19.

viant standard of rigidity leads to a disparate system of physical geometry which - at least on the face of it - is empirically equivalent to the original one in that all judgments about congruence or length relations are preserved. In other words, one is free to assume like deformations of all measuring rods, ending up in this case with a divergent but empirically equivalent system of physical geometry. This divergent system is equally able to cope with all geometric measuring results.[5]

A possible conclusion from this scenario is that the project of ascertaining rigidity empirically leads right into a circularity and that, consequently, the standard of rigidity has to be decreed by convention. This is precisely the conclusion that conventionalists drew later, but it was not Helmholtz's conclusion. In his view, the circularity problem can be overcome by drawing on the actual mechanical behavior of moving bodies. A characteristic in point is the law of inertia. According to this law, a force-free body moves along a straight line with unchanged velocity. If the motion of an inertially moved body is described using a deviant standard of rigidity, it will no longer exhibit a constant velocity; that is, the body will no longer traverse equal distances in equal time intervals. This outcome demonstrates that the length measurement was incorrect. Consequently, if the laws of mechanics are invoked in geometric measurements, one system of physical geometry is uniquely distinguished through experience (Helmholtz 1870: 29-30).

It was this last move that was contested by conventionalists. The conventionalist claim is that physical geometry can in no way be singled out by recourse to experience alone; any geometric measurement is in need of a conventionally chosen standard. This claim was first formulated by Henri Poincaré, it was then taken up by Rudolf Carnap and assumed its canonic form in Reichenbach. I focus on Carnap's Ph.D. thesis of 1922 which can be taken in its relevant parts as a direct rejoinder to Helmholtz.

Quite in the spirit of Helmholtz's project Carnap asked what are the prerequisites for measuring out spatial relations. As he argued, one could begin by identifying straight lines. But the catch emerges immediately. Any such identification requires the comparison of a given line with a standard. For example, we judge the straightness of a given edge by using light rays, thus assuming the latter to represent straight lines. The ensuing circularity (or infinite regress) makes it clear that straightness is not a fact of the matter but can only arise from a prior stipulation. Analogously for length relations. The question if a

5 Helmholtz 1870: 24-25. However, Helmholtz's claim that such a deformation would go entirely unnoticed is mistaken; it was later corrected by Reichenbach; Carrier 1994: 146, 259.

given rod has retained its length under transportation can only be answered by comparing the measuring results obtained by its use with those furnished by another rod and so on and so forth. Again, the conclusion is that distance relations are not empirically given; rather, their determination demands a convention (Carnap 1922: 33-37).

The point where Carnap parts company with Helmholtz emerges when it comes to judging the adequacy of Helmholtz's mechanical ploy. Carnap argued that the laws of mechanics are not a fact of the matter either. That is, they can be adapted to deviant standards of straightness and rigidity. This claim was supported by picking a divergent congruence standard to the effect that the surface of the earth comes out as a Euclidean plane. Carnap demonstrated that the laws of mechanics can be adjusted to the altered congruence standard in such a way that all results of length measurements can still be accounted for mechanically (Carnap 1922: 47-54).

The gist of the conventionalist reply to Helmholtz thus consists in the contention that it is not a fact of the matter that inertially moved bodies traverse equal distances in equal times. Rather, as conventionalists insist, the law of inertia can readily be reformulated to the effect that the velocity of such bodies depends in a specified way on their location (Reichenbach 1928: 138). Put in more general terms, the laws of mechanics are unsuitable for defusing Helmholtz's circularity problem since they leave no less room for adaptation than the standards of straightness and rigidity do. Mechanical theory can always be adjusted so as to agree with the geometric stipulations one likes best.

In fact, if one accepts the project of grounding physical geometry on experience, without intrusion of physical theory, the free play for tampering with the law of inertia is preserved in general relativistic space-time. It is indeed possible to formulate alternative versions of the law of inertia which respectively entail different geodesic structures of space-time. What appears as the straightest possible spatiotemporal line on one of these versions comes out as curved on an alternative one (Carrier 1994: 152-153). It follows that Helmholtz's project of determining the facts that underlie geometry is much harder to implement than Helmholtz believed.

4. Rigidity as a Transcendental Notion

The third type of philosophical construal of geometry that can be traced back to Helmholtz's writings is *metrogenic apriorism* as exemplified by the positions of the early Bertrand Russell and of Hugo Dingler. This line of thought is intimated by the following Helmholtzian argumentation. Every geometric theory that is supposed to be applicable to reality entails claims about length relations. Such claims acquire an empirical bearing only if lengths can be measured. Length measurements demand that spatial intervals can be compared to one another and this in turn necessitates that length standards or rigid rods can be transported without thereby changing their lengths. Therefore, all spatial measurements presuppose the free mobility of bodies. Free mobility is the condition of the possibility of geometric measurements and thus acquires transcendental significance (Helmholtz 1868: 36, 60; Helmholtz 1870: 29-30).

Helmholtz mistakenly identified this transcendental aspect of rigidity with a variant of Kantian transcendentalism. In fact, however, Helmholtz's argument does not involve anything akin to a Kantian transcendental form of intuition (Torretti 1978: 168). Rather, it outlines a quite distinct kind of transcendentalism. Helmholtz's argument is concerned with the necessary preconditions for measurement, not for the organization of sensory impressions. The question is which properties have to be ascribed to measuring instruments if the results they furnish are supposed to be suitable as evidential basis of a scientific theory. What emerges as the answer to this question is a metrogenic transcendentalism of the kind just sketched. It is true, Helmholtz eventually qualified this transcendental line of reasoning by invoking the laws of motion and by insisting that geometry supplemented with mechanics is empirical. Accordingly, he did not actually endorse the transcendental view. Still, his corresponding remarks paved the way to a novel philosophical interpretation of physical geometry.

Helmholtz's remarks can be taken to suggest the following approach to physical geometry which was first outlined by Russell in 1897. The circularity problem involved in ascertaining rigidity entails the breakdown of Helmholtz's empiricist program; it demonstrates that geometry cannot possibly be obtained from experience (Russell 1897: 23, 75). There is a distinguished system of physical geometry, and since this system cannot be identified empirically, it has to be specified *a priori* by recourse to the preconditions of any spatial measurement. Accordingly, Russell endeavored to pursue and strengthen Helmholtz's transcendental project.

Russell contended that any measurement of spatial quantities presupposes the free mobility of bodies; it presupposes that a body's shape is not affected by its location. Put briefly, the argument runs as follows. The determination of length relations necessarily involves a comparison between the length to be evaluated and a measuring rod. The like variation of measured length and measuring rod would not affect the outcome of any such comparison. What appeared before to be equally long comes out as equally long afterward. Therefore, the existence of such a variation can never be confirmed by actual measurements,[6] and its assumption is meaningless for this reason. Russell's conclusion is that free mobility is among the conditions of the possibility of measurements and, consequently, that the curvature of space is necessarily constant (Russell 1897: 51, 69, 72, 152-154).

It is worth noting that Russell here develops part of the conventionalist scenario and rejects it as absurd. What is operative in Russell's argument is the verificationist theory of meaning, but it is not this theory that marks the disagreement between Russell and conventionalism. Conventionalists insisted that the right conclusion from the circularity problem, given verficationist semantics, would have been that *any claims* about length standards - including their constancy - are to be regarded as meaningless until a preferred standard is introduced conventionally. None of the conceptually disparate alternatives is distinguished on the basis of logic and experience alone.

A different project to a similar effect was pursued by Dingler. Dingler likewise urged that physical geometry had to be established on a priori grounds, and he moreover stressed that these grounds had to be metrogenic in kind. Taken in this sense, Dingler's project indeed constitutes a continuation and elaboration of Helmholtz's pertinent argumentation. On the other hand, the class of geometric systems Dingler found admissable is much more restricted than its Helmholtzian analogue. Rather than including all constantly curved geometries it is made up of one single geometric system, namely, Euclidean geometry.

Stripped to its bare essentials Dingler's approach can be reconstructed as follows. Like Russell before him, Dingler maintained that the circularity involved in the attempt to identify rigid rods empirically is inescapable and vicious. A body is qualified for geometric measurements if it is free of any deformation. The only means to ascertain that a given body fulfills this demand is to check the accordance of the results it provides with geometric theorems (Dingler 1955: 156-157). In fact, this Dinglerian claim directly echoes a Helmholtzian argument (Helmholtz 1866: 108; Helmholtz 1868: 32-33), but the use Dingler made of this observation is more in line with Russell than with Helmholtz. For Dingler, the circularity problem demonstrated that physical geometry can in no way be empirical; it is *a priori* instead.

Dingler abhorred the conventionalist nonchalance with respect to geometry and insisted that the construction of geometry must not exhibit any arbitrary features. Rather, geometry has to be based unambiguously on secure foundations. These foundations Dingler believed he had identified in basic human

6 This connection reproduces the Helmholtzian flaw; see note 5.

skills. His project was to build up geometry systematically, i.e., without gaps and circles, on this alleged foundation.

The core element of the implementation of this project is Dingler's so-called "three-plate procedure" (*Dreiplattenverfahren*). It consists in grinding three bodies alternately in pairs so that the surfaces of any two eventually match. If these three bodies fit together interchangeably their surfaces realize Euclidean planes.[7] For Dingler, this showed that the Euclidean plane can be constructed by exclusive recourse to primary human skills, and he contended, moreover, that the flat plane is the only one that can be so constructed. The construction of any determinately curved surface requires more advanced technical procedures that cannot be carried out relying on basic skills alone (Dingler 1955: 134-140). Relying on this distinction of the Euclidean plane Dingler attempted to provide a construction of full Euclidean geometry on its basis. His claim is, then, that rigid rods are necessarily characterized by the condition that they satisfy Euclidean distance relations (Dingler 1923: 126; Dingler 1964: 176-187).

The difference between Dingler's geometric apriorism and the empiricist or conventionalist approach to geometry is highlighted by considering the response to possible deviations of the measured raw-data from the Euclidean expectations. Reichenbach, for instance, explains that it is perfectly legitimate to start one's spatial measurements on the basis of Euclidean geometry. In other words, one may begin by using this geometry as observation theory; as a theory, that is, supposed to explain the behavior of the measuring instruments. However, if non-Euclidean results show up one is free to adopt an alternative geometric system as observation theory and assume that the measuring rods behave in a non-Euclidean fashion. The observation theory is adjusted in light of the measuring results (Reichenbach 1928: 41-42).

This option is entirely rejected by Dingler. Since Euclidean geometry is thought to be distinguished on epistemological grounds, prior to any measurement, adapting geometry to measurement results means going in a circle. Any deviation from the Euclidean expectation has to be attributed to the presence of physical distortions which deform the measuring rods. This procedure constitutes Dingler's "method of exhaustion" (Dingler 1923: 37-39, 135-138). Systematically speaking, the method makes use of the Duhem-Quinean room for adjusting a theoretical system to anomalous experience. Exhaustion works since it is possible to shield a given theoretical element from the impact of re-

---

7 This procedure was already mentioned by Helmholtz; however, it was merely cited as an example for the principle that if two magnitudes are equal to a third one, they are also equal to one another; Helmholtz 1887: 380.

calcitrant data - provided that drastic enough changes are made in other parts of the theoretical network involved.

What Dingler was after is a theory of geometric measurements such that the results these measurements supply no longer leave room for the conventionalist freedom of interpreting the results differently. Rather, he urged that the interpretation has to be fixed unambiguously. This is supposed to be achieved by tightening the transcendental constraints on geometric measurements, constraints that were in principle envisaged already by Helmholtz. In pursuing his project Dingler narrows down Helmholtz's epistemic requirement of free mobility - which allows for non-vanishing, if constant, curvature - to the demand for Euclidean congruence relations in finite-sized spatial regions.

This admittedly sketchy discussion of the Russellian and Dinglerian approaches should suffice to make it plausible that they indeed bear a resemblance to the transcendental line of Helmholtz's philosophy of geometry. The upshot is that Helmholtz stimulated the development of a novel, non-Kantian type of transcendentalism. He initiated metrogenic apriorism.

5. One Hundred Years After, or: Philosophy of Geometry Today

If the importance of a philosopher is to be judged in terms of the amount of discussion his work engendered, then Helmholtz doubtlessly qualifies as an important figure in the philosophy of geometry. His writings spawned three major schools of 20th-century philosophy of geometry. Empiricists and conventionalists join in regarding Helmholtz as one of their own (Ehlers/Pirani/Schild 1972: 65; Reichenbach 1928: 48). And it is not an easy matter to stimulate disparate lines of thought without being flatly inconsistent. So, as regards the philosophy of geometry, Helmholtz certainly merits a centennial celebration.

This holds true in spite of the fact that most of what he actually said about physical geometry is wrong - if judged with the benefit of hindsight. As already said, the conventionality problem cannot be solved by simple recourse to the laws of mechanics. Furthermore, whereas Helmholtz considered the free mobility of finite sized bodies - and the constant curvature that goes along with it - as a fact of nature and even as a sort of transcendental precondition of any geometric measurement, space (or space-time) is actually endowed with variable curvature, and we still manage to measure out physical geometry. It would be foolish to blame Helmholtz for the fact that he had no inkling of the possibility of geometrizing a physical interaction. Still, it might be of interest to examine how Helmholtz's overall approach fares - if judged by present lights. In the remainder I want to inquire into the fate of free mobility and to sketch the dominant line of contemporary philosophy of geometry, the one that does not have its roots in Helmholtz.

As to free mobility, at first sight it looks as if Helmholtz came out splendidly. Helmholtz's argument from free mobility actually comprises two steps. In the first step it is demonstrated that Riemann's expression for the metric can be derived from *infinitesimal* free mobility. In the second step it is assumed that free mobility extends to finite-sized bodies, and it is inferred that under these circumstances spatial curvature is constant (Helmholtz 1868: 41, 58-59).

General relativity introduces a (pseudo-)Riemannian space-time of variable curvature. It is characteristic of all these space-times that in infinitesimal regions the metric assumes its special relativistic or Minkowski shape. Minkowski space-time is distinguished by vanishing - and thus constant - curvature. Accordingly, infinitesimal free mobility is retained in general relativity.[8] The apparent upshot is that the first half of Helmholtz's double claim is preserved in the present perspective; only the extension of free mobility to finite-sized bodies has turned out to be unjustified.

But on second glance, infinitesimal free mobility loses much of its epistemic appeal. The empirical effects of variable curvature can only be made to vanish at a single point (but not in its small neighborhood) or along a one-dimensional trajectory (but not in the small region surrounding it). In particular, curvature-induced deformations of a three-dimensional body do not disappear on letting the body shrink to infinitesimal extension (Synge 1960: 115; Ohanian 1977: 903-906; Friedman 1983: 199-202). Infinitesimal free mobility holds in general only for infinitesimal one-dimensional rods; it is not guaranteed to apply to any infinitesimal body. It seems inappropriate to place heavy epistemic emphasis on this highly idealized case.

Admittedly, the alternative procedure of using freely falling particles and light rays for providing the empirical basis of geometry likewise involves idealizations. However, these idealizations are less restrictive in crucial respects. Strictly speaking, if there are tidal forces, there are no rigid bodies - no matter whether the bodies be large or small. But there are still geodesic motions under these circumstances. The particle method works in regions of variable curvature where the rod method fails. The existence of rigid bodies is contingent on the presence of certain space-time conditions; the particle

8 It is true, Minkowski space-time is endowed with a metric of indefinite or Lorentz type. But even this possibility was envisaged by Helmholtz - though rejected on empirical grounds; Helmholtz 1868: 59-60.

method is free from restrictions of this kind. Only the latter method is applicable to the entirety of general relativistic space-time structures. Accordingly, it is rather the law of inertia and the invariance and isotropy of light propagation on which physical geometry rests. These are the facts - if there are any - which lie at the foundation of geometry. Time has come to bid farewell to free mobility.

Finally, I want to give a brief description of the presently dominant line in the philosophy of physical geometry which is conspicuous by its absence in Helmholtz's writings. It rather originates from the work of Hermann Weyl in the 1920's and was taken up by Hilary Putnam in the 1960's. Since then, its popularity has increased considerably.

According to this *field interpretation* of physical geometry, geometric quantities like the metric or the geodesic structure are to be construed as physical fields like the electromagnetic field. These geometric fields are quantities in their own right, and conceptually independent of their material sources. Physical bodies interact with space-time structure; they affect the magnitude of geometric quantities and these quantities in turn determine trajectories and influence metric standards. The behavior of rigid rods is among the indications of space-time structure, but it is in no way the privileged one. Physical geometry - just like other physical fields - becomes manifest empirically in a multitude of indicator phenomena (Weyl 1923a: 219-226; Putnam 1963). The field interpretation regards physical geometry as a theoretical state rather than an observational state. Geometric relations are not obtained by simply measuring out space-time. They are specified by theory and multiply but only indirectly accessible empirically.

Philosophically speaking, the field interpretation involves a three-fold shift with respect to the earlier approaches. First, as to meaning, it is assumed that scientific concepts are specified by their integration in a nomic network. Meaning is conferred by the theoretical context, it does not accrue from a piecemeal connection between concept and fact. Second, as to methodology, it is supposed that a theory's truth can increase through non-empirical virtues. Methodologically preferred theories are more true than their inferior - although empirically equivalent - counterparts. Since all of the allegedly conventional alternatives to standard general relativity lag behind the latter methodologically, these alternatives are wrong; no conventional choice among them is called for. This last step had been emphatically rejected by empiricist and conventionalist approaches. They had insisted that the only ground for increased verisimilitude is empirical merit. Third, as to the object of scientific theory, the field interpretation is committed to scientific realism. In contrast to the conventionalist view, physical geometry is not regarded as an abstraction formed on the basis of geometric measurements. It is not just a convenient shorthand notation for measurement results. Rather, the metric and geodesic fields exist independently of any human endeavor to get hold of them empirically.

Helmholtz is part of the older empiricist tradition; the tradition that ramified into the three branches mentioned. All these approaches accept the principle that physical geometry has to emerge directly from geometric measurements. Geometry has to be built up from scratch; it has to be erected on something simple and immediately given - be it simple facts, possibly supplemented with basic decisions, or be it primary skills. This former standard approach has become a minority view now. Still, Helmholtz isn't yet in need of support through affirmative action. This is all the more true since paradigm shifts in philosophy are never as unanimous and irreversible as in the sciences. Helmholtz may well stage a comeback.

References

Carnap, Rudolf: *Der Raum. Ein Beitrag zur Wissenschaftslehre.* Berlin: Reuther & Reichard, 1922.

Carrier, Martin: *The Completeness of Scientific Theories. On the Derivation of Empirical Indicators within a Theoretical Framework: The Case of Physical Geometry.* Dordrecht: Kluwer, 1994.

Dingler, Hugo: *Die Grundlagen der Physik. Synthetische Prinzipien der mathematischen Naturphilosophie*. Berlin: de Gruyter, [2]1923.

Dingler, Hugo: *Die Ergreifung des Wirklichen* [1955]. Frankfurt 1969.

Dingler, Hugo: *Aufbau der exakten Fundamentalwissenschaft.* München: Eidos, 1964.

DiSalle, Robert: Helmholtz's Empiricist Philosophy of Mathematics. Between Laws of Perception and Laws of Nature. In: D. Cahan (ed.), *Hermann von Helmholtz and the Foundations of Nineteenth-Century Science.* Berkeley: University of California Press, 1993, 496-519.

Ehlers, Jürgen: The Nature and Structure of Spacetime. In: J. Mehra (ed.), *The Physicist's Conception of Nature.* Dordrecht: Reidel, 1973, 71-91.

Ehlers, Jürgen: Einführung in die Raum-Zeit-Struktur mittels Lichtstrahlen und Teilchen. In: J. Audretsch/K. Mainzer (eds.), *Philosophie und Physik der Raum-Zeit.* Mannheim: Bibliographisches Institut, 1988, 145-162.

Ehlers, Jürgen/Pirani, Felix A.E./Schild, Alfred: The Geometry of Free Fall and Light Propagation. In: L. O'Raifeartaigh (ed.), *General Relativity. Papers in Honour of J.L. Synge.* Oxford: Clarendon, 1972, 63-84.

Friedman, Michael: *Foundations of Space-Time Theories. Relativistic Physics and Philosophy of Science.* Princeton: Princeton University Press, 1983.

Helmholtz, Hermann von: Über die tatsächlichen Grundlagen der Geometrie [1866]. In: Helmholtz. *Abhandlungen zur Philosophie der Geometrie.* Cuxhaven: Junghans, 1987, 108-112.

Helmholtz, Hermann von: Ueber die Thatsachen, die der Geometrie zum Grunde liegen [1868]. In: Helmholtz, *Über Geometrie.* Darmstadt: Wissenschaftliche Buchgesellschaft, 1968, 32-60.

Helmholtz, Hermann von: Über den Ursprung und die Bedeutung der geometrischen Axiome [1870]. In: Helmholtz, *Über Geometrie.* Darmstadt: Wissenschaftliche Buchgesellschaft, 1968, 1-31.

Helmholtz, Hermann von: Zählen und Messen, erkenntnistheoretisch betrachtet [1887]. In: Helmholtz, *Wissenschaftliche Abhandlungen* III. Leipzig: Barth, 1895, 356-391.

Ohanian, Hans C.: What is the Principle of Equivalence? *American Journal of Physics* 45, 1977, 903-909.

Putnam, Hilary: An Examination of Grünbaum's Philosophy of Geometry [1963]. In: Putnam: *Mathematics, Matter and Method,* Philosophical Papers I. Cambridge: Cambridge University Press, 1975, 93-129.

Reichenbach, Hans: *Axiomatik der relativistischen Raum-Zeit-Lehre* [1924]. Braunschweig: Vieweg, 1965.

Reichenbach, Hans: *Philosophie der Raum-Zeit-Lehre* [1928]. Braunschweig: Vieweg, 1977.

Riemann, Bernhard: Über die Hypothesen, welche der Geometrie zugrunde liegen [1854]. In: H. Weyl, *Das Kontinuum und andere Monographien.* New York: Chelsea.

Russell, Bertrand A.W.: *An Essay on the Foundations of Geometry* [1897]. New York: Dover, 1956.

Synge, John L.: *Relativity: The General Theory* [1960]. Amsterdam: North-Holland, [4]1971.

Torretti, Roberto: *Philosophy of Geometry from Riemann to Poincaré.* Dordrecht: Reidel, 1978.

Weyl, Hermann: *Raum, Zeit, Materie. Vorlesungen über allgemeine Relativitätstheorie* [1923a]. Darmstadt: Wissenschaftliche Buchgesellschaft, [6]1961.

Weyl, Hermann: Mathematische Analyse des Raumproblems [1923b]. In: Weyl: *Mathematische Analyse des Raumproblems/Was ist Materie?* Darmstadt: Wissenschaftliche Buchgesellschaft, 1977, 1-117.

Weyl, Hermann: Review of Reichenbach's Axiomatik der relativistischen Raum-Zeit-Lehre. *Deutsche Literaturzeitung* 30, 1924, 2122-2128.

V
Helmholtz und die physischen Grundlagen der Musik

Musical Thought and Practice: Links to Helmholtz's Tonempfindungen

Erwin Hiebert
Elfrieda Hiebert

I

There is a great deal of evidence by way of written record and artifact to show that no matter how far one may reach back into history, there is music. Evidently human beings are music makers in the same way that they also may be said to be tool- or instrument makers. It is appropriate to claim that wherever music is made it is an expression of sensations, inner nature, emotions, ideas - in short, a declaration of the human soul. The question that almost immediately thrusts itself upon the student of music is whether this characteristic human expression called music has an identifiable natural or scientific basis - external, that is, to the inner self of its maker. Historically, a dominant concern in studies on music - and we are restricting ourselves here to Western music - has been to discover or establish, if possible, the firm foundation and the principles on which a near-universal theory of harmony can be constructed. Most and perhaps all of the theories in which it has been claimed that this is accomplished, from early times until our day, have been challenged in one way or another by later music theorists. It goes without saying that the theories that have been proposed inevitably have generated complexities that only the well-trained music theorist will claim to manage. We may mention for example: the generation and progression of scales, chords and key systems; the rationalization of the major and minor triad problem; the identification and resolution, in context, of consonance and dissonance; the explanation and interpretation of harmonic (vertical) and melodic (horizontal) musical structures among others.

As a corollary to the historical search for a universal theory of harmony, and moving in parallel with the prospect of reaching that goal, have gone attempts to determine whether, or in what measure, that which is designated as harmony can be said to rest upon a physical or natural foundation amenable to analysis by way of mathematics and the laws of nature. Since the time of the Pythagoreans, series of tones and intervals, that are recognized as uniquely agreeable or disagreeable to the ear, have been seen to be analyzable in some measure by means of the mathematics that enters into describing the manner of physical movement of material bodies. There always have been music theorists who have contested the view that the fundamental principles of harmony are deeply rooted in mathematics or physical law. That critical position is fundamentally at odds with Helmholtz's analysis of any theory of music.

In 1863, at the height of his scientific career as Professor of Physiology at Heidelberg, Helmholtz published his *Lehre von den Tonempfindungen als physiologische Grundlage für die Theorie der Musik.*[1] As a comprehensive analysis of the state of musical acoustics at mid-century, *Tonempfindungen* incorporated the concentrated thought and experimental research in acoustics that Helmholtz had been actively engaged in for a decade. His investigations had included research on combination tones (1856), the timbre of vowel sounds (1857 and 1859), the physiological and physical causes of musical harmony and disharmony (1857/8), the theory of open and closed organ pipes (1859), musical tuning (1860), motions of the strings of the violin (1860), the analysis of timbre (1860), the theory of reed pipes (1861), and Arabic-Persian scales (1862).[2] In a lecture at Bonn, on the physiological causes of musical harmony in 1857 Helmholtz set the stage for an expansive treatment of the

1 The Preface is dated October 1862. The publication of *Tonempfindungen* in 1863 falls between volumes 2 and 3 of Helmholtz's most important physiological treatise, viz. *Handbuch der physiologischen Optik;* Leipzig 1856, 1860 and 1867. *Tonempfindungen* was substantially revised and updated for the 3rd and 4th editions of 1870 and 1877. The French edition of G. Guérault, from the German 2nd (1865), appeared in 1868; the English edition of Alexander J. Ellis, from the German 3rd, appeared in 1875, 2nd ed. in 1885.

2 Fundamental research papers on acoustics continued to appear after 1863: On the mechanism of the ossicles (Gehörknöchelchen) 1867; On the vibration of sound in the cochlea (Schnecke des Ohrs) 1869; On the mechanics of the ossicles and the typanum (Trommelfell) 1869; On the telephone and timbre (Klangfarbe) 1878; and on Lord Rayleigh's treatise on "Theory of Sound," 1878.

sensations of tone that was historically conceived and seen to be analyzable with the help of physics, physiology, psychology and the aesthetics of different cultures:

> "Es hat mich immer als ein wunderbares und besonders interessantes Geheimnis angezogen, dass gerade in der Lehre von den Tönen, in den physikalischen und technischen Fundamenten der Musik, die unter allen Künsten in ihrer Wirkung auf das Gemüth als die stoffloseste, flüchtigste und zarteste Urheberinn unberechenbarer und unbeschreiblicher Stimmungen erscheint, sich die Wissenschaft des reinsten und consequentesten Denkens, die Mathematik, so fruchtbar erwies. Der Generalbass is ja eine Art angewandter Mathematik; in der Abtheilung der Tonintervalle, der Tacttheile u.s.w. spielen die Verhältnisse ganzer Zahlen - zuweilen sogar Logarithmen - eine hervorragende Rolle. Mathematik und Musik, der schärfste Gegensatz geistiger Thätigkeit, den man auffinden kann, und doch verbunden, sich unterstützend, als wollten sie die geheime Consequenz nachweisen, die sich durch alle Thätigkeiten unseres Geistes hinzieht und die uns auch in den Offenbarungen des künstlerischen Genius unbewusste Aeusserungen, geheimnissvoll wirkender Vernunftmässigkeit, ahnen lässt" (Helmholtz 1884: 81/2).

Recognizing the enormous breadth and commanding quality of Helmholtz's scientific output, along with the fact that his life's work is not readily periodized by discipline, it nonetheless becomes pertinent to explore the way in which he balanced his interests and cultivated them either simultaneously or preferentially at various times during an active 50-year research career. The predominant and most innovative component of Helmholtz's corpus, the *Schwerpunkt* of his scientific work, we want to suggest, is anchored in physiology. More precisely, Helmholtz's research programme derives from a perspective that attributes unique importance to the study of the physiology of sensations (*Sinnesempfindungen*). Broadly conceived this includes medicine, optics and acoustics.

Helmholtz's career was launched on this terrain at age 20. The investigation of sensations and the invention of ingeneous instrumental means to study sensations commanded Helmholtz's attention early on and reached its zenith in the 1850s and 1860s. Subsequently, except for a never-ceasing fascination with optics, the focus on physiology tapered off as physically-oriented problems were taken up. Except for *Erhaltung der Kraft* (1847) thermodynamics did not loom large in Helmholtz's output until the 1850s and then it reached brilliance in connection with chemistry in 1882/3. Electrodynamics, mechanics, hydrodynamics, and mathematics came late in Helmholtz's career, mostly after the age of 50.

There are two points we want to make while ranging so freely over the many disciplines that Helmholtz mastered in one way or another. The first is that from the start of his career he had a solid command of physiology, optics, acoustics, mathematics, energetics and traditional mechanics. Mastery of electromagnetism, chemistry, hydrodynamics and meteorology were acquired several decades later. Secondly, it is evident, we believe, that for most of his life Helmholtz's perspective on the natural sciences was much more closely and more deeply aligned with physiology than with physics. It is true that during the last two decades of his life he occupied the most celebrated chair in theoretical physics in Germany. He was respected and honoured in this capacity by the public, colleagues, foreign associates, and the German industrial-government establishment. He was recognized shortly after his death as one who, as comprehensively as any living scientist, had mastered and substantively contributed to the old physics that within a decade was to be turned on its head. Physicists were quick to claim him as their own. But perhaps, in retrospect, his more innovative and path-breaking ideas and contributions belong equally, and even more, to the physiological sciences. In 1857, shortly after he had conceived of the idea of taking up the study of musical acoustics in a systematic and comprehensive way, he mentioned that by approaching the field of physical acoustics from a physiological point of view, the subject underwent substantial clarification.

> "Indem ich die physikalische Akustik vom physiologischen Standpunkte aus betrachtete, d. h. näher der Rolle nachging, welche dem Ohr in der Wahrnehmung der Töne zuertheilt ist, schien sich manches in seinem Zusammenhange klarer darzustellen; und so will ich denn versuchen, ob ich Ihnen einiges von dem Interesse mittheilen kann, welches diese Fragen in mir erregt haben, indem ich Ihnen einige Ergebnisse der physikalischen und physiologischen Akustik anschaulich zu machen suche" (Helmholtz 1884: 82).

The publication of *Tonempfindungen* in 1863 elicited both an immediate and a continuing response on the part of musicians, i.e. music critics, historians of music, composers, and performers, as well as instrument builders and scientists at home and abroad. It was a treatise that one had to come to grips with, a treatise that has continued into our times to merit reflections commendatory and critical. In his volume *Die Musik im 19. Jahrhundert*, the late Carl Dahlhaus of Berlin commented about the status of musical thought at mid-century in relation to the expansiveness in technology, transportation, economics and the new thrust in the natural sciences that characterizes the period following the revolution of 1848:

> "Musical thought was beholden no less to the natural sciences than to the age's characteristic fondness for the grandiose, and to the increased pace of evolution that became noticeable in virtually all areas of culture. The only work that actually proved "epoch-making" in nineteenth-century music theory, influencing practically everyone from the philosopher pondering the problems of aesthetics to the humble musician teaching how to write chord progressions, was Hermann von Helmholtz's *Lehre von den Tonempfindungen* (1863)" (Dahlhaus 1989: 192/3).

Others have asserted that Helmholtz's treatise was mainly a monumental synthesis of what was known at the time about musical acoustics; that in many of its details it was either no more than a beginning for refined analysis, or else downright inadequate (for instance, A. von Oettingen (1866) in Shirlaw 1913: 385 and 387; Donington 1982: 39).

The response to *Tonempfindungen* in music journals was immediate in 1863. A steady stream of commentaries and reviews in the *Allgemeine Musikalische Zeitung* (Leipzig, Selmar Bagge, ed.) ran until around 1870 when they ceased, the readership apparently saturated. In the *Neue Musikalische Zeitung* (Düsseldorf, Franz Brendel, ed.) of which Robert Schumann had been the first editor, four articles and reviews appeared in the 1860s. During this period *Dwight's Journal* in Boston carried four items. From these prompt appraisals and notices one only can conclude that musicians were fishing around in the sciences, hoping to find validation for their theoretical views about music, at times adapting Helmholtz's analyses to fit their own agendas in music theory (for example, Riemann's synthesis). Some musicians were genuinely anxious to discover that science could inform, illuminate, and assist in the interpretation of music. Although it was agreed, in general, that Helmholtz's studies in the physiology of sound were novel and significant, there were critics who rejected the essential relevance of his findings for music.

There was little response on the part of Helmholtz to either positive or negative (critical) evaluations of his work. Two alternatives may explain his reticence. The musicians should be left to battle it out among themselves. He was confident that his scientific analyses rested on solid ground and would speak for themselves.

On this question of reactions to Helmholtz as scientist invading the domain of music we may well ask: Who is outsider? Who is insider? Analysis of sensations was for Helmholtz *terra firma*; besides, he was not an outsider to music: It may have appeared that way to some musicians especially in Germany where musicology and music theory had become highly professionalized disciplines. Several points merit comment: Helmholtz recognized his limitations and the boundaries of his research. He had studied music and was well informed about the literature of music (McKendrick 1899: 134). He obviously was a genuine *Kulturträger* who recognized that his work in the sciences would speak for itself. Nevertheless, he also was firmly wedded to an historical perspective on science and other disciplines, and was ready to admit that what he had accomplished was a product of his own culture-conditioning. This reveals itself in the way in which he refrains from looking upon himself as a great innovator of new ideas - a genius somehow ahead of his times. This attitude also characterizes his position on the reception to *Tonempfindungen*.

The most concerted effort to challenge Helmholtz's *Tonempfindungen* at its very roots came within months of its publication from Moritz Hauptmann (1792-1868). Hauptmann conceded that Helmholtz appeared to be an exemplary physiologist, but called into question the significance of his contributions to the art of music:

> "... [da] scheint mir alles Physiologische sehr bedeutend, ja von grossem Werth zu sein, auch da noch, wo von Tonempfindung die Rede ist; das Psychologische, im Sinne des Tonverständnisses, das, wodurch in der Musik ein musikalisch bestimmter Sinn auszudrücken ist, was Musik zur Sprache macht, wird im Buche nicht erklärt; das muss selbst etwas Bestimmteres sein als Helmholtz es angiebt [...]. Ich kann aber in alledem, was er als vortrefflicher Physiolog vorbringt, etwas Aufklärendes für das Tonverständniss immer nicht finden [...]. Ich sehe, wie die Königin vor des alten Hamlet Geist, alles was da ist, und sehe doch nichts [...]. Ich habe aber so viel nachgedacht über diese Dinge, habe an allen den Punkten, die ich von Andern als erreichtes Ziel behalten sehe, auch gestanden und bin weiter gegangen, weil mich die Ergebnisse an jenen Punkten nicht befriedigen konnten: interessante Einblicke in die Natur, merkwürdige Krystallformationen gingen auf, aber keine Architektur; nur das Elementar-Natürliche, nicht die vernünftige Menschennatur und die Bedingungen ihres freien Kunstgebildes" (Hauptmann 1863: 669-670).

Hauptmann's *Natur der Harmonik und der Metric* had appeared ten years earlier (1853) and had been hailed by many scholars in the musical world as the most valuable work on harmony ever published. Over the years Hauptmann had become convinced of the inadequacy of linking acoustical phenomena, mathematical proportions and progressions to music theory. For example, he disregarded theoretical explanations based on partial-tone theory. Contrary to Helmholtz he renounced empirical investigations, advocated deductive reasoning and dialectial discourse and declared that music theories be derived solely from natural logic (Jorgenson 1986: 103, 189; Rummenhöller 1963: 36; Shirlaw 1913: 352-354).

Beyond the various criticisms levelled against *Tonempfindungen*, especially in the 1860s, the work itself, but more often other textbooks built on Helmholtz's theory of music, came to be listed as required reading for a musical career. Such was the case with a work on musical acoustics by John Broadhouse. It was designated as "The Student's Helmholtz". Broadhouse reported in 1881 that "the Universities of Cambridge and London set papers in Acoustics at their examinations for musical degrees" and that "there is [...] little doubt that the other [British] Universities will in due course follow the example [...] by adding Acoustics to their curriculum for musical degrees" (Broadhouse 1881: Preface). Yet, we must remember there had been previous textbooks modelled on Helmholtzian acoustics in England.[3]

Throughout the 19th century there were attempts on the part of others besides Hauptmann to assert that matters such as arithmetical series, the analysis of partial and combination tones, the aural interpretation of consonance and dissonance, and the role of the natural sciences in general had little to do with music itself. We mention here the most noteworthy and influential critics, some more moderate than others: the Belgian music theorist and critic François Fétis (1784-1871), the theoretical biologist, metaphysician and music aesthetician Hermann Lotze (1817-1881), the physicist and music theorist Arthur von Oettingen (1836-1920), the psychoacoustician and philosopher Carl Stumpf (1848-1936) and the musicologist and theorist Hugo Riemann (1849-1919) who nevertheless made efforts to combine divergent viewpoints.

II

Our topic is one that lies on the borderland where science and music, or science and culture, intersect. How, we have asked ourselves, might we become engaged in sorting out the various links that connect Helmholtz's contributions to the study of acoustics with the world of music? Did his work (or how did his work) affect the thinking about music, the making of music and the listening to

3 The foremost among these were: John Tyndall, *Sound. A Course of Eight Lectures Delivered at the Royal Institution*, London in 1867; Sedley Taylor (Trinity College, Cambridge), *Sound and Music: An Elementary Treatise on the Physical Constitution of Musical Sounds and Harmony* (London, 1873); W.H. Stone (lecturer in physics at London) *The Scientific Basis of Music* (London, 1878); and William Pole (civil engineer and music scholar) *The Philosophy of Music*. Being the substance of a Course of Lectures Delivered at the Royal Institution of London in 1877 (London, 1879).

music? Such a broad topic inevitably raises more questions than one can possibly answer.

Helmholtz's *Tonempfindungen* stimulated the writing of books and the establishment of lecture demonstrations. Much was written on the physics of sound, the psychology of hearing, the design of "sound producing" instruments, the aesthetics of music and music theory.

Did *Tonempfindungen* invade the world of practice? The British physicist and acoustician Llewellyn Lloyd, in an article on "Helmholtz and the Musical Ear", observed in 1939, "the importance of acoustics [on composers] [...] naturally fell on deaf ears" (Lloyd 1939: 171). This is too simplistic an evaluation and turns out not to be true as one pursues the question in more depth. It already has been mentioned that musicians who wished to verify theory turned to Helmholtz knowing fully well that scientific analysis only partially explains the art of music. Helmholtz himself realized the limitations of empirical findings in understanding the "character of musical systems" and "aesthetic norms" in various cultural contexts. His concepts and experiments nevertheless set the stage and showed the way for novel musical practice.

The enthusiasm, yet often mixed with controversial responses that Helmholtz's *Tonempfindungen* incited, rang through the musical world. In many cases it was an indirect response, such as the music written for the blending, enriched sounds of the piano - Brahms, Liszt, Debussy, for instance. Increased range, improved damping, sustained homogenized sounds, bigger and richer (thicker) sounds due to the possibilities of the 19th century piano spurred new ideals in piano writing. We postpone for the moment how Helmholtz's work was related to piano construction.

There were also direct responses to *Tonempfindungen*. They merit attention, but we only can refer briefly to a few of them in this paper. Rimsky-Korsakov explicitly turned to Helmholtz in a work of instruction on the principles of orchestration. In the memoirs of his musical life he wrote:

> "After reading the works of Tyndall and Helmholtz, I framed an introduction to my work, in which I endeavoured to expound the laws of acoustics as applied to the principles governing the construction of musical instruments" (Rimsky-Korsakov 1913: VII).

This work never was completed. The task, as he saw it, was too mind-boggling. George Ives, father and teacher of Charles Ives, in an essay in 1880 on the theory of music, stressed Helmholtz's idea that consonance and dissonance were a matter of degree rather than of kind. This emphasis shows up in a new dimension of freedom in the music of his son Charles Ives.[4]

Similarly Schönberg probably gained support from Helmholtz in developing freedom in the use of tonality. Dahlhaus observes:

> "One can base an attempt at a more precise definition of what Schönberg had in mind on Helmholtz's *Die Lehre von den Tonempfindungen*, a book which Schönberg, for all his pretended lack of erudition, had presumably studied before writing his *Harmonielehre*; for he shares with Helmholtz two striking ideas that depart from the general run of music theory: First the view that the difference between consonance and dissonance is one of degree, not of kind, and secondly the conviction that tonality is not a natural law of music but merely a formal principle. Thus, if one assumes that Schönberg had at least a cursory acquaintance with *Die Lehre von den Tonempfindungen*, it is clear that he could have deduced the idea of a 'logic' of *Klangfarbenmelodie* from what Helmholtz says about the 'logic' of the melody of pitches. According to Helmholtz, the 'feeling for the melodic relationship between consecutive notes' is based on the unconscious 'sensation of similar overtones in the sounds in question" (Dahlhaus 1987: 142/3).

The links between Helmholtz and Hindemith show up in their common emphasis on the ear as a determining agent for what we learn to hear, become conditioned to hear, want to hear, and are permitted or forced by natural law to hear. To examine these potential Helmholtzian links in the music of composers such as Rimsky-Korsakov, Ives, Schönberg, Hindemith and others would require a greatly expanded study of their compositions and written statements.

Let us examine a little more closely the work of Leos Janácek (1854-1928). In the Czechoslovakian territories Helmholtz's *Tonempfindungen* was taken up immediately, especially by lecturer/theorist Frantisek Skuhersky (1830-1892) and aesthetician Josef Durdík (1837-1920). In 1874, as a 20-year-old student in Prague, Janácek buried himself in studies at the Prague Organ School and in independent reading. His personal, profusely annotated and underlined copies of the works of Durdík, Zimmermann, Helmholtz, Wundt and others are located in the library of the Janácek Archive of the Moravian Museum in Brno.[5] He devoured these works, dating them so one can see how he absorbed or refuted readings along the way. Janácek became fired up to read the *Tonempfindungen* from his teacher, Skuhersky and from Durdík's references to Helmholtz

4 George Ives's copy of Helmholtz's *Tonempfindungen* (English ed.) is located at the Ives House in East Redding, according to the book list (p. 47b) in the Yale University Library. Baron 1992: 279 n.17; see also Boatwright 1965: 25.

5 Beckerman 1994: 20. The information on Janácek is mainly based upon the work of Michael Beckerman. The authors offer grateful recognition.

in his *General Aesthetics* (1874). Having been trained in mathematics and the natural sciences, Durdík had turned to these disciplines for an underpinning to his aesthetics arguments. Janácek writes: "I plunged diligently into his [Helmholtz's] work from February 16, 1876 to January 22, 1879" (Beckerman 1983: 397).

Mulling over the Helmholtzian idea that consonance and dissonance are related to the degree of "roughness" of sound and alert to the influence of experiential and cultural factors, Janácek developed the principle of chord connection - "antecedent relations", as he referred to it - in which less consonant chords resolve to the nearest more consonant ones. Janácek classified the connections according to the effect (Beckerman 1983: 396). Step by step Helmholtz's ideas were adapted to Janácek's thinking when he explained that "antecedent relations" also are produced by memory. This he later rationalized by referring to Helmholtz's theory "that the transverse fibers of the basilar membrane of the inner ear vibrate sympathetically with music tones, such that we continue to perceive an entity for a fragment of a second after its frequency has ceased". Thus, he believed that an "after image" of the first interval or chord and a "sonic shadow" is formed on the second chord. There also is a "chaotic moment" when the two interpenetrate - "percolate".[6]

Max Brod expresses this aspect of Janácek's philosophy of music succinctly:

> "Seit 1876 beschäftigte sich Janácek mit Helmholtz's *Lehre von den Tonempfindungen* und entnahm ihr die Hauptstütze der eigenen Theorie, die er gegen Helmholtzens Theorie von den Obertönen entwickelte. Die Theorie von den Obertönen genügte ihm nicht zur Erklärung der Ton-Verwandtschaften, auf die Helmholtz seine Lehre von den Akkordverbindungen aufbaut. Denn die Obertöne sind bei verschiedenen Instrumenten verschieden. Da kam ihm die andere Feststellung Helmholtzens zuhilf: dass der Ton nach seinem Aufhören noch ein Zehntel Sekunde lang in einem Zehntel seiner Stärke nachklingt. In dieses Nachklingen tönt der nachfolgende Akkord herein und verursacht (nach Janáceks Terminologie) ein Chaos von Tönen, das gesetzmässig aufgelöst werden kann und dabei bestimmte Empfindungen erweckt. Diese Theorie hat Janácek in seiner Schrift [...] "Bau und Verbindung der Akkorde" 1897 entwickelt" (Brod 1956: 23).

Janácek justifies his connecting forces, (i.e., the antecedent relation phenomenon) with what he refers to as the glue component that he discovered in Helmholtz and Wundt. Beckerman remarks: "With the use of the latest scientific proofs [the reference being to Helmholtz] Janácek, as a true positivist [much

6 Beckerman 1983: 397. "Percolate" is Beckerman's translation of the Czech "prolinani".

imbued by Herbertianism] felt that his theory of '*antecedant relations*' was rooted in the laws of nature" (Beckerman 1983: 397/8). Janácek continually modified and reformulated his theories regarding chord connection, experimenting with ideas that would generate musical practice. Broadly ranging over various disciplines, as Helmholtz had done, Janácek was genuinely interested in history and boldly attempted, as he said: "to integrate musical acoustics, philosophy, modern theory, and most notably human behaviour" (Beckerman 1994: 14).

III

Finally, we would like to demonstrate that there were important links that connected the world of musical instruments (in this instance, the piano) with Helmholtz's *Tonempfindungen*. The effect of Helmholtz's studies on acoustics rippled through the piano workshops of Western Europe as well as America. Significant and far-ranging researches, based mainly upon the determination of how partials influence the timbre of instruments, came to be of the utmost importance in the construction of the piano during the second half of the 19th century. "[...] Musical concerns brushed shoulders with a new dominant scientism in European thought [...]" (Plantinga 1990: 11-12).

As a versatile scientist, science organizer (*Wissenschaftsorganisator*) and enlisted member of the ceremonious political and social atmosphere of the Kaiserreich, it was natural for Helmholtz to be thoroughly *au courant* with the musical world. He was an accomplished musician, played the piano, studied music theory in considerable depth, and moved with his family in the most elite musical circles. He was thoroughly acquainted with musical literature and exhibited a special fondness for Bach, Beethoven, Mendelssohn and Wagner. He owned a collection of old instruments that today is part of the Neupert Collection at the Germanisches Nationalmuseum in Nürnberg.[7] Theodore Steinway, whom we shall introduce shortly, also owned an instrument collection of which the "[...] old musical instruments of every kind (except the pianofortes)" are now in the Städtisches Museum in Braunschweig.[8] The pianos, it

7 Personal communication, 3 February 1994, from Sabine Klaus, Germanisches Nationalmuseum, Nürnberg.

8 Personal communication, 29 December 1993, from Henry Z. Steinway. Information

has been said, went to a museum in Hamburg (for which, to date, there is no documentation).

While the piano factories of Blüthner, Bechstein, Bösendorfer, and Chickering all were alert to the expanding scientific world surrounding them, and experimented and incorporated new findings into the construction of their pianos, first place must be given here to the house of Steinway as a family enterprise. Henry Engelhardt Steinway moved from Germany to New York with his family in 1850, leaving his eldest son, Theodore, behind in Braunschweig until 1865 when he joined them in America.

Together with his sons, Henry gradually built up a piano firm from a small workshop to a mammoth factory, staking out a large area in New York (Astoria) where the Steinways created schools and a community for their workers. As a skilled craftsman, Henry based his expertise more on empirical judgment than on scientific pursuits. His son William was delegated to provide publicity and to run the business and manufacturing operations.

Theodore, however, is recognized as the leading force behind changes that were made in the construction of the piano during the second half of the 19th century. He was a student of physics and mathematics and for many years managed the industry in regard to scientific and technological ventures. Henry Z. Steinway, one of the remaining members of the Steinway house, in a recent communication to us observed: "Theodore was the designer - engineer - inventor who built the Steinway piano as it sits today [...]. It was Theodore who knew Helmholtz and consulted with him".[9] This association moreover is confirmed in a letter from the Steinway firm to Helmholtz in 1871:

> "Unser Herr William Steinway wird wahrscheinlich im Monat Mai in Berlin eintreffen und sich glücklich schätzen mit Ihnen Bekanntschaft zu machen.
> Unter den besten Grüssen unseres Herrn Theodor Steinway, der sich noch immer mit grossen Vergnügungen der anregenden Unterhaltungen erinnert, welche derselbe im vorigen Jahr mit Ihnen hatte [...]".[10]

Although Theodore followed his father to New York, his heart remained in Europe, and so he travelled back and forth between Germany and New York, maintaining contact with new developments on the Continent. Theodore's mind, it is said, was constantly devising new experiments for improving the

---

from Theodore Steinway's will.

9 Personal communication from Henry Z. Steinway, 16 November 1993.

10 One autograph letter, dated 11 March 1871, from the Steinways to Helmholtz exists in Berlin: Archiv der Berlin-brandenburgischen Akademie der Wissenschaften. Nachlaß Helmholtz Nr. 448. Thanks to Petra Werner for making a photo copy available to us.

sound of the piano according to the aesthetic conceptions that he perceived to prevail at the time. This resulted in an effort to construct a piano with a homogenized sound throughout the registers.

Theodore Steinway lent a hand to Helmholtz in some of his researches for the *Tonempfindungen* and, according to Fanny Morris Smith, provided pianos for his experiments on strings.[11] Theodore utilized Helmholtz's idea (although it had been attempted earlier) that the part of the string not set in vibration by the hammer could be used to enrich the vibrating string with its harmonics. He thus developed, as a "Duplex Scale", the non-struck back portion of the string so that it would vibrate, according to calculated values, in sympathy with the struck length of the string, thereby adding fullness to the sound.

Theodore Steinway learned from Helmholtz about the striking point of the hammers and the placement of the bridge (brought in from the edge with more curvature) which helped to unify and enrich the sound (see also Winter 1988: 271/2, 288/9). In addition, there was the problem of the composition, firmness and geometry of the hammer. Previous hammers, Helmholtz suggested, had been too hard, and accordingly had generated an overabundance of overtones that covered up the fundamental, and thus produced a too-brilliant and harsh tone, especially in the upper registers. Furthermore, the thickness, stiffness and elasticity of the strings were a concern of Helmholtz. In his chapter on "Unterschiede der musikalischen Klangfarbe" Helmholtz enters into considerable detail on such issues as 1. The nature of the stroke; 2. The place struck; 3. The density, rigidity, and elasticity of the string (Helmholtz/Ellis Eng. ed. from 3rd German ed. 1875: 74-80).

Steinway was a leader in developing a uniform sound, standardizing the striking point, moving the bridge, and fashioning the hammer - all in favor of a fuller, richer sound.[12] Piano music reflected these changes in the late 19th century. We cannot here go into these matters more explicitly; the topic remains for an expanded study.

---

11 Smith 1892: 112. Smith interviewed Henry Ziegler, a nephew of Theodore, who carried out the latter's ideas in the construction of pianos. She further conferred with Steinway workmen, friends and family. Preface p. XI. See also Good 1982: 196/7.

12 These changes in producing a more homogenized sound on the piano of course can be viewed as a disadvantage when interpreting piano music of the late 18th and 19th centures (at least until the last decades). Composers at that time wrote for pianos that exhibited various colors from register to register. The differences affected balance, form and structure, factors rarely considered by pianists today.

Five letters from Helmholtz to Steinway (three autographs and two published in Steinway catalogs) are extant;[13] they are informative from a technical point of view. In the letter of 9 June 1871 Helmholtz admired the refined *sostenuto* pedal in the Steinway especially as it "helped him to play Bach." In other communications Helmholtz also commended the more unified sound in the registers, due, in part, to the bridge that was moved toward the middle of the sounding board. Over the years Theodore supplied Helmholtz with pianos for pleasure and experimentation and he was grateful for this. The last letter was written after Helmholtz's visit to New York in 1893. It was an acknowledgement of the piano that the Steinway house generously had presented to him. There is one autograph letter, dated 11 March 1871, from the Steinways to Helmholtz (see Footnote 10). In it greetings are sent and appreciation is spelled out for Helmholtz's consultation and association with Theodore and William.

IV

Helmholtz's *Tonempfindungen* demonstrates his ability successfully to wander outside of the more rigorously demonstrable areas of the sciences into a world - the world of music - where a great host of subjective and complex phenomena could be integrated but within rather stringent limits. These were limits of which Helmholtz was painfully aware. It was a domain where history and culture were dominant, where a wide span of musical styles reigned, where strong views on changing aesthetics were evident and where divergent theories of harmony and harmonic practice existed. Aesthetic principles reigned - it was a domain that was unique as a Wissenschaft *viz.* a Geisteswissenschaft. On this treacherous terrain Helmholtz was able to demonstrate, from what others already had said, and what he added from his own research in musical acoustics, that music was amenable, analyzable, and illuminated with the help of: mathematics, the mechanics of vibrating sonorous bodies (strings, bars, organ

13 Personal communication from Henry Z. Steinway, 16 November 1993. The autographs dated 9 June 1871 and 6 October 1893 have been on loan to the La Guardia College Archive since December 1992. The autograph letter dated 4 December 1893 is located at the Steinway Archive. Letters dated 13 August 1873 and 16 March 1885 are published in Steinway catalogs; autographs are missing.

pipes, columns of air), anatomy, aural physiology (sensations), psycho-acoustic concepts, etc.

Immediately following the publication of *Tonempfindungen* and for several decades thereafter some critics judged that Helmholtz, although influential and versatile as a scientist, had not exhibited in his investigations into musical acoustics the level of aesthetic finesse and musical insight that was necessary for the task that he had set himself. This evaluation, we believe, is totally erroneous and could only derive from an inadequate reading of *Tonempfindungen*. Helmholtz, who knew the discipline of physics exactly from the inside, and who recognized where the boundaries between science and other disciplines rested, also was the person who was unusually careful not to extend science into areas of music where the science either did not apply or did not provide productive insight. Dahlhaus reminds us:

> "[...] among nineteenth century music theorists, Hermann von Helmholtz, a distinguished physicist and physiologist, knew the rules of rigorous scientific explanation far better than his theorizing musical colleagues, who naively turned to physics in order to create a basis for music theory that would enable it, too, to partake of the dignity and inviolability of the natural sciences. In 1863, a half century before composers were to abandon tonality, Helmholtz in his *Lehre von den Tonempfindungen* boldly proposed that 'the system of scales, keys, and their connecting harmonies does not rest on immutable natural laws, but is the consequence of aesthetic principles which have been, and will continue to be, subject to change during the continuing march of human evolution'" (Dahlhaus 1989: 322-323).

In *Tonempfindungen* Helmholtz emphasizes the fact that the inherent complexity of music and musical experience has given music theorists and analysts a wide canvas on which to construct schemes that purport to explain why, how, when and in what cultural contexts music has flourished. Fertile and insightful as such attempts had been, he nevertheless suggests that they had not and probably never will lend themselves to the degree of simplicity and universality that characterizes investigations in the natural sciences. Music seen as aesthetics is not readily reducible to principle by inductive science or deductive logic.

Let us give Helmholtz the last word about how, in his opinion, the aesthetics of music and the physiological foundations of music, each in their own ways, have contributed to the development of music theory. He ends chapter 19, "Beziehungen zur Ästhetic", as follows:

"Ich schliesse hiermit meine Arbeit. Soviel ich übersehe, habe ich sie so weit fortgeführt, als die physiologischen Eigentümlichkeiten der Gehörempfindung einen direkten Einfluss auf die Konstruktion des musikalischen Systemes ausüben, so weit, als die Arbeit hauptsächlich einem Naturforscher zufallen mußte. Denn wenn sich auch naturwissenschaftliche Fragen mit ästhetischen mischten, so waren die letzteren doch von verhältnismässig einfacher Art, die ersteren jedenfalls viel verwickelter. Dies Verhältnis muss sich notwendig umkehren, wenn man versuchen wollte, in der Ästhetik der Musik weiter vorzuschreiten, wenn man zur Lehre vom Rhythmus, von den Kompositionsformen, von den Mitteln des musikalischen Ausdruckes übergehen wollte. In allen diesen Gebieten werden die Eigentümlichkeiten der sinnlichen Empfindung noch hin und wieder einen Einfluss haben, aber doch wohl nur in sehr untergeordneter Weise. Die eigentliche Schwierigkeit wird in der Verwickelung der psychischen Motive liegen, die sich hier geltend machen. Freilich beginnt auch hier erst der interessantere Teil der musikalischen Ästhetik - handelt es sich doch darum, schliesslich die Wunder der grossen Kunstwerke zu erklären, die Äusserungen und Bewegungen der verschiedenen Seelenstimmungen kennen zu lernen. So lockend aber auch das Ziel sein möge, ziehe ich es doch vor, diese Untersuchungen, in denen ich mich zu sehr als Dilettant fühlen würde, anderen zu überlassen, und selbst auf dem Boden der Naturforschung, an den ich gewöhnt bin, stehen zu bleiben" (Helmholtz 1st ed. 1863: 358; 6th ed. 1913: 599).

Bibliography

Baron, Carol K.: George Ives's Essay in Music Theory: An Introduction and Annotated Edition. *American Music*, 10, 1992, 239-288.

Beckerman, Michael: Janácek and the Herbartians. *The Musical Quarterly*, 69, 1983, 388-407.

Beckerman, Michael: *Janácek as Theorist*. New York: Pendragon, 1994.

Boatwright, Howard: Ives' Quarter-Tone Impressions. *Perspectives of New Music*, Spring/Summer, 1965, 22-31.

Broadhouse, John: *Musical Acoustics; or, The Phenomena of Sound as Connected with Music*. London: William Reeves, 1881.

Brod, Max: *Leos Janácek: Life and Work*. Wien/Zürich/London: Universal Edition, 1956.

Dahlhaus, Carl: Hermann von Helmholtz und der Wissenschaftscharakter der Musiktheorie. In: Frieder Zaminer (Hg.): *Über Musiktheorie:* Referate der Arbeitstagung 1970 in Berlin. Köln: Arno Volk Verlag, 1970, 49-58.

Dahlhaus, Carl: *Schoenberg and the New Music: Essays by Carl Dahlhaus*. Translated [from German] by Derrick Puffett and Alfred Clayton. Cambridge/New York/New Rochelle/Melbourne/Sydney: Cambridge University Press, 1987.

Dahlhaus, Carl: *Nineteenth-Century Music*. English translation by J. Bradford Robinson from the German *Die Musik des 19. Jahrhunderts* (Neues Handbuch der Musikwissenschaft, Vol. 6. Wiesbaden: 1980). Berkeley, Los Angeles: University of California Press, 1989.

Donington, Robert: *The Instruments of Music*. London: Methuen, 1949/1951.

Good, Edwin M.: *Giraffes, Black Dragons, and other Pianos: A Technological History from Cristofori to the Modern Concert Grand*. Stanford, California: Stanford University Press, 1982.

Hauptmann, Moritz: Brief über Helmholtz's Lehre von den Tonempfindungen. *Leipziger Allgemeine Musikalische Zeitung*, Neue Folge I, 1863, 669-673.

Hauptmann, Moritz: *Die Natur der Harmonik und der Metrik: Zur Theorie der Musik*. Leipzig: Breitkopf und Härtel, 1853.

Helmholtz, Hermann von: *Die Lehre von den Tonempfindungen als Physiologische Grundlage für die Theorie der Musik*. Braunschweig: Friedrich Vieweg und Sohn, 1863. English transl. of 3rd ed. (1870) by Alexander J. Ellis, *On the Sensation of Tone as a Physiological Basis for the Theory of Music*. London: Longmans, Greer, and Co., 1875. Reprint of the 6th German ed. (1913). Hildesheim: Georg Olms Verlagsbuchhandlung, 1968.

Helmholtz, Hermann von: On the Physiological Causes of Harmony in Music. Transl. by A.J. Ellis from *Populäre wissenschaftliche Vorträge*, 2 vols. 1865 and 1871. In: *Popular Lectures on Scientific Subjects*. New York: Appleton, 1873.

Helmholtz, Hermann von: *Vorträge und Reden*. 3rd ed. of *Populäre Wissenschaftliche Vorträge*. Braunschweig: Friedrich Vieweg und Sohn, 1884.

Helmholtz, Hermann von: *Wissenschaftliche Abhandlungen*. Leipzig: J.A. Barth, 3 vols.

Jorgenson, Dale A.: *Moritz Hauptmann of Leipzig*. Lewiston/Queenston: The Edwin Mellen Press, 1986.

Koenigsberger, Leo: *Hermann von Helmholtz*. 3 vols. Braunschweig: Friedrich Vieweg und Sohn, 1902.

Lloyd, Llewelyn S.: Helmholtz and the Musical Ear. *The Musical Quarterly*, 25, 1939, 167-175.

McKendrick, John Gray: *Hermann Ludwig Ferdinand von Helmholtz*. London: T. Fisher Unwin, 1899.

Oskar, Paul: *Geschichte des Klaviers*. Leipzig: A.H. Payne, 1868. Reprint Leipzig: Zentralantiquariat der DDR, 1986.

Plantinga, Leon: The Piano and the Nineteenth Century. In: R. Larry Todd (ed.): *Nineteenth-Century Piano Music*. New York: Schirmer Books (MacMillan, Inc.), 1990, 1-15.

Rimsky-Korsakov, Nicolas: *Principles of Orchestration*. Maximilian Steinberg (ed.), English Transl. by Edward Agate. Berlin/Moscou/Leipzig/New York: Edition Russe de Musique, London: The Russian Music Agency, 1913.

Rummenhöller, Peter: *Moritz Hauptmann als Theoretiker*: Eine Studie zum erkenntniskritischen Theoriebegriff in der Musik. Wiesbaden: Breitkopf & Härtel, 1963.

Shirlaw, Matthew: *The Theory of Harmony: An Inquiry into the Natural Principles of Harmony (...)* London: Novello & Co., Ltd., 1913.

Smith, Fanny Morris: *A Noble Art: Three Lectures on the Evolution and Construction of the Piano*. New York: The De Vinne Press, 1892.

Winter, Robert: Striking It Rich: The Significance of Striking Points in the Evolution of the Romantic Piano. *The Journal of Musicology*, 6, 1988, 267-292.

VI
Helmholtz in Wissenschaft, Politik und Geschichte

Helmholtz' Vortragskunst und sein Verhältnis zur populären Wissensvermittlung

Horst Kant

In seiner akademischen Festrede anläßlich der Übernahme des Prorektorats an der Universität Heidelberg im Jahre 1862, in der er ausführliche Gedanken "Über das Verhältnis der Naturwissenschaften zur Gesamtheit der Wissenschaften" darstellte, bemerkte Hermann Helmholtz (1821-1894) u.a.:

> "Das Wissen allein ist aber nicht der Zweck des Menschen auf der Erde. Obgleich die Wissenschaften die feinsten Kräfte des menschlichen Geistes erwecken und ausbilden, so wird doch derjenige keine rechte Ausfüllung seines Daseins auf Erden finden, welcher nur studiert, um zu wissen. [...] Nur das Handeln gibt dem Manne ein würdiges Dasein; also entweder die praktische Anwendung des Gewußten oder die Vermehrung der Wissenschaft selbst muß sein Ziel sein. Denn auch das letztere ist ein Handeln für den Fortschritt der Menschheit" (Helmholtz 1971a: 102).

Waren Helmholtz' Interessen - ob nun in seiner Zeit als Mediziner/Physiologe oder in der als Physiker[1] - vornehmlich auf Einsicht in die theoretischen Zusammenhänge gerichtet, so trachtete er doch stets danach, seine Erkenntnisse auch in praktische Nutzanwendungen umzusetzen, zumindest den Weg dahin

---

1 Auch wenn diese gängige Zweiteilung des Helmholtzschen Wirkens - bis 1870 Mediziner/Physiologe, ab 1871 Physiker - sehr grob ist (und eigentlich die physikalische Forschung seinen gesamten Lebensbereich durchzieht - vgl. Kant, 1994b), soll sie hier doch weitgehend zugrunde gelegt sein, wobei insbesondere der Physiker Helmholtz im Blick ist.

aufzuzeigen. Nicht zuletzt beruhte - als ein herausragendes Beispiel - sein Engagement für die Physikalisch-Technische Reichsanstalt, das er praktisch seit seiner Berufung nach Berlin entwickelte, auf dieser Einsicht.[2]

Andererseits war Helmholtz -neben seinen Vorlesungen und Fachvorträgen - auch bestrebt, in breiter angelegten Vorträgen das neueste Wissen einem breiteren Publikum zu vermitteln. Wie sich diese populäre Wissensvermittlung bei Helmholtz darstellte, soll im folgenden näher betrachtet werden.

In der Vorrede zur von ihm herausgegebenen deutschen Übersetzung der *Fragments of Science* des englischen Naturforschers John Tyndall (1820-1893) schrieb Helmholtz im Jahre 1874:

> "Das auch in Deutschlands gebildeteren Kreisen erwachende und sich immer lebhafter äußernde Verlangen nach naturwissenschaftlicher Belehrung halte ich nicht bloß für ein Haschen nach einer neuen Art von Unterhaltung oder für leere und fruchtlose Neugier, sondern für ein wohlberechtigtes geistiges Bedürfnis, welches mit den wichtigsten Triebfedern der gegenwärtigen geistigen Entwicklungsvorgänge eng zusammenhängt. Nicht dadurch allein, daß sie gewaltige Naturkräfte den Zwecken des Menschen unterworfen und uns eine Fülle neuer Hilfsmittel zu Gebote gestellt haben, sind die Naturwissenschaften von dem allererheblichsten Einfluß auf die Gestaltung des gesellschaftlichen, industriellen und politischen Lebens der zivilisierten Nationen geworden; und doch wäre schon diese Art ihrer Wirkungen wichtig genug, daß der Staatsmann, Historiker und Philosoph ebensogut wie der Techniker und Kaufmann wenigstens an den praktisch gewordenen Ergebnissen derselben nicht teilnahmslos vorübergehen kann. Viel tiefergehend noch und weiter tragend, wenn auch viel langsamer sich entfaltend, ist eine andere Seite ihrer Wirkungen, nämlich ihr Einfluß auf die Richtung des geistigen Fortschreitens der Menschheit [...]" (Helmholtz 1971b: 365).

Diese Vorrede erschien später in Helmholtz' *Vorträge und Reden* unter dem Titel "Über das Streben nach Popularisierung der Wissenschaft" und ist das ausführlichste Bekenntnis von Helmholtz zur Popularisierung der Wissenschaft.

Seit er wissenschaftlich tätig war, hatte Helmholtz auch die Notwendigkeit der Popularisierung von Wissenschaft erkannt - wobei hier bewußt von "Popularisierung" und nicht von "Populärwissenschaft" gesprochen wird. Ohne letztere gering zu achten, sah Helmholtz darin nicht seine Aufgabe. Für nur oberflächlich Gebildete oder für Leute, die lediglich Erbauung suchten, waren

---

2 So war er - seit ihrer Gründung Präsident der PTR - bereits maßgeblich an der Erstellung der sogenannten Schellbach-Denkschrift im Juli 1872 beteiligt. Vgl. u.a. Buchheim 1986; Kant/Hoffmann 1987; Cahan 1992: 74.

seine Vorträge nicht gedacht. So schrieb er 1884 in der Vorrede zum ersten Band seiner *Vorträge und Reden*, daß allen Vorträgen

> "gemeinsam bleibt, dass sie Versuche sind, die Ergebnisse mathematischer, naturwissenschaftlicher oder speciell physikalischer Forschung einem Kreise von Zuhörern und Lesern mitzutheilen, deren Studien nicht gerade in dieser besonderen Richtung gelegen haben" (Helmholtz 1903: VII).

Den höher gebildeten Leser bzw. Hörer setzte er also voraus! Hier liegt allerdings auch der Widerstreit populärer Wissensdarstellung bis heute. Der Astronom Wilhelm Meyer (1853-1910), 1888 einer der Gründer der Berliner Urania-Gesellschaft, schrieb rückblickend:

> "[...] Ein populärwissenschaftlicher Vortrag? Nein, der ist immer noch zu lehrhaft; er will nur zum Verstande sprechen; die sogenannten populären Vorträge waren in gerechten Mißkredit beim Publikum gekommen. Der Drang zur Belehrung auch in dieser Form war zwar sehr groß gewesen, aber es hatten sich in Deutschland meist Unberufene daran gemacht, ihn zu befriedigen. Diese Unberufenen setzten sich aus zwei diametral gegenüberstehenden Klassen zusammen: die zu wenig Wissenden und die zu viel Wissenden. [...] Die zu viel Wissenden waren unsere gelehrten deutschen Professoren, die meinen, ein Vortrag sei allgemeinverständlich, wenn er nur keine Fachausdrücke enthält. Es gibt in der ganzen Welt keinen schlechteren Vortragenden als den deutschen Gelehrten, und das hat seinen Grund in der vorzüglichsten Eigenschaft desselben, der unbedingten Gründlichkeit. Gründlichkeit und Langeweile aber sind zwei nahe Verwandte. Von einer gelehrten Abhandlung, die die Wissenschaft fördern soll, verlangt man ja nicht, daß sie auch interessant zu lesen sei, obgleich dies auch nichts schaden kann [...]" (Meyer 1908: 74f).

Zu ergänzen wäre, daß die meisten deutschen Gelehrten damals sowieso nicht allzuviel von Popularisierungsbemühungen hielten, und so resümierte Meyer mit Blick auf die ersten Jahre nach der Urania-Gründung: "[...] Nur die gestrengen Herren Gelehrten kamen nicht, mit Ausnahme einiger universeller Köpfe wie Helmholtz und du Bois-Reymond. Der praktische Siemens gehörte natürlich zu unseren hervorragendsten Aktionären [...]" (Meyer 1908: 83).[3]

Bestätigte Helmholtz einerseits in der genannten Vorrede zu Tyndalls *Fragments of Science* indirekt die Meyersche Kritik an den diesbezüglichen deutschen Verhältnissen, wenn er schrieb: "In England besteht die Sitte der populären naturwissenschaftlichen Vorlesungen seit viel längerer Zeit als in Deutschland. [...] [Das dortige - H.K.] Verhältnis hat aber auch andererseits eine engere Berührung der Arbeiter für die Wissenschaft mit allen anderen Krei-

3 Aus der Erkenntnis, daß die Tätigkeit der Gesellschaft Urania der Weiterbildung seiner Facharbeiterschaft dienlich sein könne, unterstützte Siemens deren Arbeit auch finanziell.

sen ihres Volkes unterhalten und dazu getrieben, für die Möglichkeit des Unterrichts der nicht regelrecht vorgebildeten Schüler ausgiebiger zu sorgen" (Helmholtz 1971b: 371f), - so war er doch andererseits kein Anhänger einer "allzu populären" Darstellung und schrieb beispielsweise bereits 1864 nach seiner Rückkehr aus London an den Freund Emil du Bois-Reymond (1818-1896):

> "Was die populären Vorlesungen in der Royal Institution betrifft, so muß ich Dir aber in der Tat recht geben, daß man sich immer mehr wird besinnen müssen, ehe man dergleichen wieder übernimmt. Ich habe durchaus keine Ursache, mit dem äußeren Erfolg der meinigen unzufrieden zu sein, da ich fortlaufend ein Publikum von 300 Leuten und darunter eine Menge wissenschaftlich gebildeter Männer hatte. Aber die Konkurrenz der populären Vorlesungen in London ist so groß, daß sie im Begriff sind, in die allergemeinste Effekthascherei herabzusinken. Die englischen Physiker beschuldigen namentlich Tyndall, daß er in dieser Beziehung den schlechten Neigungen des Publikums zu viel nachgebe, und die R[oyal] Institution auf das Niveau der Polytechnic[al] Inst[itution] bringe, [...]" (Dokumente 1986: 209).

Und bei anderer Gelegenheit schrieb er dem Freunde:

> "[...] Es ist unmöglich bei solchen Gelegenheiten, es allen recht zu machen, jedenfalls aber wohl dankbarer, es den Zuhörern nicht zu leicht zu machen und für den großen Haufen einige Rätsel stehen zu lassen, deren Verständnis vielleicht nur einer kleinen Zahl der Zuhörerschaft aufgeht. Für jene anderen ist es im Grunde immer besser, ihre Verwunderung als ihr Verständnis anzuregen" (Dokumente 1986: 112).

Zu dieser Auffassung von Popularisierung paßt auch, daß Helmholtz den Unternehmungen der Berliner Gesellschaft Urania zwar positiv gegenüberstand, sich aber nicht entschließen konnte, für die von dieser herausgegebenen populärwissenschaftlichen Zeitschrift einen Beitrag zu verfassen, wenngleich er nicht grundsätzlich ausschließen wollte, "[...] daß ich schließlich einmal, wenn sich eine bisher noch nicht abzusehende Gelegenheit findet, [...] einen Aufsatz dazu liefere. Aber zur Zeit sehe ich wirklich noch nicht, woher für mich Zeit und ein Thema kommen sollten [...]".[4]

Ob Helmholtz zu geeigneten Vortragenden im Meyerschen Sinne zu zählen war, darf dabei sicher angezweifelt werden. Man darf sich wohl nicht von seinen brillanten schriftlichen Darstellungen der "Populären Vorträge" leiten lassen, von denen Max von Laue (1879-1960) einmal etwas euphorisch sagte:

---

4 Brief von H. v. Helmholtz an W. Meyer (Gesellschaft Urania) vom 22. Mai 1888. Staatsbibliothek Preußischer Kulturbesitz, Handschriftenabteilung, Sammlung Darmstaedter, Helmholtz-Mappe Blatt 209.

> "Der Form nach sind alle diese Vorträge von einer Vollendung, daß man sie den Meisterwerken der deutschen Literatur an die Seite setzen darf" (Laue 1944: 207).

Offenbar ist Helmholtz jedoch die schriftliche Ausarbeitung nicht leicht von der Hand gegangen und erforderte mehrfache Umarbeitung.[5] Sicher spiegelt sich hierin nicht nur sein hoher Anspruch wider, sondern auch die Tatsache, daß er offenbar kein besonders guter Redner war, wenngleich es dazu auch unterschiedliche Auffassungen gab, wohl je nachdem, mit welchen eigenen Voraussetzungen und Erwartungen man in seine Vorlesungen und Vorträge ging.[6] Es dürfte aber nicht aus der Luft gegriffen sein, was ein ungenannter Autor 1870 im Rahmen der Berufungsdiskussion um den durch den Tod von Gustav Heinrich Magnus (1802-1870) freigewordenen Berliner Physiklehrstuhl in der Nationalzeitung schrieb:

> "Helmholtz besitzt nicht, was man einen glänzenden Vortrag nennt, und [wir glauben], daß er in dieser Beziehung, sowie in der Kunst, die Mittelmäßigkeit anzuregen, von Vielen übertroffen wird. [...] Es gibt für Helmholtz in Preußen nur eine einzige Stellung, die eines Mitgliedes und Pensionärs der Akademie" (Dokumente 1986: 305).

Du Bois-Reymond - zu dieser Zeit Rektor der Berliner Universität - hatte diesen "zwar äußerst schmeichelhaften, im Grunde aber perfiden Artikel über Dich" (Dokumente 1986: 238) in seinem Brief an Helmholtz nach Heidelberg erwähnt, in dem er ihm über den Stand der Berufungsüberlegungen berichtete und die Gründe nannte, die das Ministerium im Augenblick noch gegen Helmholtz ins Feld führte.[7]

Die Tatsache seines für die Mehrheit der Zuhörer wenig begeisternd wirkenden Vortrages wurde in späteren Zeiten meist dadurch umschrieben, daß man betonte, daß sein Nachfolger im Berliner Ordinariat, August Kundt

---

5 "Die schriftliche Ausarbeitung [...] ist ja meist ein mühsames Werk; mir war sie es wenigstens in hohem Grade. Ich habe viele Teile meiner Abhandlungen vier bis sechsmal umgeschrieben [...]." Zudem fielen ihm beim Korrekturlesen Punkte auf, die er "[...] hätte besser oder vollständiger machen können" (Helmholtz 1966: 30).

6 Beispielsweise meint Reiner bezüglich seiner Heidelberger Physiologie-Vorlesungen: "Seine Vorlesungen zeichneten sich durch eine besonders klare und interessante Darstellung aus [...]" (Reiner 1905: 19); ähnlich äußerte sich dazu beispielsweise Scheremetevskij (zit. bei Lebedinskij et al. 1966: 104). - Wohlgemerkt ist hier und im folgenden von Helmholtz' Qualitäten als Vortragender die Rede und nicht von denen als Lehrer im umfassenderen Sinne (wobei anzumerken ist, daß er - sowohl in der Physiologie als auch in der Physik - zahlreiche Schüler hatte, ohne aber eine wissenschaftliche Schule im eigentlichen Sinne auszubilden).

7 Vgl. dazu auch Koenigsberger 1903 Bd. II: 179f.

(1839-1894), ebenso wie der auf Helmholtz' Betreiben für Berlin gewonnene und mit dem neugeschaffenen Lehrstuhl für Theoretische Physik betraute Gustav Robert Kirchhoff (1824-1887) glänzende Lehrer gewesen seien (vgl. u.a. Rubens 1910: 587f).

In der Gedächtnisrede, die Theodor Engelmann (1843-1909) an der Universität Utrecht hielt, ist dies dann so umschrieben:

> "Damit hingen auch die Eigenthümlichkeiten zusammen, welche ihn als Lehrer, im Colleg, kennzeichneten. Da er die Form des Vortrags nie im Einzelnen ausgearbeitet hatte, sondern immer frei producierte, sprach er langsam, abgemessen, gelegentlich ein wenig stockend. Seine Augen waren dabei über die Zuhörer hinweg gerichtet, wie in unendlicher Ferne die Lösung eines Problems suchend. Kein lebendiger Rapport bestand zwischen ihm und dem einzelnen Zuhörer. Er sprach wie zu einem abstracten Auditorium, durchaus unpersönlich. [...] Nur die Sache sollte wirken. [...] Wenn man ihm vorgeworfen hat, dass er zu hoch gewesen und darum immer nur von einer kleinen Zahl verstanden worden sei, so muss ich dies bestimmt bestreiten [...]" (Engelmann 1894: 32)[8].

Wilhelm Ostwald (1853-1932) berichtete:

> "Ich hörte Helmholtz einige Male in der gewöhnlichen Vorlesung für Mediziner und andere Anfänger [gemeint ist die Experimentalphysik-Vorlesung - H.K.] und kann nur sagen, daß das geringe Vergnügen, welches ihm diese Arbeit machte, aus dem Ton seiner Stimme, wie aus jeder seiner müden und gleichgültigen Bewegungen zutage trat. Dabei war inhaltlich der Vortrag so streng und klar, daß er ohne jede Änderung in ein Lehrbuch hätte aufgenommen werden können" (Ostwald 1919: 291).

Emil Fischer (1852-1919) äußerte sich ähnlich:

> "Die Wirkung von Helmholtz' Vortrag auf die Studirenden ist nach dem Grade ihrer Begabung und Vorbildung recht verschieden gewesen, da er in Folge seiner eigenen Geisteslage vielleicht unbewusst an den Zuhörer höhere Ansprüche stellte, als es in elementaren Vorlesungen üblich ist" (Fischer 1894: 2650f).

Eilhard Wiedemann (1852-1928) schließlich - um die Auswahl von Zeugnissen damit abzuschließen - gab folgende Einschätzung:

> "Der direkte Einfluss des Vortrages von Helmholtz auf die grosse Zahl der Studierenden ist ein beschränkter gewesen; sein Vortrag ist für sie zu hoch; er sieht die Schwierigkeiten nicht, die ihnen ein Problem fast unlösbar erscheinen lassen, weil er sie spielend überwindet. Für Fortgeschrittenere waren aber stets seine Vorlesungen von grosser Anregung. Da, wo er sich Zeit gelassen hat, und lassen konnte, Vorträge oder Werke vollkommen auszuarbeiten, da entstehen Meisterwerke der Darstellung und des Stiles, Werke, in denen Form und Inhalt einander

8 Man bedenke bei diesem und nachfolgenden Zitaten, daß in Nachrufen u. dgl. zumeist wohlwollend formuliert wird.

> vollkommen entsprechen, so in seinen öffentlichen Vorlesungen [...]" (Wiedemann 1893: 59).

Helmholtz hat seine Vortragsschwäche wohl auch selbst gesehen. So schrieb er Anfang 1873 im Zusammenhang mit der Ablehnung einer Einladung nach Amerika: "Und was meine Vorlesungen betrifft, so habe ich mich doch überzeugt, daß ich wohl sachverständigen Leuten wissenschaftliche Dinge in trockener sachlicher Weise auseinandersetzen kann, aber ich habe nicht Herrschaft genug über die Sprache, um dasselbe so zu tun, daß ich ein größeres Auditorium von nicht fachmäßig Gebildeten fesseln könnte". Hinzu fügte er noch, wieviel schwerer es ihm dann in einer fremden Sprache falle (Koenigsberger Bd. II 1903: 219).

Werfen wir einen kurzen Blick auf die populären Vorträge und Schriften von Helmholtz. Umfaßt sein Veröffentlichungsverzeichnis insgesamt rund 220 Titel[9], so liegt die Zahl der veröffentlichten populärwissenschaftlichen Arbeiten darin - je nach Art der Einordnung bzw. Abgrenzung - etwa zwischen 25 und 35,[10] wobei diese zumeist mehrfach vorgetragen wurden. Es spricht für Helmholtz' Bemühen, das neueste Wissen wirklich populär zu vermitteln, daß er diese Vorträge zumeist vorher in der Familie gewissermaßen "testete" - sowohl seine erste Frau Olga von Velten (1826-1859) als auch seine zweite Frau Anna von Mohl (1834-1899) wurden von ihm umfassend einbezogen. So vermerkte Helmholtz' Schwägerin in späterer Erinnerung:

> "Seine Frau [Olga - H.K.] wurde ihm in der That alles was er von ihr erhofft und vorausgesetzt hatte. Sein treusorgendes Weib und seine ebenbürtige Gefährtin. Sie arbeitete und schrieb für ihn - er las ihr seine Vorträge, die für die Oeffentlichkeit bestimmt waren, ehe er sie hielt, um an ihrem Verständniss das allgemeine gebildeter Menschen zu messen" (Johannes 1990: 297).

Seine zweite Frau Anna schrieb in einem Brief an ihren Onkel Julius am 1. Januar 1867:

> "Diese Stille hat mir Muße gegeben, in Gemeinschaft mit Professor Wiedemann, dem Physiker am Karlsruher Polytechnikum, oder vielmehr mit seiner Frau, Tyndalls Buch "Heat considered as a mode of motion" zu übersetzen. Natürlich trage ich mein Opus immer meinem Manne vor, der die letzte Hand anlegt. Nun wird bereits lustig daran gedruckt, was umso befriedigender für unser Übersetzerge-

---

9 Vgl. Auswahlbibliographie Hermann von Helmholtz, hrsg. von D. Hoffmann, H. Kant, H. Reddner. Forschungsschwerpunkt Wissenschaftsgeschichte und -theorie, Preprint Nr. 1, Berlin 1993.

10 In die *Vorträge und Reden*, wie die *Populären wissenschaftlichen Vorträge* ab der dritten Auflage (1884) hießen, wurden schließlich in 2 Bänden (vgl. 4. Auflage 1896) 26 Artikel nebst mehrerer Anhänge und Zusätze aufgenommen.

> wissen ist, als eine Übersetzung desselben Buches im Auftrage von Autor und Verleger gemacht worden war, und zwar von einem Physiker vom Fach, wegen gänzlicher Unbrauchbarkeit aber wieder eingestampft wurde. Ohne die Aufsicht meines Mannes hätte ich es natürlich nicht machen können, denn der Gegenstand ist mir nicht aller Orten verständlich; es fehlt mir dazu bis zu den Elementarbegriffen dieser Dinge. So aber war es eine sehr nützliche und lehrreiche Arbeit, die mir nach überstandener Anstrengung auch viele Freude macht. Auch unsere Tochter Käte hat daran gearbeitet und der Fratello Ottmar, dem der Gegenstand zwar äußerst langweilig war, half uns mit schwungvollen Worten von der Stelle, so daß das Ganze etwas den Charakter eines Familienstrickstrumpfes annahm" (Anna von Helmholtz 1929: 139).

Und Helmholtz bemerkte bezüglich seines Anteils an diesen Übersetzungen:

> "Ich habe meine Mitwirkung trotz großer Überhäufung mit anderen amtlichen und wissenschaftlichen Arbeiten nicht zurückgezogen, weil ich die Verbreitung gelungener populärer Darstellungen der wichtigeren und durchgebildeteren Teile der Naturwissenschaft für ein nützliches Werk halte" (Helmholtz 1971: 365).

Offenbar hat Helmholtz selbst nur Dinge in populären Vorträgen und Schriften behandelt, die dem Feld seiner eigenen wissenschaftlichen Betätigung entsprangen - das ja breit genug war. Das hat er auch sehr bewußt getan, denn er meinte, der Zuhörer wolle und solle nicht nur etwas berichtet bekommen, sondern in gewisser Weise an der Arbeit des Forschers teilhaben, auch etwas über das Wie dieser Arbeit erfahren. Dabei sollten in populären Darstellungen nur wohlgesicherte Tatsachen und fertige Ergebnisse der Forschung vorgestellt werden, keine unreifen Vermutungen, Hypothesen oder Träume. Der interessierte Leser oder Hörer sollte sich selbst ein Bild machen können von dem, was die Wissenschaft in Bezug auf die großen Fragen der Menschheitsentwicklung beitragen könne. In der Vorrede zu Tyndall[11] bemerkte er in diesem Sinne:

> "Mir scheint aber, daß nicht sowohl Kenntnisse der Ergebnisse naturwissenschaftlicher Forschungen an sich dasjenige ist, was die verständigsten und gebildetsten unter den Laien suchen, als vielmehr eine Anschauung von der geistigen Tätigkeit des Naturforschers, von der Eigentümlichkeit seines wissenschaftlichen Verfahrens, von den Zielen, denen er zustrebt, von den neuen Ansichten, welche seine Arbeit für die großen Rätselfragen der menschlichen Existenz bildet. Von diesem allem ist in den rein wissenschaftlichen Abhandlungen unseres Gebietes kaum je die Rede [...]" (Helmholtz 1971b: 370).

11 Nach seiner Englandreise 1864 hatte Helmholtz gegenüber du Bois-Reymond über Tyndall u.a. festgestellt, daß er "in der Tat ein ausgezeichnetes Talent für populäre Vorträge hat und im Publikum sehr anerkannt ist" (Dokumente 1986: 209).

Helmholtz erwartete von einem Popularisator durchaus ein gewisses künstlerisches Talent zur Darstellung, um auch abstrakte Gedankengänge leicht verständlich werden zu lassen, ohne dabei an Exaktheit zu verlieren. Er folgte dabei der Auffassung Tyndalls, die jener in seiner Vorrede zu dem Buch *Die Wärme* geäußert hatte, daß die Naturwissenschaft wesentlich stärker als Mittel der geistigen Erziehung dienen müsse, als dies bis dato der Fall gewesen sei (Tyndall 1867: Vorwort, XII).

Wie gesagt, wenn Helmholtz von Popularisierung sprach, dann hatte er einen relativ engen Kreis des Bildungsbürgertums im Blick; selten ging sein Trachten darüber hinaus. Dabei war ein wesentlicher Aspekt seines Bemühens auch die populäre Darstellung im Rahmen der Wissenschaftlergemeinschaft selbst, denn auch hier müssen ja die Erkenntnisse der einzelnen Disziplinen einem allgemeinwissenschaftlichen Verständnis zugänglich gemacht werden. Aber auch in der Lehre versuchte er, diesen Gedanken umzusetzen, und hielt in Heidelberg ab 1862 im Wintersemester regelmäßig einen Vorlesungszyklus unter dem Titel "Allgemeine Resultate der Naturwissenschaften, dargestellt für Studirende aller Fakultäten".[12]

Von seinen veröffentlichten populären Vorträgen fallen nur sieben in das Gebiet der Physik (im weiteren Sinne, einschließlich Meteorologie).[13] Hier sei abschließend am Beispiel seines Wirkens im Berliner Elektrotechnischen Verein und damit im Zusammenhang in der Physikalischen Gesellschaft sein Verständnis der populären Darstellung illustriert.

Seit ihrer Gründung im Jahre 1845 war Helmholtz aktives Mitglied der Berliner Physikalischen Gesellschaft und trug dort u.a. im Jahre 1847 seine berühmte und grundlegende Abhandlung *Über die Erhaltung der Kraft* vor. In diesem "mächtig anregenden Kreis von talentvollen jungen Naturforschern" (Siemens 1966: 46) lernte er auch Werner von Siemens (1816-1892) kennen[14],

12 Vgl. Vorlesungsverzeichnisse der Universität Heidelberg. War die Zeitangabe beim erstenmal "zweimal wöchentlich", so kündigte er die Vorlesung in den Wintersemestern 1863/64 und 1864/65 für Dienstag, Donnerstag und Samstag nachm. 5-6 Uhr, und von 1865/66 bis 1870/71 für Montag, Mittwoch und Freitag 5-6 Uhr an.

13 Es sind dies: Über die Wechselwirkung der Naturkräfte und die darauf bezüglichen neuesten Ermittlungen der Physik (1854); Über die Erhaltung der Kraft (1862); Eis und Gletscher (1865); Entstehung des Planetensystems (1871); Wirbelstürme und Gewitter (1875); Faradays Ideen über Elektrizität (1881); Elektrische Maßeinheiten (1881). Dazu könnte man auch noch die Nachrufe und Gedenkreden auf G. Magnus (1871), J. Fraunhofer (1887), R. Clausius (1889), G. Wiedemann (1893) und H. Hertz (1894) zählen.

14 Siemens hielt hier u.a. seine ersten Vorträge über Telegrafie.

mit dem ihn eine lebenslange Freundschaft verband; auch der Elektrotechniker Siemens hatte grundlegende seiner Arbeiten - vor allem die stärker wissenschaftlich orientierten - in dieser Gesellschaft erstmals vorgetragen.

Bereits auf den Beginn seines Wirkens in Berlin bezogen, vermerkte Siemens in den Lebenserinnerungen: "[...] Es wurde mir klar, daß technischer Fortschritt nur durch Verbreitung naturwissenschaftlicher Kenntnisse unter den Technikern erzielt werden könnte. Es herrschte damals noch zwischen Wissenschaft und Technik eine unüberbrückbare Kluft" (Siemens 1966: 47). Diese Kluft zu überwinden, war ein wesentlicher Teil seines Bestrebens. Und so schien ihm Ende der 70er Jahre die elektrische Industrie einen Stand erreicht zu haben, der es sinnvoll erscheinen ließ, ihre fachlichen Belange auch in einer eigenen Gesellschaft vertreten zu sehen. So schrieb er denn am 5. Februar 1879 an den damaligen Generalpostmeister Heinrich von Stephan (1831-1897): "Ew. Exzellenz wolle das Protektorat über einen das ganze Gebiet der Elektrotechnik umfassenden deutschen Verein übernehmen" (Siemens 1966: 302). Mit diesem Vorschlag prägte Siemens zugleich den Begriff "Elektrotechnik" für diesen sich neu herausbildenden Bereich von Wissenschaft und Technik, und auch er selbst sah in der Gründung des Vereins "die Geburt der Elektrotechnik als gesonderten Zweiges der Technik" (Siemens 1966: 302). Bemerkenswert ist hierbei, daß von vornherein die technische Anwendung der Elektrizität im Zusammenwirken von Wissenschaft, Gewerbe (sprich: Industrie) und staatlicher Verwaltung erkannt wurde.[15]

Helmholtz trat dem ETV - wie zahlreiche andere Berliner Physiker - bald nach dessen Gründung bei.[16] Bemerkenswert ist, daß von den 182 Mitgliedern, die die Berliner Physikalische Gesellschaft im Jahre 1880 zählte, 43 Herren (davon 17 Auswärtige) in jenem Jahr auch dem ETV beitraten.[17] Und be-

15 Wissenschaftliche und auch technische Gesellschaften waren ja aus gutem Grunde zumeist bestrebt, insbesondere den Einfluß des Staates herauszuhalten. Zum Elektrotechnischen Verein vgl. u.a. Kant 1994a.

16 In der ersten allgemeinen Versammlung des ETV Ende Januar 1880 konnten bereits 406 bestätigte Mitglieder gezählt werden (davon 119 aus Berlin), bis zum Ende des Jahres waren es 1504 (davon 328 Berliner); bis Mitte der 90er Jahre hielt sich die Mitgliederzahl in dieser Größenordnung. Siemens wurde zum ersten Vorsitzenden und Stephan zum Ehrenpräsidenten gewählt.

17 Noch im Jahre 1880 waren unter den Berliner Mitgliedern des ETV neben Helmholtz und Siemens u.a. folgende Physiker zu finden: W. Foerster (Prof. Berliner Sternwarte), E. Hagen (Ass. Physikal. Inst. Univ.), H. Kayser (Ass. Physikal. Inst. Univ.), G. R. Kirchhoff (Prof. Th. Physik Univ. Berlin), F. Neesen (Ass. Physikal. Inst. Univ.), C. Paalzow (Prof. Physik TH Charlottenburg), A. Slaby (Prof. Elektrotechnik TH Char-

merkenswert ist auch, daß diejenigen Mitglieder, die aktiv am Vereinsleben teilnahmen, ihre Aktivitäten nun auch in beiden Vereinen entfalteten, wie u.a. ein Vergleich der in beiden Vereinen gehaltenen Vorträge zeigt (vgl.Tafel).

Bemerkenswert ist aber auch das unterschiedliche Verhalten von Helmholtz und Siemens in beiden Gesellschaften. Während Siemens im Jahre 1869 seinen letzten Vortrag in der Physikalischen Gesellschaft gehalten hatte[18], nun aber im Schnitt zwei Vorträge pro Jahr im ETV hielt, trat Helmholtz dort nur einmal mit einem Vortrag auf, nämlich am 22. November 1881 mit einem Bericht über den Elektriker-Kongreß in Paris vom September 1881.[19] Helmholtz war Mitglied der Delegation gewesen, die das deutsche Reich zu diesem Kongreß geschickt hatte, der anläßlich der Pariser Elektrizitätsausstellung von 1880 stattfand und dessen Hauptanliegen es war, Einigung über die elektrischen Einheiten und die elektrische Terminologie zu erzielen. Der ETV hatte zehn der fünfzehn Delegationsmitglieder gestellt. Berichte über Ausstellung und Kongreß waren von verschiedenen ETV-Mitgliedern gegeben worden[20], und Helmholtz sprach über die Angelegenheiten der elektrischen Maßeinheiten (Helmholtz 1881).

Auch vor der Physikalischen Gesellschaft berichtete Helmholtz am 18. November jenen Jahres über diesen Kongreß; dort trug er aber auch am 17.11.1882 und am 9.5.1884 über die weiteren Verhandlungen und schließlich die Beschlüsse der Pariser Folgekonferenzen zu den elektrischen Maßeinheiten vor.[21] Den im ETV gehaltenen Bericht - der vermutlich nur wenig von dem in

---

lottenburg). Und unter den auswärtigen Mitgliedern finden wir u.a. die Namen folgender Physiker: R. Clausius (Prof. Univ. Bonn), F. Kohlrausch (Prof. Univ. Würzburg), E. Riecke (Prof. Univ. Göttingen), A. Töpler (Prof. TH Dresden), E. Warburg (Prof. Univ. Freiburg), G. Wiedemann (Prof. Univ. Leipzig).

18 W. Siemens: Ueber die neueste Construktion der elektrodynamischen Maschinen und einige andere elektrische Apparate (16.4.1869) - vgl. Nachrichten über die Physikalische Gesellschaft in den *Fortschritten der Physik* für das Jahr 1869.

19 Man könnte fast von einer gewissen Arbeitsteilung beider in dieser Hinsicht sprechen. Vor der Berliner Akademie der Wissenschaften dagegen präsentierte Helmholtz (seit 1857 korrespondierendes, seit 1879 auswärtiges und seit 1871 ordentliches Mitglied) 44 Beiträge (davon 33 nach 1871) und Siemens (ordentliches Mitglied seit 1873) 23 Beiträge (davon 20 nach 1873). Das spiegelt in der Größenordnung eine ungefähr gleiche Aktivität wider. Vgl. A. Harnack: Geschichte der Königlich Preussischen Akademie der Wissenschaften zu Berlin Bd. III, Berlin 1900, S.123f und 258f.

20 Vgl. *Elektrotechnische Zeitschrift* 2 (1881), verschiedene Hefte der 2. Jahreshälfte.

21 H. Helmholtz am 17.11.1882: "[...] erstattet darauf Bericht über die Thätigkeit der internationalen elektrischen Commission in Paris, welche in der zweiten Hälfte des Oktober d.J. dort zusammengetreten war, um hauptsächlich die Frage über die Wider-

der Physikalischen Gesellschaft abwich - nahm Helmholtz bezeichnenderweise in die Ausgabe seiner *Populären Wissenschaftlichen Vorträge* auf.[22] In den *Wissenschaftlichen Abhandlungen* dagegen findet sich aus jener Zeit ein Aufsatz unter dem Titel "Ueber absolute Maassysteme für electrische und magnetische Grössen", der 1882 in Wiedemanns *Annalen der Physik* erschienen war (Helmholtz 1882).[23] Erläuterte er im ersten Falle die Notwendigkeit einheitlicher elektrischer Maßeinheiten und den Diskussionsstand auf dem Pariser Kongreß, so diskutierte er im zweiten Falle konkret die Vorzüge des elektrostatischen (nach Gauss) und des elektrodynamischen (nach Maxwell) Maßsystems und machte deutlich, daß man nicht ohne weiteres das eine zu Gunsten des anderen fallen lassen könne.

Man könnte diese unterschiedliche Einordnung zunächst als mehr zufällig betrachten, zumal die Unterschiede in den Anforderungen an den Leser offenbar nur geringfügig sind - hier wird der hohe wissenschaftliche Anspruch von Helmholtz gerade auch an seine populären Darstellungen deutlich. Doch hat Helmholtz wohl sehr bewußt unterschieden, was er als populär und was er als wissenschaftlich betrachtete.

---

standseinheit zu prüfen" [abgedruckt als H. Helmholtz: Bericht über die Thätigkeit der internationalen electrischen Commission. *Verhandlungen der Physikalischen Gesellschaft zu Berlin* 2 (1882)]. Des weiteren H. Helmholtz am 9.5.1884: Ueber die Beschlüsse der internationalen Conferenz für elektrische Masseinheiten in Paris [abgedruckt als H. Helmholtz: Über die Beschlüsse der internationalen Conferenz für elektrische Maasseinheiten. *Verhandlungen der Physikalischen Gesellschaft zu Berlin* 3 (1884) 2, S.26-28]. - Vgl. Sitzungsprotokolle in den Verhandlungen der Physikalischen Gesellschaft zu Berlin 1882 und 1884.

Demgegenüber spielte dieser - gewissermaßen mehr technisch motivierte Fragenkomplex - in den Akademievorträgen von Helmholtz und Siemens kaum eine Rolle; lediglich Siemens gab 1884 eine kurze Mitteilung vor der Berliner Akademie der Wissenschaften "Über eine Einrichtung zur Darstellung der von der Pariser Conferenz zur Bestimmung der elektrischen Einheit angenommenen Lichteinheit" (Siemens 1889: 397-398), und ausführlicher sprach er im gleichen Jahre vor dem ETV "Über elektrische und Lichteinheiten nach den Beschlüssen der Pariser internationalen Conferenz" (Siemens 1889: 399-403). Siemens nahm diese beiden Abhandlungen in den wissenschaftlichen Teil (und nicht in den technischen) seiner Arbeiten auf.

22 Dort bezieht er sich aber nur auf den ETV, während der Bericht in der Physikalischen Gesellschaft keine Erwähnung findet.

23 Über diesen Artikel hat er jedoch offenbar weder in der Physikalischen Gesellschaft noch vor der Akademie der Wissenschaften explicite vorgetragen, wobei allerdings der Beitrag in Wiedemanns *Annalen* in gewisser Weise eine Zusammenfassung des physikalischen Kerns seiner Vorträge in der Physikalischen Gesellschaft darstellt (vgl. Koenigsberger Bd. II 1903: 287).

Tafel

Zahl der Vorträge einiger ausgewählter Wissenschaftler, die sowohl Mitglied der Physikalischen Gesellschaft (Angaben über die Jahre 1876-1890; für 1883 keine Angaben verfügbar) als auch des Elektrotechnischen Vereins (Angaben für die Jahre 1880 - 1890) waren

	Physikalische Gesellschaft														Elektrotechnischer Verein										
	1876	77	78	79	80	81	82	84	85	86	87	88	89	90	1880	81	82	83	84	85	86	87	88	89	90
Aron	1	3	4	1						1							2	1	1		1				
Brix															1		1								
Christiani	1	1	3	1		1	6										2	1							
du Bois-Reymond	1	1			1	1		1		1	1	1		2											
Frölich	2	2	3	4		2	1	1		1	1		2		3		1	3	2	1	2	1	1		
Goldstein					2					1										1					
Helmholtz	1	3	1	2	4	3	3	2	2	2	4	5	3	1		1									
Neesen	3	5	2	2	5	3	3	8	3	2	1	2	1			4	4	1							1
Paalzow	2	1	1	1	1												1								
Siemens															3	1	3	2	2		2	2	1	2	
Σ	11	16	14	11	13	10	13	12	5	9	7	6	6	3	8	6	14	8	5	2	5	4	2	3	1
Gesamtzahl der Vorträge pro Jahr	32	32	37	34	29	33	38	30	30	41	50	41	39	36	20	24	24	24	21	25	18	19	16	20	20

Literatur

Buchheim, Gisela: Die Denkschrift vom 30.7.1872 von K.-H. Schellbach. *NTM-Schriftenreihe* 23 (1986) 1, 98-101.

Cahan, David: *Meister der Messung - Die Physikalisch-Technische Reichsanstalt im Deutschen Kaiserreich.* Weinheim : VCH, 1992.

Dokumente einer Freundschaft - Briefwechsel zwischen H. v. Helmholtz und E. du Bois-Reymond 1846-1894. Hrsg. von Ch. Kirsten et al., Berlin: Akademie-Verlag, 1986.

Engelmann, Th.W.: *Gedächtnissrede auf Hermann von Helmholtz, gehalten am 28. September 1894 in der Aula der Universität Utrecht.* Leipzig: W. Engelmann, 1894.

Fischer, Emil: Hermann von Helmholtz. *Berichte der Deutschen Chemischen Gesellschaft* 27 (1894) 3, 2643-2652.

Helmholtz, Anna von: *Ein Lebensbild in Briefen.* Hrsg. von Ellen von Siemens-Helmholtz. Berlin: Verlag für Kulturpolitik, 1929.

Helmholtz, Hermann von: Ueber die Berathungen des Pariser Congresses betreffend die elektrischen Masseinheiten. ETZ 2 (1881) 12, 482-489. (Auch in ders.: *Populäre Wissenschaftliche Vorträge* Bd.II, Braunschweig: Vieweg und Sohn, 1903, 293-309).

Helmholtz, Hermann von: Ueber absolute Maassysteme für electrische und magnetische Grössen. *Wiedemanns Annalen* XVII (1882) 42-54. (Auch in ders.: *Wissenschaftliche Abhandlungen* Bd. 2, Leipzig: Barth, 1883, 993-1005).

Helmholtz, Hermann von: Vorrede zum ersten Bande der dritten Auflage 1884. In: Ders.: *Vorträge und Reden* Bd. I. Braunschweig: Vieweg und Sohn, 1903 (5.Aufl.), VII.

Helmholtz, Hermann von: *Hermann von Helmholtz über sich selbst.* Rede zu seinem 70. Geburtstag, eingeleitet und mit Anmerkungen versehen von Dorothea Goetz. Leipzig: Teubner, 1966.

Helmholtz, Hermann von: Ueber das Verhältnis der Naturwissenschaften zur Gesamtheit der Wissenschaften. In: Ders.: *Populäre wissenschaftliche Vorträge.* Braunschweig, Vieweg und Sohn 1865, 1-30 - Zit. nach: Ders.: *Philosophische Vorträge und Aufsätze.* Hrsg. von H. Hörz und S. Wollgast, Berlin: Akademie-Verlag, 1971, 79-108. (1971a).

Helmholtz, Hermann von: Ueber das Streben nach Popularisierung der Wissenschaft. In: *Vorträge und Reden,* Bd. II. Verlag von Fr. Vieweg, Braunschweig 1903 (5. Aufl.) 422f - Zit. nach: Ders.: *Philosophische Vorträge und Aufsätze.* Hrsg. von H. Hörz und S. Wollgast, Berlin, Akademie-Verlag 1971, 365-378. (1971b).

Johannes, Betty: Biographie von Olga und Hermann Helmholtz. Anhang zu: *Letters of Hermann von Helmholtz to his Wife 1847-1859.* Hrsg. von R.L. Kremer. Stuttgart: Steiner, 1990.

Kant, Horst und Hoffmann, Dieter: Vor 100 Jahren in Berlin gegründet - Physikalisch-Technische Reichsanstalt. *Wissenschaft und Fortschritt* 37 (1987) 12, 312-315.

Kant, Horst: Werner Siemens und sein Wirken im Berliner Elektrotechnischen Verein sowie in der Preußischen Akademie der Wissenschaften. In: *Studien zu Leben und Werk von Werner von Siemens.* Hrsg. von Dieter Hoffmann und Wolfgang Schreier. Braunschweig: 1994a (im Druck).

Kant, Horst: Hermann von Helmholtz' Bedeutung für die theoretische Physik des 19. Jahrhunderts. Wissenschaftshistorisches Symposium aus Anlaß des 100sten Todesjahres, Heidelberg Februar 1994. Leipzig: Barth 1994b (im Druck).

Koenigsberger, Leo: *Hermann von Helmholtz*. Bd. I-III. Braunschweig: Vieweg und Sohn, 1903.

Laue, Max von: Zum 50. Todestage von Hermann von Helmholtz. *Die Naturwissenschaften* 32 (1944) 27/39, 206-207.

Lebedinskij, A.V., Frankfurt, U.I., Frenk, A.M.: *Гельмгольц* (1821-1894). Moskau: Izdatel'stvo "Nauka", 1966.

Meyer, Wilhelm: *Wie ich der Urania-Meyer wurde*. Hamburg 1908.

Ostwald, Wilhelm: *Große Männer - Studien zur Biologie des Genies*. Bd. 1. Leipzig: Akademische Verlagsgesellschaft, 1919.

Reiner, Julius: *Hermann von Helmholtz* (= Klassiker der Naturwissenschaften Bd.6). Leipzig: Thomas, 1905.

Rubens, Heinrich: Das physikalische Institut. In: Max Lenz, *Geschichte der Königlichen Friedrich-Wilhelms-Universität zu Berlin*, Bd. 3. Halle: Waisenhaus, 1910. - Wiederabgedruckt in: *Wissenschaftliche Zeitschrift der Humboldt-Universität zu Berlin*, M-N-Reihe 32 (1983) 5, 583-591.

Siemens, Werner: *Wissenschaftliche und technische Arbeiten* Bd. I. Berlin: Springer, 1889 (2. Aufl.).

Siemens, Werner v.: *Lebenserinnerungen*. München: Prestel, 1966 (1.Aufl. 1892).

Tyndall, J.: *Die Wärme*. Deutsche Ausgabe, hrsg. von H. Helmholtz und G. Wiedemann. Braunschweig: Vieweg und Sohn, 1867.

Wiedemann, Eilhard: Hermann von Helmholtz; Rede zur Feier seines 70. Geburtstages. *Sitzungsberichte der Physikalisch-medicinischen Societät in Erlangen* 25 (1893) 54-67.

Anti-Helmholtz, Anti-Zöllner, Anti-Dühring: The Freedom of Science in Germany during the 1870s

David Cahan

1. Introduction

In October 1877, Hermann Helmholtz delivered his inaugural address as the rector of the University of Berlin. His address - "On Academic Freedom in German Universities" - was, as this essay seeks to show, highly political. If his audience, the faculty and students of Germany's premier university, and many others among the country's cultural elite could scarcely fail to miss the political nature of his address, a modern reader, unaware of the context in which Helmholtz spoke, may well fail to do so. For it was political not merely in the sense of *res publica*, that the rector is governor of the entire university and is its representative to other institutions and individuals. It was also political, as this essay argues, insofar as it was a warning to the students of Berlin and to others about proper conduct within the university, a warning given in the form of an explanation of the nature and function of the academic system in Germany and elsewhere, as well as in the form of explaining some of the ground rules of proper behavior of academics within the German system.

This essay thus seeks to portray the specific context that made Helmholtz's address so political. In does so, in part, by providing answers to two highly specific questions: why was Helmholtz elected the rector of Berlin in the late summer of 1877? And: why did he deliver an address less than three months later on the subject of academic freedom in German universities? The answers to these questions originate in the activities of two discontented German academics - Johann Karl Friedrich Zöllner (1834-82) and Eugen Dühring (1833-1921) - who during the 1870s challenged the German academic system and became Helmholtz's bitterest enemies. While space limitations prevent discussion of several aspects of Helmholtz's dual and extended conflicts with Zöllner and Dühring, above all the post-1877 events concerning these three figures - including such otherwise pertinent topics as spiritualism among German scientists and the anti-Semitic tirades of both Zöllner and Dühring (on which, for Zöllner, see, Stromberg 1989; and Meinel 1991) - the essay nonetheless attempts to show the essence of the antagonistic relationship between Helmholtz, on the one hand, and Zöllner and Dühring, on the other. In so doing, it tries to indicate the deeply political context in which Helmholtz became the rector and delivered his address. Although a good deal of the essay is concerned with Zöllner and Dühring (sections 2 and 3, respectively), its main focus is their effect on Helmholtz, and the final section (4) concludes with a partial analysis of Helmholtz's address, concentrating on Helmholtz's understanding of the issue of freedom of science in Germany during the 1870s.

2. Anti-Zöllner

Zöllner began his university studies at Berlin in the mid-1850s, where his physical science teachers included Gustav Magnus and H.W. Dove, two of Helmholtz's own former teachers. In 1859, Zöllner received his doctoral degree from the University of Basel, where among his teachers was Gustav Wiedemann, one of Helmholtz's close associates. As a student at Basel, Zöllner greatly improved the astrophotometer, an invention that shaped the course of his scientific career. First at Basel, and thereafter at Leipzig, Zöllner used the astrophotometer to make unprecedentedly precise studies of stars and other heavenly objects. He also made other important instrumentational advances (the reversion spectroscope and the horizontal pendulum) and became known for his highly speculative solar and comet theories (Herrmann 1976; Herrmann 1982). In addition, as his extremely respectful if not obsequious letters of 1862 to Helmholtz show, he had a keen if amateurish interest in problems of physiological optics, in particular the relationship of eye movements to problems of perception as manifested in the use of scientific instruments (Zöllner 1862).

Zöllner's astrophysical work brought him swift professional advancement: between 1862 and 1872 he climbed the academic ladder at the University of Leipzig, moving from astronomical researcher (1862) to *Privatdozent* (1865) to extraordinary professor (1866) and finally to ordinary professor of physical astronomy (1872), and turned down a series of calls elsewhere (Hamel 1983). Along the way, he became Germany's first established academic astrophysicist and advisor to Wilhelm Foerster, professor of astronomy at Berlin, and others as they planned the establishment of an Astrophysikalisches Observatorium in Potsdam between 1865 and 1872 (Herrmann 1975). In short, during the 1860s Zöllner became the leading figure in the emerging field of astrophysics.

Then in 1872 Zöllner's very positive relationship with the German academic system suddenly turned negative. This reversal of fortune was occasioned by his book *Über die Natur der Cometen. Beiträge zur Geschichte und Theorie der Erkenntniss*, a book which is not so much about comets or epistemology as it is a grossly polemical attack on the personal motives, scientific methods, and scientific results of several British physicists and their German patron and co-translator, Helmholtz (Zöllner 1872). The heart of Zöllner's diatribe came in the 72-page preface and the 270-page fourth section of the book, a total of 342 pages, which he wrote in less than 90 days (Zöllner 1872: lxxii, xcviii-xcix). Zöllner there unrelentingly scorned, disparaged, and treated with sarcasm the work and persons of William Thomson, Peter Guthrie Tait, John Tyndall, and Helmholtz. In Zöllner's eyes, the major sin of Thomson and Tait lay in the publication of their book of 1867, *Treatise on Natural Philosophy*, which Helmholtz co-translated and briefly introduced under the title *Handbuch der Theoretischen Physik*, part one of which appeared shortly after May 1871. For pages on end Zöllner attacked Thomson - he knew Tait was only small potatoes - for, among other things, claiming that the British physicists George Gabriel Stokes and Balfour Stewart had discovered the heat-radiation law of emission and absorption before the German physicist Gustav Robert Kirchhoff did; for claiming that those British physicists had discovered spectral analysis before Kirchhoff and the German chemist Robert Bunsen did; and for what Zöllner claimed to be Thomson's mystical approach to nature, in particular to comet theory (Zöllner 1872: xxxi-xlvii). Moreover, he especially attacked Helmholtz's three-page preface to the book, time and again citing a passage wherein Helmholtz spoke of the book's "high scientific importance" for "the German natural-scientific and mathematical public", and where Helmholtz praised Thomson "as one of the most penetrating and ingenious thinkers who have ever applied themselves to our science". These flattering remarks infuriated Zöllner, who cited them sarcastically (and repeatedly), and declared that the principles of physics underlying Thomson and Tait's book were "absolute nonsense" (Zöllner 1872: xlvi-xlvii, lii-liv; Thomson and Tait 1871: x-xii).

Zöllner was particularly infuriated that a German professor had the temerity to praise British physicists. He claimed that British physics, with its supposed stress on the inductive rather than the deductive approach to physics, was inferior to German physics and in decline. Above all, he was absolutely outraged at the British rejection of Wilhelm Weber's electrodynamic theory, a theory that Helmholtz had also recently called into question. Without giving any proof, Zöllner maintained that Helmholtz's claim that Weber's theory stood in contradiction to the principle of conservation of energy was false, and he further maintained that the very translation of Thomson and Tait's volume, with its anti-Weberian electrodynamics, showed Helmholtz's lack of tact and national solidarity (Zöllner 1872: xlvii-liv, xlix-l, lxii-lxiii, 329-35; Molella 1972: 138-237; Buchwald 1993, esp. 351, 363, 370-72, 386). Indeed, by translating and above all by supporting and praising a British author against his own countryman, Helmholtz had supposedly betrayed German science; he was allegedly unpatriotic, and this in a hyperpatriotic era. Zöllner proclaimed: "Germany alone has been called to be the carrier and the center of this epoch, for thanks to the depths of the German mind it alone contains that profusion of deductive needs and abilities for successfully mastering the accumulated, inductive material gained through the exact sciences" (Zöllner 1872: lxii-lxiii, lxx).

Zöllner advanced still other charges or made other insinuations against Helmholtz: Helmholtz had either stolen or failed to note the similar ideas or views of Schopenhauer on color theory, on the a priori character of thc law of causality, and on "unconscious inferences"; that he, Zöllner, had developed his own idea of "unconscious inference" and an associated theory of sense perceptions, and that Helmholtz had failed to cite or understand his work (Zöllner 1872: 321, 344-50, 378-425). Moreover, Zöllner sarcastically portrayed Helmholtz as one of those "simple professors" from Berlin who imagined themselves to be "elegant" men, who had superb institutional facilities at their disposal, and who gave well-paid popular lectures before society ladies and gentlemen while neglecting the scientist's true vocation, the pursuit of truth. In Zöllner's view, the alleged English troubles in science, the reason for its decline, was in part due to the popularization of sciencc by and among Englishman who did not understand science; and those who, like Helmholtz, translated the British into German and who popularized science, were now in turn leading Germany into scientific decline (Zöllner 1872: lvii-lviii, lix-lx).

For these and his other polemics, Zöllner paid a price. In 1873, a five-member commission, which included Helmholtz and four close associates and fellow members of the Berlin Akademie der Wissenschaften, had been empowered to outline the Observatorium's specific purposes, to project its institutional needs and costs, and to recommend a director. Zöllner had worked for several years with Foerster on the plans for the future Astrophysikalisches Observatorium, and until 1872 (i.e., until the publication of the *Cometen*) it seemed certain that Zöllner would be appointed the founding director. The new commission, however, did not even consider Zöllner's name, even though he was Germany's foremost authority in astrophysics; his polemical attack on Helmholtz apparently cost him this appointment (Herrmann 1975: 253-57).

The failure to receive the directorship was not the only price Zöllner paid for his polemics. Following the publication of his *Cometen*, there was talk among German scientists that Zöllner was mentally unbalanced, possibly insane. Rudolf Clausius now wrote Tyndall, who, like Thomson and Tait, was one of the objects of Zöllner's venom:

> "I am absolutely outraged by this book, and furthermore I can tell you that everyone with whom I have spoken about it condemns and decidedly disapproves the personal attacks contained therein. The manner in which he [i.e., Zöllner] speaks is not only unheard of in a scientific work but is also so unexplainable and peculiar that one cannot even understand how he came to speak that way. As a result, there are even rumors about his mental state, which, if they should turn out to be true, will be very sad for him" (Clausius 1872).

And two months later, in June 1872, Helmholtz wrote in a similar vein to Tyndall:

> "Here too in Germany, Zöllner's book has extremely shocked us. The first impression was that, in general, he may be insane, all the more so as one knew that several cases of mental illness had already occurred in his family, and with his mother's brother the illness began with the publication of a similar book. In the meantime it is, at the moment, not so far advanced that one might have to lock him up in a lunatic asylum".

Still, Zöllner had many friends in Leipzig, which cultivated, as Helmholtz recognized, "a kind of rivalry between the little minds of the University of Leipzig and Berlin". Helmholtz continued his judgment of Zöllner and his book:

> "Nonetheless, I must say that to me, despite a certain connection of his thoughts in that wild book, the hypothesis that he may be mentally ill is still the most probable. The poisonous envy towards the outward success of others and the morose rage may also be explicable without that [i.e., mental illness] in a man who

> has buried himself in his work, [and who] feels a certain receptive talent yet is struck with the curse of unproductivity and cannot move ahead".

Yet Helmholtz saw more than alleged mental illness at work here, for he also thought that there were philosophical, personal, and political dimensions for the success of Zöllner's "wild book". He continued:

> "Furthermore, the heart of Zöllner's opposition to us is really the animosity of the metaphysician against natural science, and unfortunately it is evident in this affair that the old natural-scientific (the metaphysical) enthusiasms still lay hidden in many German souls, and which really haven't dared to come out and be known. Such people are disposed to believing Zöllner. I myself have already had to experience how much concealed envy there is towards my (truly very modest) outward success in life. Zöllner is valued by philosophers as a knowledgeable authority about natural science, and they gladly use this to ventilate their oppressed hearts.
> Then we have among us, since the [Franco-Prussian] war, a multitude of chauvinists, precisely among those people who did nothing and suffered nothing, but who awoke one fine morning as the victors of Sedan and who now do not let themselves speak [enough] about national pride. Zöllner has very skillfully constructed his book for these types of people" (Helmholtz 1872).

In sum, Helmholtz saw a set of conscious and unconscious factors at play in Zöllner's attack on him.

Helmholtz did not restrict himself to the private responses of denying Zöllner the Potsdam directorship or writing letters to friends. He also planned, as he explained to his publisher, Friedrich Vieweg, a written public response in the form of a preface to the German translation of Tyndall's *Fragments of Science* (Helmholtz 1873). He considered Zöllner's bitter attack on Tyndall particularly deplorable because it was done "in the name of German national feeling against the infiltration of foreign scientific directions" and because it degenerated from being a legitimate scientific critique into being a deep personal attack. And he condemned Zöllner's politicized philosophy of science:

> "Mr. Zöllner recommends the "deductive" method, which he himself follows or at least intends to follow in his astrophysical speculations, as the original German method and seeks to seal off Germany's intellectual horizon by [erecting] a Chinese wall against the foreign, inductive method. At present, when Faraday has been dead only a few years and when the entire intellectual atmosphere of Europe is saturated with and stimulated by Darwin's ideas, he doesn't hesitate to pronounce English science as decrepit and dying, as poisoned and poisoning".

Indeed, he thought it simply bad philosophy of science: "The natural sciences", he wrote, "have made better and quicker progress precisely insofar as they have freed themselves from the influence of supposed a priori deductions". He condemned the pernicious influence of metaphysics and declared that it had done much harm to the German intelligentsia: it was, he said, their opium (Helmholtz 1874a: xx-xiv). Furthermore, in 1874, in his preface to Part Two of the German edition of Thomson and Tait's book, and in the pages of *Nature*, Helmholtz made many of the same points, now declaring that Zöllner's remarks were "sophistical", "libellous", and "intolerant"; Zöllner had acted, Helmholtz said, as if he were infallible, while hastily passing intellectual and moralistic judgments upon his opponents. Moreover, Helmholtz accused Zöllner of having made elementary mistakes in epistemology and logic, and again scorned him for being "the genuine metaphysician. In view of a presumed necessity of thought, he looks down with an air of superiority on those who labor to investigate the facts" (Helmholtz 1874b). With the publication of the scandalous *Cometen*, Zöllner had abused his academic freedom and his scientific career now effectively came to an end. His nationalistic and *ad hominem* attacks and associated anti-scientific attitudes and pronouncements were symptomatic, as Christoph Meinel has argued, of an emerging crisis within the culture of German science (Meinel 1991).

3. Anti-Dühring

This bitter anti-Helmholtz, anti-Zöllner exchange was soon followed and paralleled by an even more venomous anti-Helmholtz, anti-Dühring exchange.

In the early to mid-1850s, the young Dühring studied jurisprudence and a variety of sciences at Berlin. By the late 1850s, the onset of blindness led him to abandon his law practice for an academic career in philosophy. From 1859 to 1861 he studied philosophy at Berlin, and promoted there in the latter year with a dissertation entitled *De Tempore, Spatio, Causalitate Atque de Analysis Infinitesimalis Logica*. In 1863, he habilitated there, and spent the next fourteen years as a *Privatdozent* for political economy and philosophy. From the start of his new career he showed himself to be a prolific author. In addition, his lectures were reportedly extremely well attended. Yet in 1866 he was passed over for promotion to extraordinary professor (Dühring 1903: 58-127). This rebuff became the initial source of his bitterness towards his senior colleagues, and he would spend the next half-century using his polemical skills to lash out at his academic and other (supposed) enemies through numerous public lectures and in books aimed principally at non-academic audiences.

In 1869, Dühring won a first prize from the University of Göttingen's philosophical faculty for a study which he published in 1873 under the title of *Kri-*

tische Geschichte der allgemeinen Principien der Mechanik, and then again in subsequent editions. Although the work concerned the history of rational mechanics, it was written in a popular style. Dühring now declared that Julius Robert Mayer's discovery of the mechanical equivalent of heat meant that he was the discoverer of the law of conservation of energy, and he attacked Helmholtz's 1847 treatise on the subject for failing to mention Mayer's work on the equivalence of heat while discussing James Prescott Joule's work on that subject. Furthermore, he judged Helmholtz's treatise as simply "the dressing up in new words of things that had been known for centuries" (Dühring 1873: 447). Dühring and Helmholtz had no prior involvement with one another, and it seems not unlikely that Dühring may have then attacked Helmholtz because the Prussian Kultusministerium had spared no expense to bring Helmholtz to Berlin - he first joined the faculty there in the spring of 1871 - as part of its efforts to transform Berlin into the new Germany's cultural capital. By attacking Helmholtz in 1873, Dühring was attacking the establishment that had previously decided not to promote him.

Following his unprovoked attack on Helmholtz, in December 1874 and early 1875 Dühring further vented his bitterness on another of his senior academic colleagues, the leading political economist Adolf Wagner, who had come to Berlin in 1870 as the professor of political economy, and so was Dühring's rival. Dühring attacked Wagner in the pages of a Berlin financial newspaper and in his *Kritische Geschichte der Nationalökonomie und des Socialismus*; at the same time, he attacked the German universities and their professoriates as a whole. As a result, the ministry warned Dühring of remotion should another such event occur (Ministerium 1877: 7-10).

The attack on Wagner and the university system only whetted Dühring's appetite to attack again. In September 1876, he completed a polemical pamphlet - the written version of a speech he held at the Berlin Rathaus in March 1876 - on higher professional education for women and university teaching: *Der Weg zur höheren Berufsbildung der Frauen und die Lehrweise der Universitäten*, which appeared the next spring (Dühring 1877a). His authority to speak on this topic issued from the fact that, in addition to teaching at the university, in 1872 he also began teaching literature and philosophy at the Victoria-Lyceum, a Berlin educational institute for young women. In his pamphlet, Dühring spoke venomously of "the envy and intrigues in the province of the little authorities of science", and of "pseudo-science." He claimed that the universities were no more than medieval "learned cliques" that promoted "clanish subjugation"; that they constituted a "learned monopoly and exclusivities"; that they were filled with "intellectual corruption" and were run by a "feudatory, authoritarian administration". In modern times, he further fumed, universities "everywhere have increasingly deteriorated and have essentially stopped the progress of the sciences". "The clanish structures and their effects", he added, "can also be sufficiently viewed at today's German universities". Professors were allegedly nepotists and "monopolist chiefs", who earned a lot of money while simultaneously seeking to injure the economic well-being of mere lecturers. "The professorial estate", he averred, "is a type of caste that above all reproduces itself through inbreeding" (Dühring 1877a: Vorbemerkung, 36-40). In short, he declared that German universities stood for the very opposite of academic freedom.

In May 1876, the Lyceum board - whose members included several Berlin professors (among them Helmholtz's dear friend Emil du Bois-Reymond) as well as Anna von Helmholtz, Hermann's wife - decided not to renew Dühring's contract. Dühring maintained that the occasion (though not the real reason) for his dismissal was his lecture of March, 1876. It was this caste on the Lyceum board - supposedly spurred by the Jews among or behind them, as well as by Anna von Helmholtz - that led, he claimed, to his dismissal and, hence, loss of income (Dühring 1877a: 60-76).

The Rathaus speech and pamphlet on higher education for women were only two of Dühring's recent attacks. For in the second edition of his *Kritische Geschichte der allgemeinen Principien der Mechanik* (1877), Dühring, who was still a *Privatdozent* at Berlin, now extended his attack on Helmholtz and attacked professors in general. He also attacked what he claimed as the absurd, metaphysical, and confusing subject of non-Euclidean geometry. "It is not surprising", he said, "that Mr. Helmholtz, the unclear, a little philosophizing physiological physics professor, cannot refrain from taking part in the discussion and from commenting approvingly, in an essay 'Ueber die Thatsachen, welche der Geometrie zu Grunde liegen [...]', on the piquant nonsense". Furthermore, he declared that Helmholtz's ideas on acoustics were mere popularizations and extensions of Georg Simon Ohm's work, thereby insinuating still further plagiarism on Helmholtz's part. He proclaimed that, except for Gustave Lejeune Dirichlet, all the mathematicians at the University of Berlin and the Berlin Akademie der Wissenschaften had been or were secondary figures. He lambasted every aspect of university life, including scientific journals - summarily dismissing virtually all articles and university lectures as worthless. Students, he pronounced, could learn better from books than from lecturers (Dühring 1877b: 444-45, 459-60, 474, 529-30, 550-51, 554-56).

Dühring's pamphlet and the second edition of his mechanics book became the *touches finales* that led to his formal dismissal from the university. Sometime shortly before 12 May 1877, the university and ministry began remotion proceedings against him. The charges included his unfounded allegations against Wagner as well as his numerous defamatory, denigratory, and completely unsubstantiated statements against the German universities as a whole, against Berlin university mathematicians as a group, and against Helmholtz in particular, who allegedly failed to acknowledge Mayer's work. The university, quoting from Dühring's own writings, found that his charges against Helmholtz were without factual basis and were marked by "personal invective" on Dühring's part (Ministerium 1877: 10-12, 20-6). Helmholtz, for his part, informed the ministry that neither he nor his wife had ever sought Dühring's dismissal from the Lyceum; that in 1854, over twenty years before Dühring's charges, he had been the first to note Mayer's work publicly; and that he did not dispute Mayer's priority (Ministerium 1877: 27-9). Dühring denied that he had wronged anyone; claimed that the university did not even permit "a moderate freedom of scientific criticism" of its own members; maintained that he argued on the basis of facts, not personalities; and, pointing to his lack of advancement at the university since 1863, considered himself the injured party (Ministerium 1877: 13-20, esp. 16). On 7 July the ministry remoted Dühring: in mid-semester, he lost his right to teach. The ministry rejected all his charges. Minister Falk wrote Dühring: "The scientific freedom that you also rightly claim for the *Privatdozenten*, and which I would not be inclined to infringe upon, has nothing to do with such utterances [as yours]" (Ministerium 1877: 29-36, esp. 36). The case immediately became a *cause célèbre*: on 11 July, the ministry, concerned about the public discussion of the case, felt forced to publish its proceedings against Dühring (Ministerium 1877: 3-4). The next day, a protest rally took place in Berlin at which, if Dühring is to be believed, some 2,500 individuals, including over 1,500 students, protested in his favor; some compared him, he said, to Socrates and Giordano Bruno! Dühring later claimed that his real offense was his mere "presence at the university". Somewhat more specifically, he said that he was remoted because he had revealed to the world that Helmholtz had plagiarized Mayer's discovery of the conservation of force. With the aid of the Kultusministerium and the Jewish, especially the liberal, press, Helmholtz had allegedly managed to silence Dühring by having him remoted and by preventing him from defending himself in the newspapers while at the same time arranging for derogatory news paper articles about Dühring and adulatory ones about himself (Dühring 1903: 179-211, esp. 182, 206).

4. The Freedom of Science in Germany during the 1870s

On 24 July 1877, two weeks after Dühring's dismissal and the protest rally in his favor, Helmholtz conceded to a friend that the Dühring Affair had very much affected him. Seen philosophically, it was another instance, he thought, of the metaphysicians at work: "Throughout my life", he wrote, "I've sought to cut through their sophisms with facts. In the end, people are intelligent enough to believe the facts more than the subtlest theories". He thought Dühring's charges, like Zöllner's, were so patently false or frustratingly vague that he simply could not understand the source of Dühring's animosity against him. "With Zöllner, there was at least [the fact] that I did not answer a letter of his and that in my *Physiologische Optik* I did not mention an explanation that he had given for an experiment and that I did not understand, matters which could well excite a man of exaggerated vanity". But even such trivial slights as these, if such they were, had never occurred between himself and Dühring. Helmholtz told his friend: "For I have never done anything to him, never named him either in letters or in lectures, while in the Wagner Affair I spoke in general for his forbearance and, with this most recent action (concerning what was written against me), for treating it as non-existent". Nor could he understand the student support for Dühring: "I cannot imagine that students in my university days would believe such plain and transparent lies and distortions". He had wistfully hoped that Dühring's student supporters would leave the university - "we could not get a more beautiful compliment" - and he disparagingly noted their cowardice in signing a petition without providing their place of residence. Helmholtz was in a miserable mood when he wrote his friend on 24 July, even though on that very day his colleagues, doubtless in part wanting to show their solidarity and further rejection of Dühring and his insidious charges against the German university system, had elected him rector of Berlin for 1877-78. "There's not much pleasure to be had in Berlin just now", he told his friend in closing, "almost all friends are away, the Spree stinks, and it's miserably hot" (Helmholtz 1877a).

Helmholtz's election left no doubt about where the Berlin and Prussian educational establishments stood vis-à-vis Dühring's and Zöllner's charges against Helmholtz. Less than ten days later, on 2 August, he delivered an invited address before his alma mater, the Friedrich-Wilhelms-Institut in Berlin, "On Thought in Medicine". His principal aim in that address was to argue for the rise and importance of a factual (observational), experimental (empirical), non-metaphysical approach to medicine and science; to stress the importance of the inductive and the dangers of the deductive approach to medicine. And throughout his address he warned of the consequent perils when metaphysical systems were introduced into medicine and science, even noting how he had gotten involved in a number of controversies with metaphysicians (Helmholtz 1877b: passim, esp. 234-35). His address was thus in part another response to Zöllner and Dühring.

Two and a half months later, on 15 October, Helmholtz delivered his rectoral address "On Academic Freedom in German Universities": the timing and choice of subject matter were, here too, in part a response to the charges by Zöllner and, above all, Dühring, including the latter's pamphlet *Die Lehrweise der Universitäten* and, to Helmholtz at least, his surprisingly strong support among students. In his address, Helmholtz asserted that German universities had "conquered a position of honor not confined to their fatherland; the eyes of the civilized world", he told his Berlin audience, "are upon them". He warned the students that "[s]uch a position would be easily lost by a false step, but would be difficult to regain" (Helmholtz 1877c: 238-39). He portrayed the German universities, especially in comparison to those of France and Britain, as virtually flawless. It was a claim that bordered on the chauvinistic, yet one that, if nothing else, at least implicitly undermined Zöllner's and Dühring's ridiculous charges of Helmholtz's disloyalty to the state.

Helmholtz thought that the German universities had become great because their essence was the freedom to learn (*Lernfreiheit*) and to teach (*Lehrfreiheit*), that they provided a degree of freedom unmatched by any other national university system. His address read in good measure like a homily on the conditions that made such freedom possible, that allowed German universities to flourish scientifically, and that the students in particular would have to meet if German universities were to maintain their standing. German students, he said, were "responsible to themselves, striving after science of their own free will". They could migrate freely from university to university, choosing the teachers whom they preferred to study with, "and [were] free to acquire any part of their instruction from books". Though Helmholtz warned that "the guardians of public order" and "the disciplinary powers of the University authorities" could and would control the students if necessary, he averred that fundamental control came from "the sense of honor of the students", "a vivid sense of corporate union", and the associated "requirement of honorable behavior in the individual". With the large demonstration in Dühring's favor still fresh in everyone's memory, he further warned the students that "freedom necessarily implies responsibility". If the students failed to maintain their inherited academic freedom, he threatened that another, "more rigid system of supervision and control" might well come their way (Helmholtz 1877c: 249-52, 254-55; cf. 263-64).

As for university teachers, they had to be not only scientifically competent; they also had to instruct in a non-rhetorical, non-inflammatory (i.e., an academic) manner. He claimed that "the advanced political freedom of the new German Empire has brought a cure" against "outside political or ecclesiastical interference with the universities". He declared proudly that German university teachers could freely teach any point of view, but added, perhaps with Zöllner and Dühring in mind, that "it is forbidden to suspect motives or indulge in abuse of the personal qualities of our opponents, ... [and] any incitement to such acts as are legally forbidden". He portrayed the *Privatdozenten* as having "exactly the same" legal standing as ordinary professors, and maintained that, apart from having control over laboratory resources and apart from the burden of having to conduct examinations, the *Privatdozenten* were the equals of the *Ordinarien*, a point that few if any *Privatdozenten* would have shared. He further maintained that the *Privatdozenten* were really the friendly competitors of the *Ordinarien*, a point, he further noted, that astonished foreigners - and would certainly have astonished many a *Privatdozent*, not least Dühring. In extraordinary detail Helmholtz explained that ordinary professors were appointed largely by a given faculty, but that appointments were always made in consultation with the government, which, he said, only rarely opposed a faculty's choice. He portrayed the two groups as working together closely and harmoniously, perhaps thereby intending to imply that the judgment as to Dühring's remotion was widely shared (Helmholtz 1877c: 253, 255-59).

By the 1870s, Helmholtz had become the very epitome of the academic establishment, and Zöllner's and Dühring's attacks on him were fundamentally attacks by outsiders against an insider. Both Zöllner and Dühring were filled with *ressentiment* against the scientific establishment, and both, especially Dühring, were determined not to play by the normal academic rules. Instead, they chose to question their opponents' motives; to use vague metaphysical points against them instead of presenting concrete evidence; and to employ chauvinism, demagoguery, and slander in their diatribes against Helmholtz and other establishment figures. For doing so, Zöllner was essentially eliminated from the German scientific community and Dühring from the academic community entirely. Theirs was a fundamental challenge to the values and rules of academic freedom. Not for nothing did Dühring and, to a much lesser extent, Zöllner become one of the leading anti-Semites and national chauvinists in Bismarckian Germany. As one of the pillars of the German establishment, Helmholtz meant his address of 1877 in part as a warning to those who failed to uphold those values and rules, and so further contribute to the political and social discriminations of Bismarckian Germany.

Literature

Buchwald, Jed Z: Electrodynamics in Context: Object States, Laboratory Practice, and Anti-Romanticism. In: David Cahan, ed., *Hermann von Helmholtz and the Foundations of Nineteenth-Century Science*. Berkeley, Los Angeles, London: University of California Press, 1993. 334-73.

Clausius, Rudolf: Letter to John Tyndall, 4 April 1872, copy in the Eidgenössische Technische Hochschule Zürich, Hs 227: 5-163 (original in the Royal Institution, London).

Dühring, Eugen: *Kritische Geschichte der allgemeinen Principien der Mechanik*. Berlin: Theobald Grieben, 1873.

Dühring, Eugen: *Der Weg zur höheren Berufsbildung der Frauen und die Lehrweise der Universitäten*. Leipzig: Fues's Verlag, 1877. [1877a]

Dühring, Eugen: *Kritische Geschichte der allgemeinen Principien der Mechanik*. 2nd ed. Leipzig: Fues's Verlag, 1877. [1877b]

Dühring, Eugen: *Sache, Leben und Feinde. Als Hauptwerk und Schlüssel zu seinen sämmtlichen Schriften*. 2nd enl. ed. Leipzig: C.G. Naumann, 1903.

Hamel, Jürgen: Karl Friedrich Zöllners Tätigkeit als Hochschullehrer an der Universität Leipzig: Ein Beitrag zur Geschichte der Institutionalisierung der Astrophysik. *NTM-Schriftenreihe für Geschichte der Naturwissenschaften, Technik und Medizin 20:1* (1983): 35-43.

Helmholtz, Hermann von: Letter to John Tyndall, 23 June 1872. Tyndall Collection. Royal Institution, London, Case 9, Paket 8, 30/D1.9.

Helmholtz, Hermann von: Letters to Friedrich Vieweg, 22 October and 17 November 1873. Friedrich Vieweg und Sohn, Archiv, Braunschweig.

Helmholtz, Hermann von: Vorrede von Herrn Professor H. Helmholtz. In: John Tyndall, *Fragmente aus den Naturwissenschaften*. Trans. A[nna] H[elmholtz]. Braunschweig: Friedrich Vieweg und Sohn, 1874. Pp. v-xxv. [1874a]

Helmholtz, Hermann von: Helmholtz on the Use and Abuse of the Deductive Method in Physical Science. *Nature 11* (1874-75): 149-51, 211-12. [1874b]

Helmholtz, Hermann von: Letter to "Verehrte Frau," 24 July 1877. Autogr. VA, Handschriftenabteilung, Bayerische Staatsbibliothek, Munich. [1877a]

Helmholtz, Hermann von: On Thought in Medicine. In his: *Popular Lectures on Scientific Subjects*, Second Series. London, New York, and Bombay: Longmans, Green, 1898. 199-236. [1877b]

Helmholtz, Hermann von: On Academic Freedom in German Universities. In his: *Popular Lectures on Scientific Subjects*, Second Series. London, New York, and Bombay: Longmans, Green, 1898. 237-65. [1877c]

Herrmann, Dieter B.: Zur Vorgeschichte des Astrophysikalischen Observatorium Potsdam (1865 bis 1874). *Astronomische Nachrichten 296:6* (1975): 245-59.

Herrmann, Dieter B.: Zöllner, Johann Karl Friedrich. In: Charles C. Gillispie, et al., eds., *Dictionary of Scientific Biography*, 18 vols. New York: Charles Scribner's Sons, 1970-90. *14* (1976): 627-30.

Herrmann, Dieter B.: *Karl Friedrich Zöllner*. Leipzig: BSB B.G. Teubner, 1982.

Meinel, Christoph: *Karl Friedrich Zöllner und die Wissenschaftskultur der Gründerzeit: Eine Fallstudie zur Genese konservativer Zivilisationskritik*. Berlin: SIGMA-Verlag, 1991.

Ministerium der geistlichen, Unterrichts- und Medicinal-Angelegenheiten 1877: *Aktenstücke in der Angelegenheit des Privatdocenten Dr. Dühring veröffentlicht durch die Philosophische Fakultät der Königl. Universität zu Berlin*. Berlin: G. Reimer, 1877.

Molella, Arthur Philip. *Philosophy and Nineteenth-Century German Electrodynamaics: The Problem of Action at a Distance*. Ph.D. dissertation, Cornell University, 1972.

Stromberg, Wayne H.: Helmholtz and Zoellner: Nineteenth-Century Empiricism, Spiritism, and the Theory of Space Perception. *Journal of the History of the Behavioral Sciences 25* (1989): 371-83.

Thomson, William, and Peter Guthrie Tait: *Handbuch der theoretischen Physik*. Trans. H. Helmholtz and G. Wertheim. Vol. 1, Part 1. Preface to the German edition by Helmholtz. Braunschweig: Friedrich Vieweg und Sohn, 1871; Vol. 1, Part 2. Preface to the German edition by Helmholtz. Braunschweig: Friedrich Vieweg und Sohn, 1874.

Zöllner, Johann Karl Friedrich: Letters to Hermann von Helmholtz of 1862?, 21 October 1862, and 20 December 1862. In: Helmholtz-Nachlaß, Archiv, Akademie der Wissenschaften, Berlin.

Zöllner, Johann Karl Friedrich: *Über die Natur der Cometen. Beiträge zur Geschichte und Theorie der Erkenntniss*. Leipzig: Wilhelm Engelmann, 1872.

Hermann von Helmholtz: Aspekte einer Wissenschaftlerkarriere im deutschen Kaiserreich

Walter Kaiser

1. Einleitung

In dieser Fallstudie zur Wissenschaftsgeschichte des 19. Jahrhunderts sollen zunächst einige historische Abläufe im Umkreis der Biographie von Hermann von Helmholtz geschildert werden. Der Schwerpunkt soll aber dort liegen, wo es um die Beziehung zur Allgemeingeschichte geht. Ein historiographisches Problem, das sich hier stellt und an dem sich zugleich die tiefer liegende Wechselwirkung von politischer Geschichte und Wissenschaftsgeschichte aufweisen läßt, ist die Mentalität der Eliten des deutschen Kaiserreichs. Zentral für den vorliegenden historischen Essay werden insofern Aspekte der Biographie von Hermann von Helmholtz sein, die etwas über die Eigenart, die Rolle und die Gefährdung der Bedeutung der deutschen Bildungseliten aussagen, wobei - sozusagen als Fernwirkung des Abstiegs - auch das spätere Versagen der deutschen Bildungseliten in die Betrachtung einbezogen wird. Der allgemeinhistorische Gesichtspunkt, also der folgenschwere Wandel im Verhalten einer führenden Schicht, wird demnach nicht im Sinne einer abstrakten Diskussion einer historiographischen Theorie angegangen. Vielmehr soll gerade die wissenschaftshistorische Episode als Teil des größeren historischen Kontextes ihre besondere Bedeutung für das allgemeinhistorische Problem "zeigen". Es wird dabei deutlich werden, daß trotz des scheinbar singulären

Falls und ungeachtet der geringen Zahl von Stützstellen damit erstaunlich viel vom komplizierten Raum der deutschen Geschichte seit 1870 aufgespannt wird.

2. Hochschullehrer als Leistungs- *und* Werteeliten?

Helmholtz' universale und herausragende Begabung zeichnete sich bereits während seiner Potsdamer Schulzeit ab (Koenigsberger 1902-1903, Kirsten 1986). Eine gewisse Härte im späteren Umgang mit Menschen mag damit zusammenhängen, daß die Atmosphäre des Elternhauses stark davon bestimmt war, daß der Vater als preußischer Beamter sein aus dem deutschen Idealismus genährtes liberales politisches Denken mit Rücksicht auf die Familie unterdrücken mußte. Hinzu kam, daß der Lehrerhaushalt, aus dem Helmholtz kam, finanziell alles andere als auf Rosen gebettet war. Der Vater war deshalb bestrebt, den Sohn am Königlichen Medizinisch-Chirurgischen Friedrich Wilhelms-Institut in Berlin unterzubringen. Über diese militärärztliche Ausbildung bekam Helmholtz die Möglichkeit, seinen naturwissenschaftlichen Neigungen entsprechend zu studieren; allerdings mit der Verpflichtung, die weitgehend auf "königliche Kosten" gewährte Ausbildung durch eine achtjährige Tätigkeit als "Compagnie"- oder "Escadron-Chirurgus" abzugelten. Obwohl Helmholtz aufgrund seiner Herkunft also keinesfalls privilegiert war, bedeutete die Ausbildung und die durch den Staat gewährte Förderung eine tiefreichende Bindung an das Leistungsdenken des preußischen Erziehungssystems. Charakteristisch ist es insofern, daß Helmholtz neben der militärärztlichen Tätigkeit seine Kenntnisse in mathematischer Physik vertiefte, daß er dieses Selbststudium durch die von Gustav Magnus in Berlin angebotenen Experimentiermöglichkeiten und durch dessen physikalisches "Colloquium" ergänzte, und - nicht zuletzt -, daß er sich in die 1845 von jungen Naturwissenschaftlern (den Physiologen Emil du Bois-Reymond und Ernst Brücke, aber auch Werner Siemens) gegründete Berliner Physikalische Gesellschaft einführen ließ.

Die anatomischen Untersuchungen zur mikroskopischen Struktur des Nervensystems, die physiologisch-physikalischen Überlegungen im Zusammenhang mit dem Energieerhaltungssatz, die Untersuchung der Reizleitung in Nerven und die von ihm durchgesetzte Methode der Spiegelung des Augenhintergrunds haben Helmholtz rasch bekannt gemacht. Bereits 1851, also mit 30 Jahren, wurde er zum ordentlichen Professor der Physiologie an der Universität Königsberg ernannt.

Nachdem Helmholtz einmal in die mit dem Staat verflochtene Bildungselite eingebunden war, profitierte er nicht nur von deren Förderung, sondern nutzte sie auch selbst zielbewußt für seine weitere Karriere. Ein abgelehnter Ruf nach Kiel brachte schon 1853 in Königsberg eine Gehaltserhöhung auf 1000 Taler im Jahr. Aufgrund eines Verzichts von Emil du Bois-Reymond - der finanziell unabhängiger war und auch an Berlin gebunden war - und durch eine briefliche Intervention des einflußreichen Alexander von Humboldt beim Preußischen Kultusminister wurde Helmholtz 1855 auf eine wiederzubesetzende Professur für Anatomie und Physiologie nach Bonn berufen.

Helmholtz' Vorlesungen über Anatomie waren aber - wie vom Ministerium offenbar befürchtet - nicht über jeden Zweifel erhaben (Ebert 1949: 41, 56; Koenigsberger 1902: Bd. 1, 257, 263, 283; Kirsten 1986: 238). Der preußische Staat war Helmholtz insofern bei der Berufung nach Bonn durchaus entgegengekommen. Er wollte ihn auch in Bonn halten. Trotzdem folgte Helmholtz schon 1858 nach einem langen Gezerre zwischen den Ministerien in Berlin und Karlsruhe einem Ruf auf die Professur für Physiologie im badischen Heidelberg. In wissenschaftlicher Hinsicht war Helmholtz' Heidelberger Zeit eine Zeit des Ausbaus und der großen handbuchartigen Zusammenfassung seiner Arbeiten zur physiologischen Optik und zur physiologischen Akustik. Zunehmend trat aber die theoretische Physik in den Mittelpunkt des Interesses von Helmholtz. Vor allem faszinierte ihn die kritische Auseinandersetzung mit den beiden um die geeignete Modellierung der elektrodynamischen Phänomene konkurrierenden Theorien von Wilhelm Weber und James Clerk Maxwell.

Das Problem, das in dieser Akzentverschiebung in Helmholtz' Forschungsinteressen lag, schien 1868 lösbar zu werden. Nach dem Tode von Julius Plücker bot sich die Möglichkeit, - nun als Physiker - nach Bonn zurückzukehren. Allerdings waren die Berufungsverhandlungen erneut unerfreulich, wohl auch deshalb, weil Helmholtz seine kaufmännischen Talente immer besser entfaltete. Offenbar ohne großes Verständnis zu wecken, schrieb er (am 7. 6. 1868) an seinen Bonner Freund Rudolf Lipschitz, wie schwer es ihm angesichts seiner "4½ Kinder" falle, "sich von der gut nährenden Milchkuh der medicinischen Facultät zu trennen, wenn man einmal an ihren Brüsten" liege, "und zu der keuschen Muse der philosophischen Facultät sich hinzuwenden" (Scharlau 1986: 124 f.).

1870, zwei Jahre später, ist Helmholtz ihr doch erlegen, und zwar als Nachfolger von Gustav Magnus auf dem Lehrstuhl für Physik an der Berliner Universität. Eigentlich war der Heidelberger Kollege Gustav Kirchhoff als der "geschultere Physiker und der bewährtere Lehrer" der Wunschkandidat der

Berliner Philosophischen Fakultät gewesen. Kirchhoff konnte sich aber zunächst nicht von Heidelberg lösen. Anstelle von Kirchhoff ging der zweitplazierte Helmholtz, der "genialere umfassendere Forscher", nach Berlin (Koenigsberger 1903: Bd. 2, 179 f, 187-189; Kirsten 1986: 238-242). Die Tatsache, daß das Berufungsverfahren aus preußischer Sicht nun so erfolgreich verlief, hat auch damit zu tun, daß Helmholtz' alter Freund Emil du Bois-Reymond als Rektor die Verhandlungen führte. Jedenfalls konnte Helmholtz in Berlin nun sein erstes Königsberger Professorengehalt verfünffachen (auf 4.000 Taler im Jahr). Außerdem bekam er den Neubau eines physikalischen Instituts inklusive Dienstwohnung zugesagt. "So geschah" - wie es du Bois-Reymond ausdrückte - "das Unerhörte, dass ein Mediciner und Professor der Physiologie den vornehmsten physikalischen Lehrstuhl in Deutschland erhielt" (Koenigsberger 1903: Bd. 2, 187). Den überragenden Rang des 50jährigen Helmholtz und die pompöse Atmosphäre der 1870er Jahre spiegelt aber auch ein einziger gedrängter Satz Leo Koenigsbergers über Helmholtz' Abschied in Heidelberg: "[...] alle beherrschte [...] das Gefühl, dass der grösste Denker und Forscher Deutschlands dorthin gehöre, wo dem Gründer des Deutschen Reiches der gewaltigste Staatsmann und der genialste Feldherr zur Seite standen" (Koenigsberger 1903: Bd. 2, 189 f.). Helmholtz also als "Reichskanzler der Wissenschaften" (Koenigsberger 1903: Bd. 3, 97) in einer Reihe mit Wilhelm I., mit Bismarck und mit dem Generalfeldmarschall Helmuth Graf von Moltke.

Betrachtet man die Eliten des Kaiserreichs aus dem historischen Abstand, so kommt man jedoch nicht umhin, den Lobeshymnen des Helmholtz-Biographen Koenigsberger einige weniger harmonische Töne hinzuzufügen. Denn das 19. Jahrhundert brachte nicht nur diese penetranten Elogen hervor, sondern gleichzeitig auch unglaubliche Grobheiten in den wissenschaftlichen Auseinandersetzungen. Und für beides bietet eben auch die Biographie Helmholtz' markante Beispiele.

Auf der einen Seite stilisierte Koenigsberger Helmholtz zum "gottbegnadeten Fürsten im Reiche geistiger und sittlicher Macht [...]" (Koenigsberger 1903: Bd. 3, Vorwort). Oder, um eine etwas erträglichere Passage zu zitieren:

> "Schon seit mehr als zehn Jahren war das Helmholtz'sche Haus der Sammelpunkt der auserlesensten Geister der neuen Reichshauptstadt gewesen; hier fanden sich - nicht auf neutralem, sondern auf einem für alles Gute und Schöne empfänglichen Boden - die ernstesten Denker mit den genialsten Künstlern zusammen und befruchteten gegenseitig Verstand und Gemüth; die Schüler *Moltke's* und die Eingeweihten *Bismarck's* fanden Fühlung unter der olympischen Ruhe des grossen Naturforschers und an der Hand der ausgezeichneten Frau, welche hier ihre

> glänzende Gabe entfaltete, die verschiedensten Geister mit einander in Berührung zu bringen; äussere Stellung allein bedeutete wenig für sie, wenn nicht Vorzüge des Geistes oder vornehmer ästhetischer Gesinnung kenntlich waren" (Koenigsberger 1903: Bd. 2, 383).

Und in Helmholtz' eigenen Worten:

> "Ich habe mich mein Leben lang gegen ein niedriges Niveau von Umgang gewehrt, und wo es mir nicht octroyirt ward, es mir auch ferne gehalten. Gute Lebensformen und einen geistigen Inhalt, der nach irgend einer Richtung hin mir überlegen oder doch interessant ist, habe ich als erstes Erforderniss zum Verkehr stets empfunden. Hierin *darf* man nicht bescheiden sein, wenn man nicht in der Mittelsorte untergehen will" (Koenigsberger 1903: Bd. 2, 383).

Also die Vorstellung von "Leistungselite" *und* "Werteelite" par excellence, so, wie sie das Denken der deutschen Hochschullehrer seit Fichtes Vorlesungen über die "Bestimmung des Gelehrten" von 1811 prägte (Schwabe 1988: 16 f.).

Auf der anderen Seite hatte Helmholtz seinem Freund Emil du Bois-Reymond in einer auch in einem privaten Brief wenig geschmackvollen Weise zum Ableben eines unbequemen Künstler-Kollegen förmlich gratuliert. Und etwas später brachte Helmholtz in einem weiteren Brief an du Bois-Reymond den "maßlosen Eigendünkel" und die "pekuniären Interessen" eines eigenen Bonner Mediziner-Kollegen unverblümt mit den "schwachen Seiten der semitischen Nationalität" in Verbindung. Voller Verachtung äußerte sich Helmholtz auch über den großen französischen Physiologen Claude Bernard. Weil die französischen Kollegen einen "solchen Lump wie Bernard wegen seiner liederlichen Arbeit über den pankreatischen Saft [also über das Sekret der Bauchspeicheldrüse!] bekränzen", wollte er "diese Bande in ihrer Nichtigkeit vorläufig ganz [...] ignorieren [...]" (Briefzitate in Kirsten 1986: 89, 92; 160 f; 105). Die "Philosophen von Fach" bezeichnete Helmholtz schließlich in einem Brief an Rudolf Lipschitz (vom 2.3.1881) als "impotente Bücherwürmer, die nie ein neues Wissen erzeugt haben" (Scharlau 1986: Bd. 2, 130). Angesichts solcher Fehlleistungen ist man zumindest gehalten, die Zugehörigkeit der deutschen Hochschullehrer im 19. Jahrhundert zu einer "Leistungselite" *und* zu einer "Werteelite" (Schwabe 1988: 16) strikt historiographisch zu verstehen und sie ausschließlich dem eigenen Selbstverständnis und dem Selbstverständnis der Zeit zuzuschreiben - wenn man nicht sogar eine betont kritische Haltung gegenüber Helmholtz einnehmen will.

Aus einer solchen kritischen Distanz erscheint dann die wissenschaftsorganisatorische Führungsposition auch als eine Folge geschickt genutzter Freundschaften und einer natürlichen - und zeittypischen - Fähigkeit, eigene materielle Interessen durchzusetzen. Selbst die wissenschaftliche Leistung scheint eben immer auch verbunden mit einem merkwürdigen Ignorieren oder Verflachen fremder Leistungen. Unverkennbar ist dieser Charakterzug Hermann von Helmholtz' beispielsweise in seinen Beiträgen zur Elektrodynamik (Kaiser 1993): So konnte er die Schwächen seiner Kritik an der Elektrodynamik Wilhelm Webers letztlich nur mit der druckvollen Rhetorik seiner Publikationen überspielen. Das Lob der Maxwellschen Feldtheorie gab ihm umgekehrt Gelegenheit, mit eigenen elektrodynamischen Feldgleichungen (Buchwald 1985: 177-186) in das Geschehen einzugreifen, wobei der schiere Umfang der Arbeiten und das von Helmholtz suggerierte Enthalten-Sein der Maxwellschen Theorie in seinem eigenen, umfassenderen Gleichungssystem eher zur Verschleierung der Konkurrenzsituation beitrugen.

3. Helmholtz, Siemens und die Gründung der Physikalisch-Technischen Reichsanstalt

Abschluß und Krönung der beruflichen Laufbahn sollte für den bereits 66jährigen Hermann von Helmholtz die Präsidentschaft der 1887 in Berlin gegründeten Physikalisch-Technischen Reichsanstalt werden. Mit ihrer Kombination von Grundlagenforschung und technischer Anwendung wurde die PTR als Produkt der Aktivität von Interessengruppen und von staatlicher Wissenschaftspolitik (Pfetsch 1974: 109-127; von Weiher 1990) ein frühes Beispiel für "gelenkte Forschung". Gleichzeitig markiert die Gründung der PTR einen ersten Höhepunkt in der tiefgehenden Verwissenschaftlichung aller anwendungsorientierten und technischen Disziplinen in Deutschland. Vorreiter waren die chemische Industrie und die frühe Elektrotechnik gewesen, und in charakteristischer Weise hatte Werner von Siemens mit dem ihm schon in jungen Jahren eigenen Selbstbewußtsein von der Telegraphie als einer eigenen "Branche der wissenschaftlichen Technik" gesprochen, in der er sich "einigermaßen berufen" fühle, "organisierend [...] aufzutreten [...]" (von Weiher/Goetzeler: 1981: 8).

Werner von Siemens und der ihm seit 1845 aus der Berliner Physikalischen Gesellschaft bekannte Hermann von Helmholtz gehörten dann auch zu einer kleinen Gruppe von mathematisch-naturwissenschaftlich ausgerichteten Lehrern, Hochschullehrern und Technikern, die 1872 eine Denkschrift zur Errichtung eines Instituts zur Förderung der Präzisionsmechanik an das Preußische Kultusministerium richteten (Bortfeldt, Hauser, Rechenberg 1987: 27-48;

Cahan 1992). Die zu einem Gutachten herangezogene Preußische Akademie ließ den Antrag jedoch ins Leere laufen. Generalstabschef Helmuth von Moltke, der Verantwortlicher für die Landesvermessung - und Freund der Familie Helmholtz - war, setzte daraufhin einen neuen Ausschuß ein, der mit den schwerpunktmäßig noch auf die Technik ausgerichteten Plänen an den Preußischen Landtag und an das Kultusministerium herantrat. Die Zustimmung des Kultusministeriums wurde durch die Auseinandersetzungen um Inhalt und Finanzierung der gleichzeitig entstehenden Technischen Hochschule Berlin-Charlottenburg aber weitgehend wieder entwertet.

Werner von Siemens war es dann, der 1883/1884 im Zusammenwirken mit Helmholtz den nun schon seit mehr als zehn Jahren auf der politischen, der akademischen und der technisch-industriellen Ebene ablaufenden Diskussionsprozeß den entscheidenden Schritt voranbrachte. Zusammen mit aus früheren Kommissionen verbliebenen Teilnehmern arbeiteten sie am 23. Mai 1883 und am 16. Juni 1883 Denkschriften zugunsten der Begründung eines "Institutes für die experimentelle Förderung der exacten Naturforschung und der Präzisionstechnik" aus. Helmholtz hatte in seinem Sondervotum den Part übernommen, neben einer mechanisch-technischen Abteilung auch auf die Schaffung einer wissenschaftlichen Abteilung zu drängen, und Siemens den Part, mit Blick auf die internationale Konkurrenzfähigkeit der Industrie eine außeruniversitäre institutionalisierte Form der Erzeugung und des Transfers von physikalischem Grundlagenwissen zu fordern (Koenigsberger 1903: Bd. 2, 347, Siemens an Helmholtz, 26.4.1883, 29.4.1883; Siemens Archiv München; Cahan 1992: 80-82). Siemens machte außerdem das Angebot, durch die Schenkung eines entsprechenden Grundstücks und durch einen Geldbetrag von 300.000 Mark, der zum größeren Teil aus dem Erbe des Bruders William zur Verfügung stand (Cahan 1992: 92), das Startkapital für die PTR bereitzustellen.

Wie stark Werner von Siemens mit der Gründung der Physikalisch-Technischen Reichsanstalt befaßt war, geht schließlich daraus hervor, daß er auf Bismarcks Ersuchen hin 1884 einen Organisationsplan konzipierte. Demnach sollte die PTR zwei Abteilungen bekommen: eine physikalische Abteilung (die spätere wissenschaftliche Abteilung I), die Grundlagenforschung mit besonders hohem Aufwand an Zeit, Gerät, Material und Dienstleistungen betreiben sollte, sowie eine technische Abteilung, die vor allem der Förderung der Präzisionsmechanik dienen sollte (was ja der Ausgangspunkt der ganzen Bestrebungen gewesen war), weiter der Eichung und Überprüfung von Meß- und Regelgeräten und dem Bau von Spezialapparaten.

Erst im März 1887 bewilligte der Reichstag die nötigen staatlichen Mittel von 700.000 Mark, und dies offenbar nur nach massiven Appellen an das Nationalbewußtsein und unter Hinweis auf den scharfen wirtschaftlichen Wettbewerb der Nationen. Rudolf Virchow verknüpfte diese Argumentation vor allem mit dem Namen Werner von Siemens: "Wenn nun Siemens kommt und den Vertretern der Nation gegenüber direkt erklärt: dies halte ich für die dringendste Aufgabe, wenn überhaupt die Nation in dem Konkurrenzkampf mit den anderen Nationen siegreich bestehen [...] soll [...] dann, muß ich sagen, sagt der deutsche Reichstag nur Ja und Amen" (Pfetsch 1974: S. 116, Anm. 17). Träger der Entscheidung war letztlich, um hier Frank Pfetsch zu folgen, ein kleiner elitärer Kreis führender Persönlichkeiten aus Wissenschaft, Industriebürgertum und Politik (Pfetsch 1974: 112-124).

Die nachhaltige Unterstützung der Gründung der PTR durch Werner von Siemens war zunächst eine gewisse Selbstbestätigung des erfolgreichen Unternehmers. Über diese Rolle als Mäzen hinaus ging es Siemens um eine Stärkung des preußisch geführten Deutschen Reiches durch eine hochstehende Wissenschaft, durch eine Förderung von Technik und Industrie. Hinzu kamen angesichts der 1881 bis 1884 auf den internationalen Pariser Konferenzen über die Festlegung elektrischer Maßeinheiten offenkundig gewordenen wissenschaftsorganisatorischen Schwächen konkrete industrielle Wünsche an die Errichtung eines außeruniversitären meßtechnischen Institutes (W. v. Siemens: 1893: 285; Koenigsberger 1903: Bd. 2, 285-287, W. v. Siemens an Helmholtz, 28. 7. 1892, Siemens Archiv München). Nicht zuletzt war die Gründung der PTR aber auch Ausdruck seiner denkbar engen persönlichen Beziehungen zu Helmholtz. Und in diese Zeit - genau auf den 10. November 1884 - fällt auch die Heirat von Helmholtz' Tochter Ellen aus erster Ehe mit Arnold Wilhelm von Siemens, dem ältesten Sohn von Werner von Siemens (Koenigsberger 1903: Bd. 2, 315).

Die Mehrzahl der deutschen Physiker hat sich im Vorfeld der Gründung der PTR weitgehend passiv verhalten, abgesehen von Ernst Abbé, mit dem Helmholtz wissenschaftlich durch seine Abhandlung über "Die theoretischen Grenzen für die Leistungsfähigkeit des Mikroskops" in Berührung gekommen war. In Süddeutschland unterstellte man sogar offen, daß Werner von Siemens die Reichsanstalt, und hier wiederum die physikalische Abteilung, seinem alten Freund Helmholtz regelrecht zum Geschenk gemacht habe (Cahan 1987: 35). Die Betrauung mit dem Präsidentenamt an der PTR war insofern gleichermaßen Ausdruck für den hohen wissenschaftlichen Rang des "ersten Physikers" Helmholtz (Koenigsberger 1903: Bd. 3, 100; W. v. Siemens 1893: 286) und Beweis für seine Zugehörigkeit zu den dominierenden Berliner Eliten des Deutschen Kaiserreichs. Und Helmholtz war nicht zuletzt auch deshalb der geeignete Präsident, weil er durch zahlreiche Ehrungen - 1882 war er geadelt worden - die geeignete personelle Brücke zu den politischen und militärischen Akteuren bot.

Allerdings wurde die Konzentration der Kräfte in der Metropole immer erdrückender. Gerade im wissenschaftlich-technischen Bereich, und seit 1884 sinnfällig geworden durch den neuen Prachtbau der nun im Deutschen Reich führenden TH Berlin-Charlottenburg, erwies sich die überragende Rolle der preußischen und insbesondere der Berliner Hochschulen (König 1988, 193-194). Hinzu kam der Boom der "Berliner Elektro-Großindustrie" und, damit eng verwoben, die Macht der "Berliner Banken" (Barth: 1980: 286-293). Und nicht umsonst sah sich Werner von Siemens veranlaßt, auf Helmholtz im Sinne einer Öffnung der PTR einzuwirken. In einem Brief vom September 1885 schrieb Siemens:

> "Lieber Freund!
> Am Sonnabend haben wir bei p. Weymann[1] die Vorlage betreffs der physik.-techn. Reichsanstalt durchberaten und [es] wird Ihnen das Protokoll wohl demnächst zugehen. Es machte nur ein Punkt Bedenken und ich habe es übernommen Ihnen darüber zu schreiben. Es betrifft die allgemeine Organisation wie sie aus Ihrer Denkschrift und dem Organisationsplan *ersichtlich* ist! Wir stimmten ja bei früheren Beratungen schon darüber überein, dass es absolut notwendig sei, dem Institut den spezifischen Berliner Charakter dadurch zu nehmen, dass es durch seine Organisation als ein, den Gelehrten von ganz Deutschland zugängliches, als ein Arbeitsplatz für tüchtige, deutsche Gelehrte überhaupt erscheint. Findet dies nicht statt, so wird der schon bedenklich erhobene Ruf - es soll der wissenschaftliche Fortschritt für ein Berliner Institut und die Angestellten desselben monopolisiert werden - überall, namentlich aber im Bundesrat und Reichstage in bedenklichem Masse erhoben! Das muß jedenfalls vermieden werden [...]" (W. v. Siemens an Helmholtz, 7. 9. 1885, Siemens Archiv München).

Helmholtz hatte zwar mit dem Neuaufbau des Berliner Physikalischen Instituts in den 1870er Jahren bereits wichtige administrative Erfahrungen gesammelt. Trotzdem dauerte es bis 1892, bis die wissenschaftliche Abteilung der PTR

---

1 Es handelte sich hier um den "Geh. Ober-Regierungsrath" Weymann, der Helmholtz im April 1887 im Auftrage des "Staatssecretärs des Innern" das Angebot überbrachte, das Präsidium der PTR zu übernehmen. Vgl. Koenigsberger 1903, Bd. 2, 351; Helmholtz sagte am 4. April zu, unter der Voraussetzung, daß er finanziell für seine beiden Lehrämter an der Berliner Universität und am Medicinisch-Chirurgischen Friedrich-Wilhelms-Institut entschädigt würde und seine Stellung an der Berliner Akademie der Wissenschaften unangetastet bliebe, a. a. O.

bezüglich Personal- und Sachmittel wirklich arbeitsfähig war. Auch war die von Helmholtz gewünschte reine Grundlagenforschung zunächst weitgehend zugunsten technischer Problemlösungen zurückgestellt worden. Schließlich waren die Aufgaben, die Helmholtz übernommen hatte, kaum mehr zu übersehen: Aufbau und Verwaltung der PTR samt ihrer Publikationen, öffentliche Ämter wie an der Akademie oder in der Prüfungskommission der Internationalen Elektrizitätsausstellung 1891 in Frankfurt, Fortsetzung der Lehrtätigkeit an der Universität und vor allem auch Weiterführung der Forschung im Umkreis der Elektrodynamik.

Helmholtz hat zwar immer versucht, seinen Mitarbeitern an der PTR seine ungebrochene Begeisterung für die Wissenschaft zu vermitteln. Allerdings gibt es Anzeichen, daß Helmholtz im Alter keinesfalls mehr der belastbare und kommunikationsfreundliche Organisator war, den die Aufbausituation der PTR im Grunde erfordert hätte; er sei "hartnäckig und rücksichtslos gegen die Menschen geworden, die [...] [seine] Zeit in Anspruch zu nehmen" gesucht hätten, wenn er müde gewesen sei (Koenigsberger 1903: Bd. 2, 263). Ganz abgesehen davon, daß eine Kette von Todesfällen aus dem engsten Kreis seine Amtszeit als Präsident der PTR verdüsterte: Erst starb sein Sohn Robert 1889, eben noch - wie es bei Koenigsberger heißt - "ohne Wissen des Vaters zu dessen freudigster Ueberraschung zum Assistenten an der Reichsanstalt" (Koenigsberger 1903: Bd. 3, 23) ernannt worden, dann Werner von Siemens 1892, Heinrich Hertz 1894. 1894 sollte auch Helmholtz' Todesjahr werden.

Manche der biographisch und wissenschaftsorganisatorisch problematischen Aspekte des Aufbaus der PTR durch den alternden Hermann von Helmholtz sind in eine vorwiegend auf Helmholtz gezielte Kritik des bedeutenden Göttinger Kristallphysikers Woldemar Voigt eingeflossen. 1912 sagte er in einer Rede:

> "Im allgemeinen besitzt die [wissenschaftliche] Abteilung [der PTR] aber nicht die ungeteilte Anerkennung der Physiker. Die ganz ungewöhnlichen Dimensionen ihrer Gesamtanlage erfordern zu einer zielbewußten und erfolgreichen Leitung einen Präsidenten von gleichfalls ganz ungewöhnlicher Begabung und Frische, und die an sich ausgezeichneten Männer, die in die bez[ügliche] Stellung berufen worden sind, waren sämtlich in einem Alter, wo sie gegenüber ihrem früheren Amte Erleichterung wünschen mußten. Keiner von ihnen hat bisher der Abteilung den Stempel seines Geistes aufgedrückt; die bedeutendsten aus ihr hervorgegangenen Untersuchungen lagen vielmehr dem Arbeitsgebiet des bez[üglichen] Leiters fern. Damit entfällt aber im Grunde die Rechtfertigung für die enorme Häufung der Hilfsmittel an einer Stelle, und man darf vermuten, daß eine verständnisvolle und elastische Verwendung derselben an wechselnden Stellen, wo immer ein wichtiges Problem und eine treibende Persönlichkeit sich finden,

> segensreicher wirken würde. Bei Errichtung und Betrieb solcher Rieseninstitute, deren leider bereits neue [die Kaiser-Wilhelm-Institute!] im Werke sind, drängt sich leicht ein *dekoratives* Element ein, das im Widerspruch mit wahrer Wissenschaftlichkeit steht" (Warburg 1912: 1092).

Emil Warburg als derzeit amtierender Präsident der PTR hielt Voigt in seiner geharnischten Antwort entgegen, Helmholtz habe ganz bewußt die wissenschaftliche Abteilung am langen Zügel geführt (Warburg 1912: 1093). Und eine der allgemeinen Aufgabenstellungen, moderne Meßmethoden für die Thermodynamik zu entwickeln, hat tatsächlich zu den Arbeiten von Willy Wien über die Energieverteilung in der Wärmestrahlung geführt, zu den Messungen der Energieverteilung der Hohlraumstrahlung durch Otto Lummer und Ernst Pringsheim, durch Heinrich Rubens und Ferdinand Kurlbaum. Neben den Messungen der spezifischen Wärmen von Festkörpern bei tiefen Temperaturen bei Walther Nernst ist so an der PTR eine der beiden experimentellen Säulen der frühen Quantentheorie entstanden (Hermann 1969: 146 f.), wobei die Präsidentschaft allerdings in der entscheidenden Phase bereits bei Friedrich Kohlrausch lag und einer der Genannten, nämlich Willy Wien, durchaus zu den Kritikern der PTR zu rechnen ist.[2]

4. Decline of the Mandarins

Diese Vorgänge im Umkreis der Entstehung der PTR sind nun insbesondere auch für den *Wandel* der Rolle der Hochschullehrer als eine der - nach Max Weber "ständisch" geprägten - Eliten des Deutschen Kaiserreichs signifikant. In zweifacher Hinsicht: Obwohl Helmholtz 1877 in seiner Antrittsrede als Rektor der Berliner Universität noch das Hohe Lied der akademischen Freiheit der deutschen Universitäten gesungen hatte - sie reiche von den "extremsten Consequenzen materialistischer Metaphysik" über die "kühnsten Speculationen auf dem Boden von Darwin's Evolutionstheorie" bis zur "extremsten Vergötterung päpstlicher Unfehlbarkeit" (Helmholtz 1896: 205) - rückte Helmholtz nun an die Spitze einer Institution, die, bewußt losgelöst von der - von Helmholtz zunehmend als Last empfundenen - Lehre (Koenigsberger 1903: Bd. 2, 345; W. v. Siemens an Helmholtz, 7. 9. 1885, Siemens Archiv München; W. v. Siemens 1893: 285), zu einem guten Teil gelenkte oder zumindest

2 Wien konstatierte 1918, daß die Verbindung von Wissenschaft und Technik nicht geglückt sei und überhaupt die PTR sich anders, als es das ursprüngliche Konzept besagte, entwickelt habe. Vgl. Wien 1921: 236 f.

anwendungsorientierte Forschung betrieb. Und obwohl Helmholtz ersichtlich und unbestreitbar auf dem Höhepunkt seines wissenschaftlichen und gesellschaftlichen Einflusses war, markiert das Zusammenwirken mit Werner von Siemens, oder sogar die dominierende Rolle von Werner von Siemens, nun den Aufstieg einer neuen Elite, nämlich der des Industriebürgertums, des wirtschaftlichen Großbürgertums.

Fritz Ringer hat diese sich zwischen 1890 und 1920 vollziehenden Verschiebungen mit der wuchtigen, aber eindrucksvollen Metapher "The Decline of the German Mandarins" belegt, wobei er die noch 1969 vorgenommene Beschränkung auf die Geisteswissenschaften später gelockert hat (Ringer 1969/1987; Ringer 1988: 103 f.; Burchardt 1977: 40 f.). Zu dem durch Konkurrenz verursachten Bedeutungsverlust der alten Bildungseliten kam zwangsläufig und nur zeitlich phasenverschoben die soziologische Öffnung. Sichtbar an der sich wandelnden Herkunft der deutschen Ordinarien drangen seit 1890 in die sich ursprünglich weitgehend aus sich selbst heraus regenerierenden akademischen Eliten Kaufmanns- und Industriellenfamilien ein (Burchardt 1988: 152 f., 163). Auch umgekehrt wurden die Grenzen durchlässiger. So absolvierte Helmholtz' Sohn Richard das Polytechnicum in München, ging in die Industrie und wirkte bis 1918 als Leiter des Konstruktionsbüros und als Oberingenieur der Münchner Lokomotivenfabrik Krauss (Koenigsberger 1903: Bd. 3, 245), eine der Vorgängerfirmen der Krauss-Maffei AG.

Der ungeachtet der neuen Rangfolge wachsende Austausch zwischen den Eliten ging einher mit einem Wandel in der politischen Orientierung. Zwei Entwicklungslinien im Umkreis von Werner von Siemens und Physikalisch-Technischer Reichsanstalt mögen beispielhaft für die offenbar nicht mehr aufzuhaltenden historischen Weiterungen stehen: Auf den patriotischen, aber linksliberalen Werner von Siemens folgte der nationalliberale, von sozialdarwinistischem Gedankengut beeinflußte Wilhelm von Siemens (Kocka 1969: 386-389; Barth 1980: 173-178e, 201, 294-296). Und schließlich war die PTR nicht davor gefeit, nach in der Regel gemäßigt konservativen, aber politisch unauffälligen Präsidentschaften (Helmholtz, Friedrich Wilhelm Kohlrausch, Emil Warburg, Walther Nernst[3], Friedrich Paschen) Johannes Stark, den herausragenden Vertreter der "Deutschen Physik", an ihre Spitze berufen zu bekommen (Kern 1987: 106), wodurch der wissenschaftliche Abstieg der

3 Nernst war allerdings Unterzeichner des berüchtigten "Aufrufs an die Kulturwelt"; Vgl. dazu Sudermann 1914: 2.

Reichsanstalt seit Kohlrausch nun sein Pendant in der Führung bekommen hatte (Rompe 1972: 59).

Einige wichtige historiographische Themen der deutschen Geschichte des 19. und 20. Jahrhunderts werden also in diesem Essay angesprochen, nämlich der Rang und der Charakter der deutschen Bildungseliten, die Grenzen, an die sie stießen, und der zunehmende Druck, der wegen des außergewöhnlich hohen Industrialisierungstempos auf diesen Eliten lastete. Gleichzeitig sollte damit der Blick auf den fatalen Gang der Dinge gelenkt werden, auf den langsamen Abstieg und das Versagen dieser Eliten, insbesondere seit dem 1. Weltkrieg (vgl. dazu Sudermann 1914: 2), und auf die Eröffnung eines deutschen "Sonderweges" infolge des Versagens der alten Eliten. Zwar wird die Vorstellung von einer sich überschlagenden deutschen Industrialisierung durch ökonomische Vergleiche mit anderen europäischen Ländern nicht unbedingt gestützt. Die deutsche Industrialisierung fällt also nicht so aus dem Rahmen, daß sie den in die Katastrophe von 1933 führenden deutschen "Sonderweg" allein erklären kann. Unverkennbar ist aber die mangelnde Fähigkeit bei der Bewältigung der Folgen der Industrialisierung und ein "Ausweichen in rückwärtsgewandte, vorindustrielle Leitbilder" (Kaelble 1983: 106-118).[4] Jedenfalls zeigt der durch Hermann von Helmholtz' und Werner von Siemens' Biographien und durch die Gründung der Physikalisch-Technischen Reichsanstalt markierte Schnittpunkt des nach oben weisenden Wegs des Wirtschaftsbürgertums und des eher nach unten weisenden Wegs der Bildungseliten[5] ein über die Familien Siemens' und Helmholtz' hinausreichende politisch-historische Bedeutung.

4 Man ist geneigt, die etwas archaischen Tendenzen, wie sie etwa im Besitz der Rittergüter von Werner und Wilhelm von Siemens aufscheinen, gerade in diesem Zusammenhang zu sehen. Vgl. Kocka 1969: 386; Barth 1980: S. 178b, S. 294.

5 Hierher gehören auch weitere Details der Familiengeschichte von Werner von Siemens: Werner von Siemens heiratete in erster Ehe die Tochter des Königsberger Althistorikers Wilhelm Drumann, seine Schwester heiratete den Göttinger Physiker August Friedrich Karl Himly. Vgl. zu den Details des sozialgeschichtlichen Kontextes Barth 1980: 173-178e. - Offenbar konnten Privatdozenten in den naturwissenschaftlichen Fächern ihre sowohl im persönlichen wie im universitären Bereich schlechten finanziellen Verhältnisse gelegentlich nur durch die Eheschließung verbessern. So heiratete der Chemiker Carl Harries Werner von Siemens' Tochter Hertha, der Physiker und Privatdozent der Philosophie Leo Arons, dem wegen seiner Vortragstätigkeit für die SPD im Jahre 1900 die Lehrbefugnis entzogen wurde, ging mit der Tochter eines Berliner Privatbankiers eine Ehe ein. Vgl. dazu Burchardt 1988: 163; vgl. wegen Leo Arons auch Schwabe 1988: 13.

Literatur

Barth, Hans Martin: *Berliner Elektro-Grossindustrie in der deutschen Politik. Elektroindustrie-Verbände-Parteien (1862-1920)*. Philosphische Dissertation FU Berlin: 1980.

Bortfeldt, J., Hauser, W., Rechenberg, H. (Hrsg.): *Forschen - Messen - Prüfen. 100 Jahre Physikalisch-Technische Reichsanstalt/Bundesanstalt 1887 - 1987*. Weinheim: Physik-Verlag, 1987, 27-48.

Buchwald, Jed Z.: *From Maxwell to Microphysics, Aspects of Electromagnetic Theory in the Last Quarter of the Nineteenth Century*. Chicago/London: The University of Chicago Press, 1985.

Burchardt, Lothar: Halbstaatliche Wissenschaftsförderung im Kaiserreich und in der frühen Weimarer Republik. In: Mann, Gunter/Winau, Rolf (Hrsg.): *Medizin, Naturwissenschaft, Technik und das Zweite Kaiserreich*. Göttingen: Vandenhoeck & Ruprecht, 1977, 35-51.

Burchardt, Lothar: Naturwissenschaftliche Universitätslehrer im Kaiserreich. In: Schwabe, Klaus (Hrsg.): *Deutsche Hochschullehrer als Elite 1815-1945*. Boppard: Harald Boldt Verlag, 1988, 151-214.

Cahan, David: Die Physikalisch-Technische Reichsanstalt 1887-1918. In: Bortfeldt, J., Hauser, W., Rechenberg, H. (Hrsg.): *Forschen - Messen - Prüfen, 100 Jahre Physikalisch-Technische Reichsanstalt/Bundesanstalt 1887-1987*. Weinheim: Physik-Verlag, 1987.

Cahan, David: *Meister der Messung. Die Physikalisch-Technische Reichsanstalt im Deutschen Kaiserreich*. Weinheim/New York/Basel/Cambridge: VCH, 1992.

Cahan, David (ed.): *Hermann von Helmholtz and the Foundations of Nineteenth-Century Science*. Berkeley/ Los Angeles/London: University of California Press, 1993.

Ebert, Hermann: *Hermann von Helmholtz*. Grosse Naturforscher, Bd. 5. Stuttgart: Wissenschaftliche Verlagsgesellschaft, 1949.

Helmholtz, Hermann von: Ueber die akademische Freiheit der deutschen Universitäten. Rede gehalten beim Antritt des Rectorats an der Friedrich-Wilhelms-Universität zu Berlin 1877. In: Ders.: *Vorträge und Reden*, 4. Auflage, Bd. 2, Braunschweig: Vieweg, 1896, 191-212.

Hermann, Armin: *Frühgeschichte der Quantentheorie (1899-1913)*. Mosbach: Physik Verlag, 1969.

Kaelble, Hartmut: Der Mythos von der rapiden Industrialisierung in Deutschland. In: *Geschichte und Gesellschaft*, 9, 1983, 106-118.

Kaiser, Walter: Helmholtz's Instrumental Role in the Formation of Classical Electrodynamics. In: Cahan, David (ed.): *Hermann von Helmholtz and the Foundations of Nineteenth-Century Science*. Berkeley/ Los Angeles/London: University of California Press, 1993, 374-402.

Kern, Ulrich: Die Physikalisch-Technische Reichsanstalt 1918 bis 1945. In: Bortfeldt, J., Hauser, W., Rechenberg, H. (Hrsg.): *Forschen - Messen - Prüfen. 100 Jahre Physikalisch-Technische Reichsanstalt/Bundesanstalt 1887 - 1987*. Weinheim: Physik-Verlag, 1987, 68-112.

Kirsten, Christa (Hrsg.): *Dokumente einer Freundschaft. Briefwechsel zwischen Hermann von Helmholtz und Emil du Bois-Reymond 1846-1894*. Berlin: Akademie-Verlag, 1986.

Kocka, Jürgen: *Unternehmensverwaltung und Angestelltenschaft am Beispiel Siemens 1847-1914*. Stuttgart: Ernst Klett Verlag, 1969.

Koenigsberger, Leo: *Hermann von Helmholtz*. 3 Bde., Braunschweig: Friedrich Vieweg und Sohn, 1902-1903.

König, Wolfgang: Höhere technische Bildung in Preußen im Kaiserreich. In: Sodan, Günter (Hrsg.): *Die Fachhochschule Berlin im Spektrum Berliner Bildungsgeschichte*. Berlin: Technische Fachhochschule Berlin, 1988, 183-213.

Pfetsch, Frank R.: *Zur Entwicklung der Wissenschaftspolitik in Deutschland 1750-1914*. Berlin: Duncker & Humblot, 1974.

Ringer, Fritz K.: Das gesellschaftliche Profil der deutschen Hochschullehrerschaft 1871-1933. In: Schwabe, Klaus (Hrsg.): *Deutsche Hochschullehrer als Elite 1815-1945*. Boppard: Harald Boldt Verlag, 1988, 93-104.

Ringer, Fritz K.: *Die Gelehrten. Der Niedergang der deutschen Mandarine 1890-1933* [1969, dtsch 1983]. Nachdruck München: Deutscher Taschenbuch Verlag, 1987.

Rompe, Robert: Einige Worte über Helmholtz und die Physikalisch-Technische Reichsanstalt. In: Gedanken von Helmholtz über schöpferische Impulse und über das Zusammenwirken verschiedener Wissenschaftszweige, *Sitzungsberichte des Plenums und der Klassen der Akademie der Wissenschaften der DDR*. Jahrgang 1972, Nr. 1, Berlin: Akademie Verlag, 1972, 57-63.

Scharlau, Winfried (Hrsg.): *Rudolf Lipschitz. Briefwechsel mit Cantor, Dedekind, Helmholtz, Kronecker, Weierstrass und anderen*. Braunschweig/Wiesbaden: Friedrich Vieweg und Sohn, 1986. = Dokumente zur Geschichte der Mathematik, Bd. 2.

Schwabe, Klaus (Hrsg.): *Deutsche Hochschullehrer als Elite 1815-1945*. Boppard: Harald Boldt Verlag, 1988, 9-25.

Siemens, Werner von: *Lebenserinnerungen*. 3. Auflage, Berlin: Julius Springer, 1893.

Sudermann, Hermann: Es ist nicht wahr... ...Eine große Anzahl hervorragender Vertreter von Kunst und Wissenschaft erläßt folgenden Aufruf: An die Kulturwelt. *Zweites Morgenblatt der Frankfurter Zeitung*, 59, Nr. 275, 4.10.1914, 2.

Warburg, Emil: Über die Ziele der Physikalisch-Technischen Reichsanstalt; zur Abwehr. *Physikalische Zeitschrift*, 13, 1912, 1091-1093.

Weiher, Sigfrid von und Goetzeler, Herbert: *Weg und Wirken der Siemens-Werke im Fortschritt der Elektrotechnik 1847-1980*. 3. Auflage, Wiesbaden: Franz Steiner Verlag, 1981.

Weiher, Sigfrid von: Vorgeschichte und Gründung der Physikalisch-Technischen Reichsanstalt in Berlin. In: Vom Bruch, Rüdiger/Müller, Rainer A. (Hrsg.): *Formen außerstaatlicher Wissenschaftsförderung im 19. und 20. Jahrhundert. Deutschland im europäischen Vergleich*. Stuttgart: Franz Steiner Verlag, 1990, 53-62.

Wien, Willy: Physik und Technik. Vortrag, gehalten am 6. Mai 1918 in Reval. In: W. Wien: *Aus der Welt der Wissenschaft, Vorträge und Aufsätze*. Leipzig: Johann Ambrosius Barth, 1921, 235-263.

Bemerkungen zu Helmholtz' Geschichtsverständnis

Wolfgang Küttler

Für die allgemeine Geschichtswissenschaft ist Helmholtz nicht nur als "Reichskanzler der deutschen Wissenschaft", das heißt wegen seiner wissenschaftspolitischen und -organisatorischen Aktivitäten und seiner exponierten Stellung in der damaligen Bildungselite interessant (Cahan 1992, 153ff.). Letztere vor allem lenkt das Interesse eines theoriegeschichtlich arbeitenden Historikers auch auf übergreifende Fragen der Stellung der Naturwissenschaften im gesellschaftlichen Diskurs der damaligen Zeit. Selbst kein Spezialist für Biographie und Werk von Helmholtz, möchte ich im folgenden zu rekonstruieren versuchen, welche Bedeutung das Historische im weiteren und ein Geschichtsverständnis im engeren Sinne für die Auffassungen und Arbeiten dieses großen Naturforschers hatte. Es soll also jener Teil des Themas Helmholtz als Gestalt seiner Zeit erörtert werden, der seine individuelle Reflexion von Geschichtlichkeit und Zeiterfahrung betrifft. Daß sich Helmholtz' eigenes Epochenbewußtsein und wissenschaftliches Selbstverständnis wesentlich von der heutigen Zuordnung unterscheiden, wäre wie für jede beliebige Biographie eine triviale Feststellung, wenn damit nicht zugleich das Problem der Prioritäten bei der Beurteilung und Periodisierung seines Schaffens aufgeworfen wäre. Dafür nämlich eröffnet sich im Vergleich zur wissenschaftsgeschichtlichen Draufsicht aus der Distanz eines weiteren Jahrhunderts eine andere Perspektive nicht nur auf Leben und Umfeld, sondern auch auf wissenschaftliche Leistungen, Methodologie und philosophische Ansichten von Helmholtz.

Als Quelle dienen hauptsächlich Helmholtz' populärwissenschaftliche Reden und Vorträge, die zu diesem Thema für einen Naturwissenschaftler überraschend viel Material enthalten, was Helmholtz selbst in der Vorrede zu seinen gesammelten Vorträgen damit begründet, daß es ihm bei "Vorlesungen vor einem überwiegend literarisch gebildeten Publikum" besonders darauf angekommen sei, "das Wesen und die Tragweite der Naturgesetze und ihre Beziehungen zu den geistigen Tätigkeiten des Menschen anschaulich zu machen" (VR I, S. VI).

Dieser Intention nachgehend, sollen im folgenden vor allem Aspekte des Zusammenhangs von Natur- und Geschichtsbild, von Natur- und Menschheitsgeschichte sowie von Natur- und Geisteswissenschaften behandelt werden. Im einzelnen werden folgende Schwerpunktfragen behandelt: 1. Welche historischen Erfahrungen prägten Helmholtz' gesellschaftspraktische Orientierungen? 2. Von welchen historisch- kulturellen Perspektiven ließ er sich leiten? 3. Welchen Platz hatte das Historische in seinem Weltbild? 4. Wie beurteilte er die mit Geschichte und Kultur befaßten Disziplinen im Verhältnis zu den Naturwissenschaften?

I

Heute gilt Helmholtz zu Recht als einer der Vollender der klassischen Physik und des von ihr geprägten Weltbildes, wobei "Vollendung" immer auch Ende und Übergang zu etwas Neuem meint. In diesem Sinne beanspruchen Vorgriffe auf den Perspektivenwechsel in der Physik, an dessen Schwelle Helmholtz' Schaffen endet, und demzufolge die Suche nach Andeutungen auf spätere Entwicklungen in seinem Werk heute häufig mehr Interesse als der Vergleich mit den Ausgangspositionen, von denen er herkam und von denen seine Intentionen verändernden Eingreifens geprägt wurden (Röseberg 1994).

So gesehen, hatte Helmholtz schon angesichts der eigenen Karriere und ihres Pro und Kontra im Kontext deutscher Zustände im zweiten Drittel des 19. Jahrhunderts auch und in vieler Beziehung gerade nach 1871, als er sich immer größere Wirkungsmöglichkeiten erschließen konnte, keineswegs das Bewußtsein, am Ende oder im Übergang wissenschaftstheoretischer Leitvorstellungen zu stehen. Er sah seine eigene Lebensleistung wie auch die Kulturaufgaben der Naturwissenschaften durchaus noch am Anfang oder im allgemeinen Aufschwung der Verwirklichung ihrer im 17. und 18. Jahrhundert begründeten Grundlagen und Ziele. Das verwundert nicht, wenn wir seine Biographie allgemeinen Daten der deutschen Geschichte zuordnen. Die üblicherweise wichtigste lebensprägende Phase zwischen einem Alter von 15 und 30 Jahren fällt bei Helmholtz in die Zeit von der Mitte der 1830er bis zum Beginn der 1850er Jahre: 1838 begann er mit dem Studium, 1849 wurde er außerordentlicher, 1855 ordentlicher Professor in Königsberg. Es ist dies die konfliktreiche Zeit des Vormärz, der Revolution von 1848/49 und der folgenden Restauration. Das Revolutionsjahr 1848 erlebte er in Potsdam und später in Berlin, ohne daß die dramatischen Zeitereignisse Spuren in seinen Briefen hinterlassen hätten. Dafür mögen, wie sein Biograph Koenigsberger vermutet, Rücksichten auf seine Stellung als Militärarzt und der Umstand ausschlaggebend gewesen sein, daß er sich mit seinem Vater zu Hause mündlich austauschen konnte (I, 106ff.). Nach dem Zeugnis seines Freundes Emil du Bois-Reymond (1912, II, 563f.) war er vom Beginn der Revolution in Berlin tief beeindruckt, was zur Tradition des Elternhauses paßt - sein Vater war den freiheitlich idealistischen, liberalen Auffassungen Fichtes verpflichtet (VR II, 314; Koenigsberger I, 1-21). Insgesamt zeugt der Duktus seiner späteren Reden nicht gerade von innerer Nähe zum 1848er Liberalismus oder gar zu noch weitergehenden demokratischen Bestrebungen. Auch du Bois-Reymond stand revolutionären Erhebungen später ablehnend gegenüber (1912, I, 603).

Die blendende Karriere nach 1871 verstärkte bei Helmholtz - hierin typisch für das deutsche Bildungsbürgertum - die Richtung einer national, macht- und kulturpolitisch begründeten Loyalität zu Kaiser und Reich.[1] Er gehörte nicht nur wegen seines wissenschaftspolitischen Einflusses, sondern auch nach dem Stil seines Hauses (Werner/Irmscher 1993, 22ff.) - er verkehrte mit den einflußreichsten Persönlichkeiten, war mit der kaiserlichen Familie vertraut und erhielt 1883 den Adelstitel - zweifellos an höchst exponierter Stelle zum damaligen Establishment.

Insgesamt war er in politisch-historischen Urteilen jedoch sehr zurückhaltend. Die ersten Kriege Bismarcks 1863/64 und 1866 kommentierte er mit Sorge um die preußischen Alleingänge (Kirsten 1986, 221f.); erst 1870 bekennt er sich deutlich zu den preußisch-deutschen Kriegszielen gegen Frankreich, allerdings - trotz bekundeter Zustimmung (ebenda, 247, 253) - auch jetzt wesentlich moderater als sein Freund du Bois-Reymond, der in Berlin leidenschaftliche, stark nationalistisch und antifranzösisch eingefärbte Reden historischen Inhalts hielt ("Der deutsche Krieg", "Das Kaiserreich und der Friede,

1 Vgl. die kritischen Beiträge von W. Kaiser und D. Cahan in diesem Band, in denen Helmholtz' Ansichten und Karriere im Lichte der Einbindung des deutschen Bürgertums in die Herrschaftsstrukturen und den geistig-politischen Diskurs des kaiserlichen Deutschlands hervorgehoben werden.

1912, I, 393-430). Nur einmal - im einleitenden Abschnitt des Vortrags "Die Tatsachen in der Wahrnehmung" von 1878 - nach den Attentaten auf Kaiser Wilhelm und kurz vor Erlaß des Sozialistengesetzes - legte er sich heftiger für seinen Monarchen, in dem "sich Alles vereinigte, was die Menschheit bisher als würdig der Verehrung und der Dankbarkeit betrachtet hat" (VR II, 216), ins Zeug - freilich auch an dieser Stelle im Ton gemäßigter als viele andere Kollegen (Köhnke 1993, 414ff.).

Die explizit historischen Textpassagen seiner Reden, in denen er gewissermaßen Bilanz erlebter und mitgestalteter Veränderungen zog, beziehen sich dagegen überwiegend auf kontrastierende Vergleiche mit Besonderheiten der deutschen Geschichte, die dem Fortschritt und der gesellschaftlichen Geltung der Naturwissenschaften im Wege standen. Durchgehend, von den ersten Reden in Königsberg bis zum Lebensrückblick 1891, bezieht er sich auf die Ausgangskonstellation, als noch kein Vater seinem Sohne empfehlen konnte, Physik zu studieren (VR I, 20), und die entsprechenden Auseinandersetzungen der Zeit zwischen 1830 und 1848 wie auch auf deren Zusammenhänge mit der klassischen Philosophie und Dichtung, dem Idealismus der Befreiungskriege und der Romantik, die für Helmholtz' Vater prägend gewesen waren. Von da aus lassen sich auch Invariablen im Pro und Kontra der historischen Orientierung seiner Reden besser begreifen: Es ging ihm um die konsequente Durchsetzung der Erfahrungswissenschaften in allen Wissens- und Bildungsbereichen, um die Verbreitung eines darauf gegründeten Weltbildes, das auch die sittlich-geistige Tätigkeit normativ prägen sollte. Das bedeutete zugleich engagierte Polemik gegen metaphysische Indoktrinierung und Deduktionen, sofern damit innerwissenschaftliche Geltungsansprüche begründet werden sollten, gegen die Einflüsse spekulativer und normativer (theologischer, juristischer, klassisch-philologischer und wie auch immer gearteter) Vorurteile. In diesem Sinne verteidigte er den Empirismus der neuen Naturwissenschaften gegenüber den Sorgen seines Vaters, er könne sich von der klassischen Philosophie vollends abwenden (Koenigsberger I, 56f.), und im Briefwechsel mit dem Freunde und Kampfgefährten Emil du Bois-Reymond mokierte er sich gelegentlich über die Borniertheit der Universitätsverhältnisse (Kirsten 1986, 168f.).

Ganz entgegengesetzt zu den Grundlagendebatten des 20. Jahrhunderts, war die Ausgangslage für Helmholtz somit dadurch gekennzeichnet, daß sich die aufstrebenden neuen Naturwissenschaften zumindest im Deutschland des Vor- und Nachmärz, das heißt in den 1830er bis 1860er Jahren, noch gegen die Vormacht von Theologie und Philosophie wie auch der von diesen teilweise noch stark dominierten Geisteswissenschaften zu rechtfertigen hatten. Dem entspricht das Bild der neueren deutschen Geschichte, das Helmholtz wiederholt und am ausführlichsten 1871 in der Rede zum Gedächtnis von Gustav Magnus (VR II, 41ff.) wie folgt skizziert: Deutschland hatte ausgangs des Mittelalters mit der Reformation einen Riesenschritt zur geistigen Befreiung getan, ihn aber mit dem hohen Blutzoll, der Ohnmacht und Rückständigkeit der Epoche der Religionskriege bezahlt - eine finstere Zeit, in deren lähmende Nachwirkungen nur das Licht des großen Preußenkönigs Friedrichs II. etwas Hoffnung gebracht habe.

Diese Skizze deutscher Geschichte und Politik, die in früheren und späteren Reden immer wieder anklingt, dient Helmholtz jedoch nicht zu einer direkten liberalen Kritik, sondern diese kommt vielmehr indirekt, als Tadel hindernder oder - in bezug auf die Zeit nach 1871 - als Lob fördernder Rahmenbedingungen für den Aufstieg und die Entfaltung der Naturwissenschaften zu Wort. Letztere wiederum erscheinen nicht einfach als Fachdisziplinen, sondern sollen in einer gewissen Stellvertreterrolle für gesamtgesellschaftliche Vorstellungen Geist und Normen der modernen Gesellschaft prägen.

II

Helmholtz war leidenschaftlicher Naturforscher und hatte auch ein vom klassisch-modernen Leitbild dieser Tätigkeit bestimmtes historisches Perspektivenbewußtsein. Schon in seinem ersten Goethevortrag in Königsberg 1853 klingt dieses Leitmotiv an: Zwar zerstöre der Blick in den Mechanismus der Natur den schönen Schein, so daß Goethe noch gegen die Physik als etwas polemisierte, "das ihn jeden Augenblick in seinem poetischen Behagen zu stören droht. [...] Wir können aber den Mechanismus der Materie nicht dadurch besiegen, daß wir ihn wegleugnen, sondern nur dadurch, daß wir ihn den Zwecken des sittlichen Geistes unterwerfen". In dieser Aufgabe sieht Helmholtz "die große Bedeutung der physikalischen Forschung für die Kultur des Menschengeschlechtes und ihre volle Berechtigung gegründet" (VR I, 45). Helmholtz' historische Perspektive selbst beruht auf einem derart explizit mechanistisch geprägten Weltbild, und zwar durchgehend. Diese Orientierung hatte im wesentlichen den Horizont eines durch Naturwissenschaften und Technik geprägten Fortschritts (Helmholtz 1903, 21), und die "strenge Methode" der exakten physikalisch-mathematischen Forschung ist darin im Sinne einer zivilisatorischen Triebkraft (Cahan 1993) zugleich Modell und Instrument geistig-sittlicher Weltbeherrschung im allgemeinen wie auch der Macht und des Einflusses der kultivierten Nationen: Neben den ungeheuren praktischen Veränderungen, die sie bewirkt hat und die für alle offenkundig sind, so daß die europäischen Nationen sehr viel Geld in ihre Förderung investieren, ist "viel tiefer gehend noch und weiter tragend, wenn auch viel langsamer sich entfaltend [...] ihr Einfluß auf die Richtung des geistigen Fortschritts der Menschheit" (VR II, 423).

Diese Grundüberzeugung ist auch maßgeblich für leitende - wir würden heute sagen: lebensweltliche - Hinsichten seiner Forschungs- und Lehrtätigkeit ebenso wie seines Interesses an praktisch wirksamer Verbreitung des neuen Welt- und Wissenschaftsverständnisses: Erstens sind für ihn die Normen der *scientific community* zugleich Kulturideale, und er macht auch bei zunehmenden Bedenken wegen einer drohenden Wertekrise keinen oder wenigstens keinen deutlich artikulierten Unterschied zwischen den Idealen selbstverleugnenden Forscherdrangs und nicht näher gekennzeichneten höchsten allgemeinen Werten, an die er die akademische Jugend erinnert (VR II, 216-219). In diesem Sinne ist es plausibel, wenn er einerseits die Wissenschaftsförderung als entscheidend für die wirtschaftliche und militärische Macht der Nation einfordert und andererseits in der Wissenschaft wie in der Kunst das letzte noch funktionierende internationale Band zwischen den europäischen Kulturvölkern sieht (VR I, 12).

Zweitens hatte Helmholtz aus diesem Interesse an der Kulturmacht Wissenschaft auch eine allgemein kulturgeschichtliche Perspektive auf die Wissenschaftsentwicklung. Er reflektierte seine eigenen Forschungen nicht nur bewußt und kenntnisreich immer auch in ihrem wissenschaftsgeschichtlichen Zusammenhang. Er kritisierte in diesem Sinne die Geschichtslosigkeit eines Typs von Naturforschern, der von seinen geisteswissenschaftlichen Kollegen nicht nur als "auffällig gleichgültig gegen literarische Schätze", sondern "sogar für die Geschichte seiner eigenen Wissenschaft gleichgültiger als recht ist" (VR I, 166) betrachtet werden müßte. Vielmehr boten ihm vergleichende Äußerungen zur Kultur- und Wissenschaftsgeschichte zusätzliche Gelegenheit, auf die Schwierigkeiten des Weges hinzuweisen, der zur damaligen Geltung der Naturwissenschaft geführt hatte. Hierher gehört auch seine hohe Wertschätzung des deutschen Universitätswesens wegen der weitgehenden Durchsetzung akademischer Freiheit, wo gleichzeitig materialistische und klerikale Anschauungen ungehindert vorgetragen werden könnten (VR II, 205) - gewiß eine Idealisierung, die aber nicht nur als Artikulation des Professorenestablishments, sondern auch aus der Distanz zu den Verhältnissen der 30er bis

50er Jahre und aus Reiseeindrücken in England und Frankreich verständlich wird.[2]

Drittens - damit eng verbunden - ergab sich aus seinen ausgeprägten philosophischen und speziell erkenntnistheoretischen Interessen die Neigung zu wissenschaftstheoretischen Erwägungen über das Verhältnis der verschiedenen Disziplinen zueinander. Was seine philosophischen Auffassungen allgemein und seine Beziehung zu Kant speziell angeht, sei in diesem Zusammenhang wiederum auf die schon skizzierte prägende Ausgangskonstellation seines Wirkens verwiesen, die er immer wieder reflektierte. Dafür waren viertens praktische bildungspolitische Gründe wichtig, da Philosophie und traditionelle geisteswissenschaftliche Fächer noch weitgehend das Bildungsideal und den Gymnasialunterricht in Deutschland beherrschten. Die Erinnerung an seine Schulzeit, in der "die Mathematik immer nur als Fach zweiten Ranges betrachtet worden" sei, während der lateinische Aufsatz "damals noch wesentlich die Siegespalme bestimmte", blieb, wenn auch durch die Förderung der Naturwissenschaften abgemildert, auch nach 1871 aktuell (VR I, 13).

Klassik und Romantik sind für ihn nicht unbedingt wie schon für viele Zeitgenossen und später für Meinecke (1965, 2) durch die Entstehung des Historismus die zweite große Tat des deutschen Geistes, sondern vielmehr bei allem freiheitlichen Aufschwung und aller Größe von Dichtung und Philosophie eher eine Fortsetzung jener Flucht in Metaphysik und Kunst, bei der mit der Wirklichkeit auch die empirische Naturforschung beiseite liegenblieb (VR II, 43). Dafür macht er vor allem auch den Einfluß der Hegelschen Philosophie verantwortlich, der die aufstrebenden empirischen Naturwissenschaften in Deutschland von der Philosophie erst eigentlich entfremdet habe (I, 164f.). Kant und - aus der väterlichen Tradition besonders - Fichte sieht Helmholtz dagegen als Zeugen eines Philosophieverständnisses, das sich noch eins weiß mit der Entwicklung der Naturwissenschaften (VR I, S. 85ff.), und partielle Kritiken an Kant gelten später wahrgenommenen Inkonsequenzen dieser Einheit (Vorrede 1884, VR I, VIII; II, 405), die für Helmholtz empiristisch und positivistisch fundiert sein soll. Diese Art der Berufung auf Kant entspricht, wie Köhnke zeigt (1986, 152ff., 412ff.), in der frühen Phase den ersten Ansätzen neukantianischer Programmatik in den 1850er Jahren, und später - allerdings nur teilweise rezipiert - der idealistischen Wende um 1878.

---

2 Cahans Beitrag in diesem Band analysiert den hochschulpolitischen und persönlichen Hintergrund des Vortrags "Über die akademische Freiheit der deutschen Universitäten" von 1872. (VR II, 193ff.)

Die schon zitierte Berliner Rektoratsrede von 1878 zeigt die zwiespältige Anpassung von Helmholtz an Wandlungen des Zeitgeistes. Zum einen zeigen sich hier in einem Moment, da innere Konflikte das neugegründete Reich zu gefährden drohen, auch skeptische Zweifel an der Wirksamkeit des einheitlich gedachten Fortschritts von Wissenschaft, Kultur und Nation. Helmholtz beklagt "zynische Verachtung aller idealen Güter des Menschengeschlechts auf den Straßen und in der Presse", Mißbrauch politischer Freiheit zu gemeinen Motiven und Schrankenlosigkeit, Tendenzen zu überzogenen Ansprüchen der unteren Stände, denen die jüngsten politischen und humanen Bestrebungen doch mehr Spielraum für sorgenfreieres und menschenwürdigeres Dasein bereitet hätten - "fast mit Mühe" müsse man sich daran erinnern, "daß erst acht Jahre verflossen seien" seit dem großen Aufbruch der Reichseinigung, und er fordert die akademische Jugend auf, "über den untergeordneten und praktisch nützlichen Aufgaben" nicht die höheren Ziele von Kultur und Wissenschaft aus dem Auge zu verlieren. In diesem Kontext finden sich erstmals auch Töne der Bewunderung für den idealistischen Geist des Jahrhundertbeginns, sogar für die Schwärmerei der Romantik, die trotz aller Eitelkeiten wenigstens primär aus der Verfolgung hoher Ideale erwachsen sei (VR II, S. 215ff.). Aber im folgenden wissenschaftlichen und philosophischen Text bleibt Helmholtz wie auch in allen späteren Äußerungen substantiell unbeirrt bei seinem Optimismus, daß der moderne Geist der Erforschung und Unterwerfung der Naturkräfte auch die Hauptrichtung zur sittlichen Vervollkommnung der Nation bilden müsse.

III

Diese natur- bzw. allgemein erfahrungswissenschaftlich fundierte Fortschrittsperspektive schließt ein, daß auch das Geschichtliche selbst als Teil einer natürlichen Ordnung aufgefaßt wird, die durch das Walten von Gesetzen bestimmt wird. In diesem Denkmuster ist es nicht nur rückschauende Betrachtung seiner individuellen Neigungen, wenn er 1891 sagt, schon früh habe sich z.B. beim Lernen von Vokabeln oder bei der Unterscheidung von rechts und links sein "schwaches Gedächtnis für unzusammenhängende Dinge gezeigt. [...] Der Geschichte vollends, wie sie damals gelehrt wurde, wußte ich kaum

Herr zu werden. Stücke in Prosa auswendig zu lernen, war mir eine Marter"[3] (VR I, 6).

Erkenntnis ist ihm stets Erfassen von Zusammenhängen; das Sammeln, Aufzeichnen, Dokumentieren der Fakten in den literarisch-philologischen Disziplinen und auch in den beschreibenden Naturwissenschaften ist dafür wichtige Vorarbeit. So wichtig und wertvoll aber die Sammlung der Kenntnisse in Nachschlagewerken ist, so wenig dürfe "unser Wissen [...] in der Form der Kataloge liegen bleiben. [...] Es ist nicht genug, die Tatsachen zu kennen; Wissenschaft entsteht erst, wenn sich ihr Gesetz und ihre Ursache enthüllen" (VR I, 169). Abgesehen von dem hier ausgedrückten Kontrast eigentlicher (Natur-)Wissenschaft zu den mit Kultur und Geschichte befaßten Disziplinen, wird damit eine ganz klare Einordnung des Historischen in den klassischen Mechanismus als Weltbild (VR I, 45, 88, 379f.) deutlich, woran Helmholtz ungeachtet aller methodischen und fachtheoretischen Relativierungen konsequent, in späteren Texten fast ängstlich beschwörend festhielt.[4]

Das Historische in Natur und Gesellschaft besitzt dementsprechend kognitive Bedeutung nur als sekundäre Erscheinung des Wirkens der Gesetze der Natur und der Geistestätigkeit des Menschen. Der beides umgreifende "Kosmos" bedeutet ähnlich der klassischen Etymologie dieses Begriffs das Weltganze in gesetzlicher Ordnung (VR I, 191; II, 89ff.), die zugleich auch in den Künsten als Schönheit tieferer Zusammenhänge der natürlichen und geistigen Welt erahnt und ausgedrückt werden kann (II, 350ff.). Geschichte der Natur, des Lebens und der Menschen stellt die vielfältige, komplizierende und komplexe Gesamtheit von Prozessen dar, in denen diese Gesetze erscheinen und wirken, d.h. auch empirisch nachweisbar sind. Die einfachen, mathematisch exakt auf den Begriff zu bringenden Gesetze werden durch immer umfangreichere Kombinationen und Zusammensetzungen verhüllt. Dadurch kompliziert sich jeweils die wissenschaftliche Erkenntnis schon in den Naturwissenschaften jenseits der Physik, setzen schließlich die Vorgänge der geistigen Tätigkeit und ihre Resultate exakter Begrifflichkeit vorerst unübersteigbare Grenzen (VR I, 169ff.; 379ff., Helmholtz 1903).

---

3 Mit der Erwähnung des Geschichtsunterrichts bezieht sich Helmholtz wohl auf seinen Unterricht bei Johann David Preuß, dem königlich-preußischen Hofhistoriographen und Verfasser einer populären, von Menzel illustrierten Biographie Friedrichs II.; vgl. Cahan 1993a, 49f., 59.

4 Am deutlichsten wird der Zwiespalt dieser Kontinuität mit neuen Fortschritten der Forschung im Vorwort zu Heinrich Hertz, Principien der Mechanik, 1894, VR II, bes. S. 376-378.

Da aber alle geistige Tätigkeit und Erkenntnis letztlich der gesetzlichen Ordnung der Naturkräfte folgt, ergibt sich gegenständlich dennoch eine Einheit von Menschen- und Naturgeschichte (VR I, 83), die Helmholtz in einer durchweg anzutreffenden nomologischen Metaphorik zum Ausdruck bringt: Jede erfolgreiche Nationalgeschichte ruht auf Gesetz und Ordnung in Staat, Politik, Recht und Sittlichkeit. Die Brandung des Meeres (VR I, 134f.), das Klangerlebnis einer Sinfonie (VR I, 154f.), das Farbenspiel der Natur und Malerei (II, 126ff.) enthüllen zugleich die innere Ordnung der Gesetze in deren Wahrnehmung und künstlerischen Umbildung. Sogar das Klima erweise sich als in kulturgeschichtlicher Zeit kaum veränderlich. Daß es in Ostpreußen im Mittelalter Weinbau gegeben habe, spreche nicht für "die Wärme des Klimas, sondern nur für die Kehlen der deutschen Herren" (VR I, 82).

Sehr plastisch läßt folgendes Zitat aus dem Vortrag "Eis und Gletscher" (1865) diese aus heutiger Sicht fast bizarre Wertschätzung gesetzmäßiger Ruhe gegenüber historisch zufälligen Veränderungen erkennen:

> "Unförmliche Steinblöcke beginnen dem aufmerksamen Beobachter ihre Geschichte zu erzählen [...], die weit über die Vergangenheit des Menschengeschlechtes hinausreicht in das Dunkel der Urzeit. Ruhiges, gesetzmäßiges und segensreiches Walten ungeheurer Naturkräfte wird da offenbar, wo beim ersten Anblick sich nur Wüsten zeigen, entweder unabsehbar hingestreckt in trostloser öder Einsamkeit, oder voll wilder gefahrdrohender Verwirrung, ein Tummelplatz zerstörender Gewalten" (VR I, 234).

Ordnung und Gesetz verbinden sich somit zu einem Ideal des sich der geistigen Tätigkeit in den empirischen Wissenschaften enthüllenden Wesens der unbelebten Natur und der sittlich-kulturellen Herrschaft des Menschen über die Natur. In Staat und Nation wirken ungezügelte materielle Interessen ähnlich störend wie der ungebändigte Zufall für die Wissenschaft. So begrüßt er einerseits begeistert die Fortschritte in der Erkenntnis von Gesetzen der Bildung komplizierter Organismen durch Darwin (VR I, 293, 386ff.). Andererseits zeigt er sich irritiert von Defiziten auf dem Wege zur Erkenntnis von Gesetzen, die sich für ihn vor allem in mangelnder Vorhersagbarkeit bestimmter natürlicher Prozesse wie z.B. des Wetters zeigen, so am Schluß des in Hamburg 1875 gehaltenen Vortrags "Wirbelstürme und Gewitter":

> "So besteht für unsern Gesichtskreis noch der Zufall; aber er ist in Wirklichkeit nur der Ausdruck für die Mannigfaltigkeit unseres Wissens und die Schwerfälligkeit unseres Kombinationsvermögens. Ein Geist, der genaue Kenntnis der Tatsachen hätte und dessen Denkoperationen schnell und präzis genug vollzogen würden, um den Ereignissen vorauszueilen, würde in der wildesten Launenhaftigkeit des Wetters nicht weniger, als im Gang der Gestirne, das harmonische Walten

> ewiger Gesetze anschauen, das wir nur voraussetzen und ahnen" (VR II, S. 162f.).

Die These, daß Helmholtz etwa seit den 1870er Jahren einen Wandel durchmachte, der von der Erkenntnisgewißheit der früheren Schaffensperioden zu wachsender Einsicht in die Grenzen der Geltungssicherheit, d.h. zur Hypothetisierung naturwissenschaftlicher Theorien führte (VR II, 393ff., 401f.; Heidelberger 1993; Schiemann 1994), ist sicher in bezug auf die Theorie- und Methodologiegeschichte der Physik plausibel. Eine Diskontinuität des Weltbildes von Helmholtz insgesamt stützt sie hingegen nicht.

IV

Aus dem bisher Skizzierten ergibt sich auch Helmholtz' Zuordnung der Geisteswissenschaften, wie er es in der Heidelberger Prorektoratsrede "Über das Verhältnis der Naturwissenschaften zur Gesamtheit der Wissenschaften" von 1862 und - im wesentlichen unverändert - in späteren Reden oder Redeabschnitten wissenschaftstheoretischen Zuschnitts dargelegt hat. Sein Interesse an diesem Problem war dreifach begründet - durch sein philosophisches Engagement, durch hochschul- und allgemein bildungspolitische Ab- und Rücksichten wie auch durch sein von Anfang an schon aufgrund des beruflichen Werdegangs von der Medizin über die Physiologie zur Physik waches interdisziplinäres Verständnis von Wissenschaft. Bei seinen Darlegungen über die Bedeutung der Medizin und bei Erläuterungen seiner physiologischen Forschungen hob er deren Funktion als Bindeglied zwischen der Physik einerseits und der Psychologie wie anderen mit den geistigen Tätigkeiten befaßten Disziplinen hervor (VR I, 267).

Die historischen Disziplinen im engeren Sinne werden kaum separat, häufig aber in diesem Zusammenhang beschrieben. Es ist auffällig, daß er Geschichte und Kultur als Gegenstände durchweg dem Bereich des Erforschens der geistigen Tätigkeiten und ihrer Ergebnisse und damit hauptsächlich der Psychologie oder - methodisch gewendet - der Induktion mittels "psychologischen Takts" (VR I, 172) zuschreibt. Wirtschaft und Gesellschaft als Gegenstandsqualitäten und die damals schon hochentwickelte Nationalökonomie als Wissenschaft kommen dagegen hier gar nicht vor. Wenn die historisch-philologischen Disziplinen mitunter auch als Hilfswissenschaften bezeichnet werden, dann für Staats-, Politik- und Moralwissenschaften, das heißt aber zugleich für Gebiete, in denen er die größten Gefahren des Mißbrauchs wissenschaftlicher Ansprüche erkennt, wenn unbestätigte Hypothesen voreilig als Gesetze ausgegeben werden, womit die ganze Wissenschaft in Mißkredit gebracht werde (Helmholtz 1903, 20).

Bei seinen Skizzen des Systems der Wissenschaften fällt auf, daß Geschichte und Geschichtserkenntnis nicht widerspruchsfrei zugeordnet werden. Akzente grundsätzlichen Kontrastes von Natur- und Geisteswissenschaften überkreuzen sich mit Hoffnungen auf wachsende Vereinheitlichung aller empirischen Wissenschaften. Dabei lassen sich in bezug auf Inhalt und Aufgaben der letzteren drei Sichtweisen unterscheiden:

1. Durchweg kommt die Abgrenzung der Geschichtserkenntnis als künstlerisch-intuitives Erfassen (I, 171ff.; II, 194, 318f., 339ff.) oder psychologisches Verstehen (I, 169ff.) geistiger Tätigkeiten der Menschen und ihrer Ergebnisse vor, wodurch sie als extremer Gegenpol von Mathematik und Physik betrachtet und damit in Gegensatz zu den Naturwissenschaften gestellt wird: Die entsprechende, auch im Alltag sehr bedeutsame "Art der Induktion, welche nicht bis zur vollendeten Form des logischen Schließens, nicht zur Aufstellung ausnahmslos geltender Gesetze durchgeführt werden kann", ist für Handlungen und Gemütsstimmungen, für den Bereich des freien Willens usw. charakteristisch (VR II, 171).

2. Versuche einer empiristisch-positivistischen Zuordnung zur allgemeinen Entwicklung der Wissenschaften beruhen zunächst darauf, daß von den historischen Disziplinen wie von den Geisteswissenschaften insgesamt eine spätere Angleichung an Theorie und Methodologie der exakten Naturwissenschaften erwartet wird: Die Wissenschaften von der unendlich komplizierten geistigen Welt des Menschen sind in der Ausformung der "echten", strengen und exakten wissenschaftlichen Methode noch weniger reif, aber in ihren Aufgaben, geistige Ordnungen in Staat und Kultur zu erforschen, von höherem Rang (VR I, 173ff.). Vor allem in Deutschland, wo diese Disziplinen allerdings auch früher entwickelt gewesen seien, habe Hegels metaphysisches System aus dem gegenständlichen Unterschied einen entfremdenden Gegensatz gemacht (1891, VR I, 11). Der Widerspruch zwischen der Negation strenger Gesetzestätigkeit in diesem Bereich und der Hoffnung, mit verbesserten Methoden dereinst diesen Mangel in den kompliziertesten Gegenständen der Natur und Kultur doch noch beheben zu können, wird dabei nicht aufgelöst.

3. Ansätze dafür bietet allerdings eine vermittelnde psychologische oder literarisch-philologische Begründung der Besonderheit der Geschichtsforschung innerhalb der "gemeinsamen Beziehungen der Wissenschaften zueinander", seitdem sich die Wissenschaftsentwicklung durch hochgradige Spezia-

lisierung und Verfachlichung auszeichnet (VR I, 159; 367-370). Künftig werde mit den bleibenden Erfolgen der Naturwissenschaften "eine volle Bildung des einzelnen Menschen, wie der Nationen, nicht mehr ohne eine Vereinigung der bisherigen literarisch-philologischen und der neuen naturwissenschaftlichen Richtung möglich sein" (VR II, 425). Der Einzug der überprüfbaren Tatsachenforschung besonders in Geschichte und Philologie habe den Gegensatz der Wissenschaftsrichtungen schon gemildert.

Da Helmholtz Quellen nicht angibt und sich zumeist namentlich nur auf Autoritäten der ersten Jahrzehnte des 19. und aus dem 18. Jahrhundert bezieht, würde es langwieriger Textvergleiche bedürfen, genauere Rezeptionslinien auch zu zeitgenössischen Entwicklungen herzustellen. Zumeist wird auf den durch seinen Vater vermittelten Einfluß von Johann Gottlieb Fichte und dessen Sohn Immanuel Hermann Fichte verwiesen, der Helmholtz' Pate war (Koenigsberger I, 6f.; Köhnke 1986, 152). Aus dem Bereich der Geschichts- und Geisteswissenschaften gehörten zu seinem gesellschaftlichen Umfeld so unterschiedliche Geister wie der Demokrat Gervinus, der Liberale Mommsen, wie Treitschke und Dilthey, aber auch Ranke als Symbolfigur des klassischen Historismus (Werner/Irmscher, 1993, 24f.). Hier läßt sich aus Namen kaum etwas für bestimmte Einflüsse ableiten.[5]

Helmholtz' Verbindungen mit dem Neukantianismus bringen, wie erwähnt, die frühe empiristisch-naturwissenschaftliche Richtung der Kantrezeption zum Ausdruck. Für die philosophische und wissenschaftstheoretische Neubegründung der Geisteswissenschaften durch die südwestdeutsche Schule trifft es jedoch keineswegs zu. Helmholtz' Schlüsseltext zum Problem der Zuordnung der Geisteswissenschaften, auf den er sich später immer wieder bezogen hat, ist die Heidelberger Prorektoratsrede von 1862, liegt zeitlich also etwa ein Jahrzehnt vor den ersten wichtigen Äußerungen dieser Richtung, die die idiographischen Kultur- von den nomothetischen Naturwissenschaften grundsätzlich abzugrenzen und als empirische Wissenschaften eigener Qualität vor allem wertphilosophisch zu begründen suchte. Obwohl sich Helmholtz 1878 bei seinen Warnungen vor Materialismus und Egoismus auf "die ewigen Ziele" bzw. "die ewigen Ideale der Menschheit" beruft (VR II, 217f.), findet sich keine Spur einer besonderen Begründung der Kulturwissenschaften aus Werten und Ideen.

5 H. Hörz (Berlin) bereitet die Herausgabe des Briefwechsels von Helmholtz mit Geisteswissenschaftlern vor.

Resümierend lassen sich folgende Gesichtspunkte auch für die weitere Forschung festhalten: Erstens legen Untersuchungen von Welt- und Geschichtsbild im Zeitkontext für Helmholtz' wissenschaftsgeschichtliche Zuordnung Schlüsse nahe, die manche spätere "Dekontextualisierung" seiner Methodologie und Erkenntnistheorie relativieren und modifizieren. Es ist ein bedeutender Unterschied, ob primär der Kontext der damaligen Zeit oder die Retrospektive von einer völlig veränderten Kultursituation und Wissenschaftsauffassung aus gefragt ist, auch wenn sich diese natürlich von der heutigen Beobachterposition aus nicht völlig ausblenden läßt. Das gilt auch hinsichtlich der in den Beiträgen dieses Bands ausführlich diskutierten Frage von Kontinuität und Diskontinuität des klassisch mechanistischen Paradigmas beim frühen und späten Helmholtz (Heidelberger 1993) mit der von G. Schiemann (in diesem Band) vorgeschlagenen Zäsur etwa um die Wende von den 1860er zu den 1870er Jahren. Dabei ist vor allem zwischen der allgemeinen Diskursfunktion des mechanistischen Weltbildes einerseits und seinen innerwissenschaftlichen Relativierungen und Wandlungen andererseits zu unterscheiden. Meine These ist, daß unter dem ersten Gesichtspunkt die bisher skizzierten Invariablen eine erstaunliche Kontinuität bewirkten, während die Argumente für einen Wandel in letzterer Hinsicht als plausibel erscheinen. Der Begriff "modern" ist hier schon zwiespältig anzuwenden, kann er doch auf Newton wie auf Planck und Einstein bezogen werden, im weiteren Sinne der empirischen Wissenschaften und im engeren der "modernen" Physik des 20. Jahrhunderts.

Zweitens zeigt sich die Fruchtbarkeit von Weltbildvergleichen, und zwar über die Persönlichkeit und das Werk von Helmholtz hinaus, in einem größeren Zusammenhang der Wechselbeziehungen zwischen Geschichts- und Naturvorstellungen in den sich verändernden Diskursen unterschiedlicher Epochen des modernen Denkens und seiner Entstehung. Dieses Problem rückt in neueren Diskussionen und Forschungen sowohl zur allgemeinen Wissenschaftsgeschichte als auch zur Entwicklung des historischen Denkens immer mehr in den Vordergrund.

Drittens zeigt sich am behandelten Beispiel die allgemeine Bedeutung einer Kontextanalyse zeitspezifischer Diskurse für die Geschichte der Naturwissenschaften, wenn deren gesamtkultureller Einfluß erhellt werden soll. Die Auffassungen von Geschichtlichkeit und spezifisches Geschichtsdenken sind besonders wichtig, wenn die jeweilige fachübergreifende Funktion naturwissenschaftlicher Erkenntnisse in den Blick genommen wird. Wie wir sahen, war gerade das eines der Hauptanliegen der populärwissenschaftlichen Bestrebungen von Helmholtz. Diese Implikationen eines naturwissenschaftlich geprägten

Weltbildes lassen wiederum wichtige geistesgeschichtliche Konsequenzen auch für die jeweilige Position der Geschichts-, Kultur- und Sozialwissenschaften zu. Beiderseits kann durch derartige Vergleiche auch wesentlich zu einem kritischen Wissenschaftsverständnis beigetragen werden.

Viertens: Wie wichtig umgekehrt die Impulse sein können, die von übergreifenden Fragen der Veränderungen des Natur- und Naturwissenschaftsverständnisses für die Erforschung von Geschichtskultur und Geschichtswissenschaft ausgehen können, demonstriert das Beispiel Helmholtz hinsichtlich der gängigen Vorstellungen von den Besonderheiten des deutschen Geisteslebens im 19. Jahrhundert. Es gilt als die Domäne des idealistischen Historismus. Wie wir sahen, steht Helmholtz in vieler Hinsicht quer zu einer solchen Generalisierung. Vielmehr vermischen sich bei ihm Elemente eines strikten, eher positivistischen Empirismus, eines fortschritts- und erkenntnisoptimistischen nomologischen Weltbildes, das durchaus noch die Tradition der Aufklärung fortsetzt, und Einflüsse des kleindeutsch-preußischen Historismus (nationaler Machtstaat, Nation als Überindividuum, Fixierung auf Staat, Politik und Recht usw.). Auch hierin zeigt sich Helmholtz als eine Persönlichkeit, die allgemeines kulturgeschichtliches Interesse beanspruchen kann.

Literatur

Cahan, David: *Meister der Messung. Die Physikalisch-Technische Reichsanstalt im Deutschen Kaiserreich.* Weinheim/New York/Basel/Cambridge 1992.

Cahan, David: Helmholtz and the Civilizing Power of Science. In: Ders. (Hg.): *Hermann von Helmholtz and the Foundations of Nineteenth Century Science.* Berkeley/Los Angeles/London 1993, 559ff.

Cahan, David (Hg.): *Letters of Hermann von Helmholtz to his Parents. The Medial Education of a German Scientist 1837-1846*, Stuttgart 1993 (1993a).

du Bois-Reymond, Emil: *Reden* in zwei Bänden. 2. vervollständigte Aufl. Mit einer Gedächtnisrede v. J. Rosenthal, hrsg. v. Estelle du Bois-Reymond, Leipzig 1912.

Heidelberger, Michael: Force, Law, and Experiment: The Evolution of Helmholtz' Philosophy of Science. In: David Cahan (Hg.): *Hermann von Helmholtz and the Foundations of Nineteenth Century Science.* Berkeley/Los Angeles/London 1993, 461ff.

Heidelberger, Michael: Helmholtz' Erkenntnis- und Wissenschaftstheorie im Kontext der Philosophie und Naturwissenschaft des 19. Jahrhunderts. In diesem Band.

Helmholtz, Hermann von: *Vorträge und Reden.* 4. Aufl., Bd. 1-2. Braunschweig 1896 (Abkürzung: VR).

Helmholtz, Hermann von: *Vorlesungen über theoretische Physik*, Bd. 1, 1: Einleitung zu den Vorlesungen über theoretische Physik. Hrsg. v. A. König u. C. Runge, Leipzig 1903.

Kirsten, Christa (Hg.): *Dokumente einer Freundschaft, Briefwechsel zwischen Hermann von Helmholtz und Emil du Bois-Reymond 1846-1894*. Berlin 1986.

Köhnke, Klaus Christian: *Entstehung und Aufstieg des Neukantianismus. Die deutsche Universitätsphilosophie zwischen Idealismus und Positivismus*. Frankfurt/Main 1993.

Koenigsberger, Leo: *Hermann von Helmholtz*, Bd. 1-3. Braunschweig 1902-1903.

Meinecke, Friedrich: *Die Entstehung des Historismus*. Hrsg. u. eingel. v. C. Hinrichs. München 1965 (= ders., *Werke*, hrsg. v. H. Herzfeld/C. Hinrichs/ W. Hofer, Bd. 3).

Werner, Petra/ Irmscher, Angelika (Hg.): *Kunst und Liebe müssen sein. Briefe von Anna von Helmholtz an Cosima Wagner 1889 bis 1899*. Bayreuth 1993.

Röseberg, Ulrich: Ontologische und erkenntnistheoretische Dimensionen des Gesetzesproblems. In diesem Band.

Schiemann, Gregor: Die Hypothetisierung des Mechanismus bei Hermann von Helmholtz. Ein Beitrag zum Wandel der Wissenschafts- und Naturauffassung im 19. Jahrhundert. In diesem Band.

VII
Helmholtzforschung heute

Gleaning from the Archives? The 'Helmholtz Industry' and Manuscript Sources

Richard L. Kremer

Darwin, Helmholtz and "Industrial" Scholarship

In 1982 and immediately following, over one-hundred symposia, exhibits and other events plus at least seventy publications of collected essays celebrated the centenary of Darwin's death (Wassersug/Rose 1984). For these publications, the prize by weight (over 2 kg) and by historical sophistication went to the 1100-page collection of thirty-one essays entitled *The Darwinian Heritage* (Kohn 1985). This book aptly reflected the "state of the art" of historical Darwin scholarship in 1982, and its shift to what might be called the third era of Darwin studies. As outlined in the volume's historiographical essay, scholars during the first era, which continued into the early 1960s, sought to portray Darwin's ideas "as the culmination of some major trends in western thought, not merely as an event in the internal history of biology" (La Vergata 1985, 904). For intellectual historians like Lovejoy, Barzun, Loewenberg, Himmelfarb or Eiseley, "Darwinism" (which could mean many things) was an important worldview, like "Newtonianism" or "Marxism", and the historian's task was to delineate that worldview, its logical precursors and its cultural influences. In the second era, which according to La Vergata opened around 1960 as Gavin de Beer began to publish Darwin's transmutation notebooks, attention shifted from "Darwin's place in history" to Darwin's "actual process of discovery" in the 1830s. The eight notebooks Darwin kept from 1836 to 1839 opened new possibilities for phenomenological studies of his private reflections on the *Beagle* materials, his reading, his conversations and the day-to-day construction of his ideas, as distinguished from the rhetorical strategies of his published work. In this era, mastery of the manuscripts epitomized the Darwin scholar, and questions of dating, transcription, concordances and word choices, and black versus brown ink were debated with talmudic thoroughness. The third era, opened by the 1982 commemorations and the appearance of the first volume of Darwin's correspondence in 1985, and marked especially by the publication of a massive new biography (Desmond/Moore 1991), seeks an "historical contextualization" of Darwin's development and tries to locate him within the shifting scientific, professional, political and social communities of Victorian England. For these scholars Darwin's correspondence (over 14,000 letters have been found) and the marginalia jotted into the books and offprints of his library have become the new sources of choice (Burkhardt/Smith 1985; Di Gregorio 1990). If it took twenty years to master the notebooks, it may well require several more decades before scholars glean all the grain from Darwin's letters and marginalia.

I cannot here begin to analyze those factors - such as the emergence of "intellectual history" and then the "history and philosophy of science" as academic disciplines or the crystallization of the modern synthetic theory of evolution - which have shaped the three eras in Darwin scholarship (see Churchill 1982). Instead, I wish to focus more narrowly on the "Darwin Industry", a name deliberately chosen by that group of Darwin scholars who in the second era described above plunged into the notebooks and correspondence. This industrial metaphor does not refer simply to the sheer bulk of scholarship, for, as one reviewer has noted, historians of science refer neither to a "Descartes Industry" nor to an "Einstein Industry", even though publishing rates on these two figures equal or even exceed those on Darwin. Rather, the industrialists "are the manuscript groupies, for whom the truth content of a statement concerning Charles Robert Darwin is measured by the degree to which it can be indisputably traced back to the characters composed by his hand, preferably in brown ink" (Lenoir 1987, 115). It is this "Darwin Industry" that will serve as a backdrop against which I want to review Helmholtz historiograpy.

In this essay, I want to consider whether a "Helmholtz Industry" has existed or might be expected to exist in the future. That is, have manuscript materials been involved (or might they sometime be involved) in any radical transformations of Helmholtz studies as they have been in Darwin studies? In particular, I shall first review the use of manuscript materials during several significant epochs of Helmholtz studies: Koenigberger's Victorian biography of 1902-3, the commemoration of the fiftieth anniversary of Helmholtz's birth in 1921, the one-hundred-fiftieth anniversary of his birth in 1971, and the recently published collection of fifteen essays edited by David Cahan to mark the centenary of his death. As seen in Fig. 1, the anniversary commemorations prompted waves of scholarship and provide convenient benchmarks to gauge general historiographic trends. Next, I shall briefly characterize three types of manuscipt materials which have yet to be fully gleaned by scholars, and shall try to predict what kind of grain (or chaff) might be found there. Finally, I shall speculate about whether in the post-modern world of the late 1990s historians of science (or anyone else) will be interested, should new grain be found in the Helmholtz manuscripts. That is, we must ask whether Helmholtz, even were arwin-like manuscripts to be found, deserves "industrial" attention.

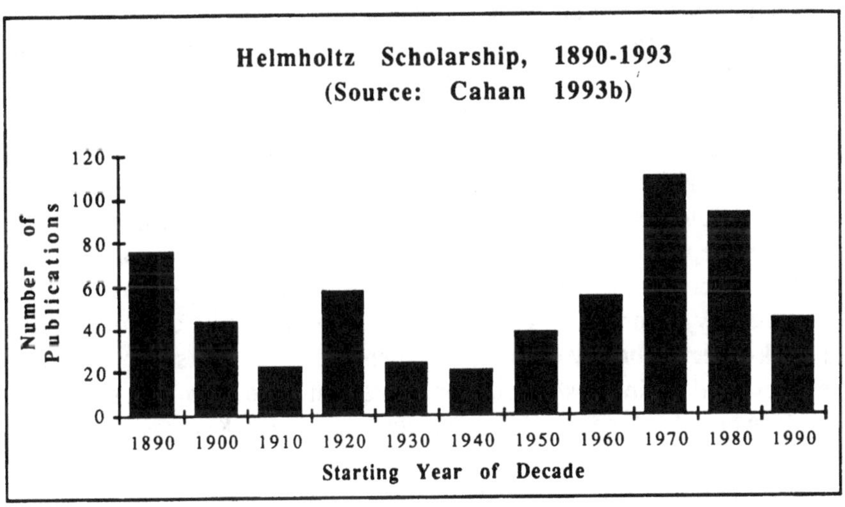

Figure 1

Some of my remarks, necessarily, must be very tentative. I cannot claim to have read all 584 items in the bibliography of the secondary literature on Helmholtz recently prepared by David Cahan (Cahan 1993, 603-36), and hence may have an incomplete understanding of the use of manuscript materials by previous scholars. Likewise, I have not identified, let alone read, all unpublished Helmholtz manuscript materials which may be extant in libraries, archives or personal collections across Europe and America. As is well known, the richest single source of unpublished Helmholtziana, the family papers held in the archive of the former Akademie der Wissenschaften der DDR in Berlin, was essentially closed to most Western scholars (myself included) from the early 1970s until 1989, and I have only begun to explore those materials. Thus some of the following conclusions may be premature, to say the least.

Helmholtz Historiography and Its Sources

In January of 1902, the Braunschweig publisher, Friedrich Vieweg und Sohn, announced the preparation of a "great Helmholtz biography" by the Heidelberg mathematician, Leo Koenigsberger. The printed *Voranzeige* indicated that Koenigsberger had been granted access to "the complete scientific *Nachlaß*, the letters of Helmholtz to his father and the latter's replies, and the comprehensive correspondence with personal and scientific friends", and that "with powerful support from the side of the family, a comprehensive representation of the life and work of the great scholar" would be produced (Vieweg 1902). As is well known, Koenigsberger's non-analytical, hagiographic volumes mostly paraphrase Helmholtz's published papers and quote, abridge or rework testimony from Helmholtz's correspondence (Koenigsberger 1902-3). However, we need not force our editorial or analytical standards on Koenigsberger. His biography represents, even to the present, the most extensive collection of manuscript materials. Indeed until very recently Koenigsberger was the sole source quoted by historians when they sought to go beyond Helmholtz's published works. Let me briefly characterize, therefore, three types of manuscript sources presented in Koenigsberger's completely unfootnoted volumes.

Most important, of course, was the correspondence Koenigsberger presented, albeit with varying degrees of editorial exactitude. According to my count, he quoted from at least 325 letters, with roughly twice as many of these written to than by Helmholtz. Since Helmholtz did not usually keep copies of letters he wrote, this means that Koenigsberger tracked down at least one-hundred letters written by Helmholtz which in 1902 were not in the family papers.[1] The most prolific correspondents included by Koenigsberger were

1 For examples of Koenigsberger's efforts to secure materials for his biography, see Koenigsberger to Felix Klein, 14 January 1902, Niedersächsische Staats- und Universitätsbibliothek, Cod. Ms. F. Klein 1/519, Göttingen; Koenigsberger to Elisa-

Helmholtz's parents (72 letters), Emil du Bois-Reymond (55 letters), Carl Ludwig (25 letters), Olga von Velten (18 letters), F. C. Donders (15 letters) and William Thomson (10 letters). As I shall describe below, I suspect that Helmholtz's total correspondence numbers somewhat less than 4000 letters; hence, Koenigsberger cited around ten percent of all the letters Helmholtz may have written or received.

Second, Koenigsberger relied heavily on written testimonials he solicited c. 1902 from still living family members and colleagues who had known Helmholtz. Such reports, from people like Thomson, Betty Johannes, Pietro Blaserna, Eduard Zeller, Julius Bernstein, Theodor Engelmann, Wilhelm von Bezold and Richard Wachsmuth reflect the public image Helmholtz had acquired by the time of his death. Since their authors undoubtedly knew that Koenigsberger would publish the testimonials, these documents if anything simply round out and occasionally enhance the many obituaries published in the late 1890s.

Finally there are the unpublished autograph manuscripts which Koenigsberger sometimes presented in full. To my knowledge, the following essays, drafts, or speeches have appeared in print nowhere except in Koenigsberger: an essay on the value of historical study, which Helmholtz wrote in 1837 as part of his application to the Friedrich-Wilhelms-Institut (1:15-16); Helmholtz's trial lecture before the Berlin art school in 1848 (1:95-105); Helmholtz's speech upon leaving Königsberg in 1855 (1:253-54); a pre-1847 essay on the concept of nature and the foundations of natural science (2:126-38); an evaluation from the 1880s of a book on Kant and geometry (2:141-42); an undated fragment on epistemology (2:158-59); official reports on air resistance and the motion of air balloons from 1878 and 1894 (2:222-23); two undated notes on electrochemical theories (2:279-84); a speech in praise of Heidelberg

beth Hertz, 6 August 1902, Deutsches Museum, 3106, Munich; Koenigsberger to George Stokes, 29 September 1902, Cambridge University Library, Add 7656, K520A; Koenigsberger to William Thomson, 4 letters, 1902, Cambridge University Library, Add 7342, K116-22; and in the Staatsbibliothek Preussischer Kulturbesitz, Berlin, see Richard Helmholtz to Koenigsberger, 4 letters, 1901-2, Darmst. R. Helmholtz 1891 K(8); Franz Richarz to Koenigsberger, 2 September 1902, Darmst. Richarz F1c 1897(5); Koenigsberger to Jeanette du Bois-Reymond, 4 letters, 1902, and Koenigsberger to [Ernst Hagen], 22 January 1903, Darmst. Koenigsberger H 1884(10); Louisa Tyndall to Koenigsberger, 18 February 1902, Darmst. Tyndall F1e 1855(2); Emil Picard to Koenigsberger, 22 April 1902, Darmst. Picard H 1883(10); Adolf Wagner to Koenigsberger, 17 April 1902, Darmst. Wagner 2g 1873(5); Conrad Dieterici to Koenigsberger, 27 June 1902, Darmst. Dieterici F1e 1882(4).

delivered in 1886 (2:338-39); an unfinished essay on the thermodynamics of chemical processes from the late 1880s (2:361-79); and finally a draft for a speech to the 1894 Versammlung deutscher Naturforscher und Aerzte, which death prevented Helmholtz from delivering (3:125-34). With the possible exception of the pre-1847 essay (see Heidelberger 1993, 471-72), none of these manuscripts have attracted much attention by subsequent historians, and they generally shed little light on the development of Helmholtz's scientific or philosophical ideas. Indeed, Koenigsberger, like Darwin scholars before 1960, remained steadfastly uninterested in the processes by which Helmholtz formulated his published work. The 1902 *Voranzeige* described Koenigsberger's desire to depict "the man in the harmonic relation of his entire activity and thought". Yet by "harmonic" Koenigsberger meant Helmholtz's published oeuvre, with all rough edges, controversies, and less successful endeavors purged from the record.

Since Koenigsberger's volumes appeared in 1902-3, publications on Helmholtz rarely if ever have gone beyond that biography in trying to exploit additional unpublished materials, from the 1921 commemoration of the 100th anniversary of Helmholtz's birth through the 1971 celebration of the 150th anniversary of that event. In 1921, leading physicists, physiologists, mathematicians and philosophers reviewed Helmholtz's contributions in light of the current state of knowledge in those fields. Although these scholars praised Helmholtz for completing the "classical" period in physics, for his logical and mathematical approaches, for generalizing and systematizing the fields in which he worked, and for rejecting "metaphysics", they also pointed out his mistakes (e.g., he misunderstood Kant's transcendental epistemology) and those theories which already had been superseded (e.g., his empiricist theory of spatial perception). Similar to the questions asked during the first era of Darwin scholarship, the focus in the 1921 commemorations was on Helmholtz's place in history and the content of his published works; his context, the development of his thought, and even biographical details were seldom if ever mentioned (Warburg/Rubner/Schlick 1921; Kries 1921; Wien 1921; Nernst 1921; Riehl 1921; Rothe 1921; Stumpf 1921). Only rarely was Koenigsberger cited, and even more rarely did the 1921 commemorators enrich their accounts by referring to their own personal experiences with Helmholtz (Lummer 1921; Goldstein 1921).

Fifty years later apparently only scholars in the German Democratic Republic celebrated the 150th anniversary of Helmholtz's birth, with the Akademie der Wissenschaften der DDR and the Humboldt-Universität each holding symposia (Scheel 1972; Helmholtz' Leistungen 1973). As in 1921, these commemorations emphasized Helmholtz's "place in history", although not the "history" of 1921. The goal, wrote an organizer of the 1971 events, was to "appreciate Helmholtz' philosophical views and scientific knowledge and discoveries in the light of dialectial materialism and current physics and psychology, and to make them fruitful for [contemporary] scientific work" (Kuchling 1973). I cannot here begin to analyze the complexity or political significance of the East German discussions of whether Helmholtz was a "spontaneous materialist" (Lenin's term) or merely a "mechanical materialist", had developed a true "copy theory" of perception, or had completely freed himself from the evil influences of Kant. Instead, I only note that these scholars, like those in 1921, dealt primarily with Helmholtz's published work and rarely if ever referred to unpublished materials, even to those presented in Koenigsberger. To the best of my knowledge, only one East German scholar in 1971 explicitly discussed unpublished materials. Describing Helmholtz's tendency to rewrite his essays many times before publication as an example of "classical" thinking (Wilhelm Ostwald's category), Friedrich Herneck reproduced a page from a revised autograph in the Akademie-Archiv of Helmholtz's 1871 "Magnus-Rede," but did not ask whether such revisions might be interesting in their own right or might offer new insights into Helmholtz's creative work (Herneck 1973, 352-53).

I should note that the East Germans were not alone in showing no interest in Helmholtz manuscripts during the 1970s. Even scholars concerned with the development of Helmholtz's thought, and not just his "place in history" - e.g., Yehuda Elkana (1970) on Helmholtz's struggle to clarify the concept of "Kraft", R. Steven Turner (1977) on Helmholtz and Fichte, Joan Richards (1977) on Helmholtz's empiricism and the foundations of geometry, or Günter Bierhalter (1981) on mechanical foundations of the second law of thermodynamics - limited themselves exclusively to Helmholtz's published papers.[2] However as noted above, during these years the rich cache of Helmholtziana located in Berlin-East was not easily accessible to Western scholars. That is, the motives behind the inattention to manuscripts by scholars on either side of the Wall may not have been symmetrical. Nonetheless, it is clear that before the 1980s Helmholtz scholarship had not entered an "industrial" mode.

2 For a rare example of an article from the 1970s referring to unpublished letters (in Berlin-West), see Heimann 1974, 207, 233.

This has begun to change only within the past decade. Klaus Klauß (1981) edited a previously unknown document from the Akademie-Archiv concerning Helmholtz's construction of the ophthalmoscope. Luigi Belloni (1982) printed letters from Helmholtz to a former student, Franz Boll. Christa Kirsten published a facsimile and transcription of a draft manuscript from the Akademie-Archiv of Helmholtz's 1847 essay *Ueber die Erhaltung der Kraft*. Yet as she herself noted, this document was a clean copy probably sent to the printer and did not vary significantly from the published version (Helmholtz 1983, 66).[3] Several years later, Kirsten (1986) and colleagues published the complete Helmholtz/du Bois-Reymond correspondence, an edition that has received wide acclaim and has been extensively used by Helmholtz scholars. Winfried Scharlau (1986) edited Helmholtz's letters to the Bonn mathematician, Rudolf Lipschitz; Herbert Hörz and Andreas Laaß (1989) presented Boltzmann's letters to Helmholtz; and David Cahan (1993a) and I (1990) have edited Helmholtz's early letters to his parents and his first wife, some of which were imperfectly presented by Koenigsberger. Helmholtz's correspondence is thus becoming more accessible, an important step if an Helmholtzian industrial revolution were to begin.

Yet a perusal of the most recent "state of the art" review of Helmholtz scholarship indicates how slowly that revolution may be emerging (Cahan 1993b). The fifteen scholars writing for Cahan's book reveal themselves quite interested in Helmholtz's intellectual development, in the disciplinary contexts in which he sought to situate his work or which he sought to transcend, and in his "place in history" relative to such movements as nativism versus empiricism, classical aesthetics, electrodynamical practices or ideologies of scientism. Yet only six of these essays refer to "new" unpublished materials (Kirsten 1986 is used more frequently), and only two propose substantially new interpretations based on such materials.[4] Timothy Lenoir (1993) cites previously unpublished letters from Donders and Alfred Volkmann to show how Helmholtz sought to bolster his theoretical claims about visual perception by establishing personal contact with leading vision experimenters. And Kathryn M. Olesko and Frederic L. Holmes (1993) examine a laboratory notebook in their important study of Helmholtz's use of error analysis and least squares for his work on the speed of nerve conduction. Yet despite these two counter-exam-

3 A recent study of the 1847 essay does not even cite Kirsten 1983. See Bevilacqua 1993.

4 See Tuchman, Olesko/Holmes, Lenoir, Kremer, Vogel and Kaiser in Cahan 1993b.

ples, I think one might safely conclude that Helmholtz scholarship to date has not been radically transformed by the use of new, unpublished materials.[5] No epic divide necessitating a "before and after the notebooks" periodization of historiography has occurred for Helmholtz, as was the case for Darwin.

Prospects for an Helmholtzian Industrial Revolution

Might, however, such a divide be coming in the future? Might a Helmholtz Industry still be possible? Might the archives contain materials with the potential to alter dramatically or at least to enhance our views of Helmholtz and his science, philosophy, career or cultural impacts? Might a Helmholtzian *Red Notebook* be hiding somewhere which could offer fundamentally new insights into Helmholtz's intellectual development (Herbert 1980)? Of course, I cannot answer these questions with any degree of certainty. If I had found a *Red Notebook*, you may be assured that I here would be discussing it rather than historiography. Rather I shall offer a speculative reconnaissance of what the extant manuscripts might hold, based on what I have seen and read. Note also that I do not consider archival sources any more "objective" or "authentic" than published sources. Manuscripts rework and recreate the past just as do any other texts, all of which require critical reading and contextual interpretation. *Wie es eigentlich gewesen ist* remains a misguided gesture, and "industrial" scholarship although more richly textured is not necessarily any more "true" than "pre-industrial" studies (see Kay 1992).

Let me begin with the official records of Helmholtz's career - the university and ministerial records and documents relating to Helmholtz's tenure at the Physikalisch-Technische Reichsanstalt. David Cahan's thorough study (1989) of the early years of the PTR, I think, nicely reveals what can be done with that set of records, and I suspect that little new will appear concerning this episode in Helmholtz's career. Koenigsberger had access to most of the ministerial records in Karlsruhe and Berlin covering Helmholtz's sojourn at the universities in Baden and Prussia and included many of the more important documents concerning negotiations over Helmholtz's moves to his various university posts. Yet a closer reading of all the appointment materials could, I think, re-

5 Of the twenty-six papers presented at the Schloß Ringberg Conference in January, 1994, only those by Werner, Hiebert, Vogt and Cahan referred to new manuscript materials.

veal much more about the changing fate of Helmholtz's reputation among colleagues and governmental officials as well as his own efforts to construct an image for himself, especially as he crossed disciplinary boundaries. Both successful and unsuccessful calls should be examined (the latter include calls to Göttingen in 1859, to Cambridge in 1861, and to Prague and Bonn in 1868).

Furthermore, close attention to the mundane records of university teaching might shed new light on Helmholtz's reportedly poor performance in the classroom and on his inability to create a "school" of disciples (Koenigsberger 1902-3, 2:179-80; Turner 1993). Only recently have records become available in Poland which reveal something of Helmholtz's interest in laboratory teaching at Königsberg.[6] Yet it is well known that Helmholtz during his years in Heidelberg increasingly removed himself from laboratory teaching. Such reports, however, might need revision, for in 1868 when Helmholtz was offered the chair for physics at Bonn, forty Heidelberg medical students petitioned the ministry, urging that Helmholtz not be allowed to leave Baden:

> "Es ist selbstredend, daß durch den Weggang dieses größten Lehrers unserer Hochschule dieser ein nicht geringerer Verlust zugeführt würde, als der Wissenschaft selbst, welche unser berühmter Lehrer in Folge jener Berufung zu verlassen im Begriffe steht. Ebenso sähen sich auch die hiesigen Studirenden der Medicin in dem Gang ihrer Studien sehr erheblich gestört durch den drohenden unersetzlichen Verlust eines Mannes, der, eine fernhinstrahlende Zierde unserer Universität, auch für den wissenschaftlichen Geist dieser von unberechenbarem Werth, vom weittragendsten Einfluß ist".[7]

Was this petition merely orchestrated by an official anxious to keep a star in Heidelberg or might Helmholtz have been truly popular as a teacher? Likewise, it would be interesting to see student *Nachschriften* of some of Helmholtz's early lectures in either physiology or physics since he never published textbooks in these subjects. I know of only one such *Nachschrift* from before 1870 (lectures on sensory physiology), but have not yet found the time to analyze it in any detail. And given Kathryn M. Olesko's (1991) and Arleen Tuchman's (1993) important studies of nineteenth-century university teaching practices, one would like to know more about how Helmholtz taught experimental

6 Archiwum Panstwowe, Acta des koenigl. Kuratorium der Albertus-Universitaet zu Koenigsberg: Die Bewilligung von 300rth. zur Anschaffung von Instrumenten und Apparaten behufs der Vorlesungen ueber experimentale Physiologie, XXVIII/2, Nr. 412, Rep. 99, H30, Olsztyn. I thank Kathryn M. Olesko for providing this information.

7 Ministerium des Innern to Staatsministerium, 19 July 1868, Generallandesarchiv 76/9939, Bl. 49r-50r, Karlsruhe.

physiology before 1871, and about how his research and teaching practices interacted, if they did at all.

What about unpublished autograph manuscripts? According to my preliminary survey, the Akademie-Archiv possesses about 190 manuscripts in Helmholtz's hand; the Deutsches Museum has another ten, and several other manuscripts can be found scattered in other collections. Some of these manuscripts may be teaching notes for Helmholtz's lectures.[8] Others, like the draft of the "Erhaltung" essay published in 1983, are probably only clean drafts prepared by Helmholtz for his publishers. Indeed, despite his autobiographical claim that he often rewrote his papers "four to six times" prior to publication (Helmholtz 1896, 1:18), I have not been able to find those reworked drafts in the one case I have explored - Section 20 of the *Handbuch der physiologischen Optik* concerning color theory (Kremer 1993). The only draft of the *Handbuch* I could locate in the Akademie-Archiv varies only trivially from the published version.[9] It may well be the case, therefore, that Helmholtz preserved only the penultimate drafts of his published works. If so, then any careful analysis, such as Holmes (1985) has done for Lavoisier, of Helmholtz's writing and revising practices would be impossible.

Yet might some of the 200 extant autograph manuscripts contain laboratory notebooks or drafts for essays which Helmholtz never published? Olesko and Holmes have identified at least three laboratory notebooks, several of which they will be analyzing in their forthcoming book on Helmholtz's early physiological experiments.[10] In my own studies of Helmholtz's work on color, I found it very difficult to deduce much about his actual experimental practices, when given only his finely-crafted, rhetorically sophisticated published papers. Clearly, the discovery of additional laboratory notebooks would greatly enrich our views of Helmholtz as an experimenter. Likewise, one wonders what might be contained in Helmholtz's lengthy manuscripts entitled "Vorgänge in Elektrolyten" or "Notizbuch Elektrodynamik", which although openly accessi-

8 See, for example, in the Akademie-Archiv the following autograph manuscripts: "Blut und Respiration", 72 Bl., AW 545; "Physiologie", ca. 100, 42, ca. 60 Bl., AW 550, 551, 555; "Allgemeine Pathologie", ca. 100 Bl., AW 553; "Physiologie der vegetativen Funktionen", ca. 40 Bl., AW 556.

9 "Ueber physiologische Optik", 993 numbered pages, AW 572. For other drafts of the *Handbuch*, see AW 571/1 and AW 573.

10 "Versuche über Muskelton", "Untersuchungen über Muskeln und Nerven", "Versuch über Gährung bei Magnus", AW 544, 547, 666. See Olesko/Holmes 1993, 51-53.

ble in Western libraries apparently have not been carefully studied.[11] I doubt whether a Helmholtzian *Red Notebook* will be found, but until a thorough survey has been conducted, we must withhold judgment about the potential significance of the extant manuscripts.

Finally, let me turn to the correspondence. Might an analysis of letters not included in Koenigsberger or the newly edited correspondences yield new insights into the development of Helmholtz's thought, his personality or working practices, his efforts to enroll supporters for his various theoretical stances, his evolving political or social views, his attempts to bridge or transcend disciplines, his role as a powerful advisor in university or science policy, or his social interactions beyond the university? Might the letters tell us more about Helmholtz as a Fichtean mandarin or a prophet of "material interests" than do his public speeches which were differently crafted for various audiences (see Heidelberger 1993; Cahan 1993c; Lenoir 1992)? Might the letters reveal more about the contemporary reception of Helmholtz's work than do the published reviews or responses (see Turner 1994)? Might the letters shed new light on Helmholtz's apparent shift, much discussed at the Schloß Ringberg Conference, from an optimistic to an agnostic epistemology (Schiemann, this volume)? As with the manuscripts, much hard work will be required before these questions can be definitively answered.

As noted above, Koenigsberger excerpted about ten percent of the total correspondence Helmholtz wrote or received. In the course of preparing my edition of letters (Kremer 1990), I surveyed, albeit not very systematically, over one-hundred libraries and archives plus auction catalogues, published materials on Helmholtz or his correspondents and information internal to the extant letters, and with the help of several other interested scholars, especially David Cahan, assembled a list of 2864 letters which either currently exist or once existed. They include 1045 letters by Helmholtz and 1819 letters to him. 413 of the letters on my list have been published in full and another 370 have been published in part; i.e., roughly twenty-five percent of the known letters have been at least partially published. For 380 of the letters on my list I do not know a current location (this includes 158 letters used by Koenigsberger). For 106 of the letters, I have been unable to identify the Helmholtz's correspondent, either writer or recipient. The remaining letters are to or from a total of 643 different persons or institutions. After moving to Heidelberg, Helmholtz wrote and

11 Deutsches Museum 1312, Munich; Staatsbibliothek Preussischer Kulturbesitz, Handschrift 233, Berlin.

received an average of about 60 to 80 letters per year, a rate of production considerably lower than Darwin's or Thomson's (see Fig. 2).[12]

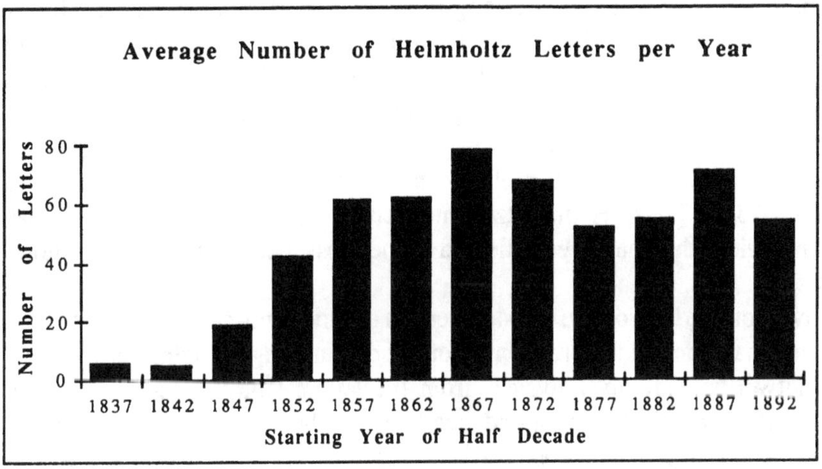

Figure 2

To his most prolific correspondents - the 101 persons on my list with whom he exchanged five or more letters - Helmholtz wrote a total of 744 letters and received another 1169 letters (see Appendix). That is, Helmholtz exchanged two-thirds of all his letters with about one seventh of his correspondents. A quick prosopography of these prolific correspondents reveals 22 physics professors, 13 physiology professors, 9 professors of other medical disciplines, 9 mathematics professors, and 4 chemistry professors, slightly more than half of the most prolific correspondents were university-based. Likewise, only half of these correspondents can be found in the *Dictionary of Scientific Biography*, a rough measure of their international scientific prominence. State and univer-

12 After some reflection, I would estimate that a maximum of perhaps 1000 additional Helmholtz letters might be extant, which would still make Helmholtz a remarkably reticent correspondent by Victorian standards. For example, the correspondence of G. G. Stokes (1819-1901) numbers about 35,000 letters, William Thomson's (1824-1907) about 13,000 letters, and John Herschel's (1792-1871) about 20,000 letters. See Wilson 1987, 181-82; *Kelvin Papers* 1977; Wilson 1976; Michael Crowe, personal communication to the author.

sity officials, family members, practicing physicians and ophthalmologists, publishers, acousticians and instrument-makers round out the list. Interestingly, among the most prolific correspondents we find only two professional philosophers (Robertson, the editor of the English journal *Mind*, and Zeller, the well-known neo-Kantian and long-time friend of Helmholtz's), one historian (Treitschke), one industrialist (Siemens) and one literary figure (Rodenberg). The list includes 11 correspondents from England, 4 from France, and several each from Italy, Holland and America; roughly 20 percent of the most prolific correspondents were non-German. And I should note that only about one-third of these prolific correspondents appear in the index to Cahan (1993b). Thus, Helmholtz's intellectual and social community was somewhat wider than has been reflected in the most recent round of Helmholtz scholarship.

I have not read all of the 2864 letters in my list and cannot predict what clues they may contain for broadening our understanding of Helmholtz and his social networks. Judging, however, from the letters I have read, I would be surprised if Helmholtz's correspondence were to shed much new light on his "process of discovery". In most cases (excepting the letters to du Bois-Reymond and to his father), Helmholtz seems to have worked out his ideas alone, and then wrote his colleagues only to report what he had completed, or to answer their questions or challenges concerning his already published work.[13] Much of the correspondence deals with logistical matters of travel and personal meetings rather than with substantive discussions per se. For example, Helmholtz's nine letters to Franz Boll, the Rome physiologist, contain almost no scientific discussions but mostly schedule personal meetings between the two men during Helmholtz's frequent vacation trips to the Alps (Belloni 1982). Yet the discussions Boll and Helmholtz had in the mountains might have been very important in shaping Helmholtz's defense of his color theory against the attacks of Ewald Hering. Thus, although the letters might enable us to map out more completely Helmholtz's social networks, they may not always provide much direct information concerning the content which passed through those networks. Still, knowing more about the networks themselves might help us better understand the contexts in which, and the reasons for which, Helmholtz moved so facilely among so many research areas throughout his life. And the letters might reveal more about Helmholtz's evolving reputation as a scientific

13 For examples of how Helmholtz, near the middle of his career, discussed his work with correspondents, see Koenigsberger 1902-3, 1:281-82, 2:11, 61-65, 123.

and cultural figure (see Kremer 1990, xix-xxvi). As such, the letters might permit some new assessments of Helmholtz's career trajectory.

In conclusion, let me return to the two questions I posed at the outset. Will a Helmholtz Industry emergy? And should a Helmholtz Industry emerge, given the extant archival materials and current trends in the history and philosophy of science? Few of Helmholtz's manuscripts will probably equal the importance of Darwin's eight transmutation notebooks. Even if scholars find another thousand Helmholtz letters, his correspondence will still be only a quarter the size of Darwin's. I have no idea whether Helmholtz's personal library still exists, and I doubt whether Helmholtz would have heavily annotated his books in any case. Any Helmholtz Industry, therefore, is guaranteed to be much smaller than Darwin's, simply by virtue of the availability and significance of manuscript materials. In addition, Helmholtz, for many reasons, was not Darwin, so that a thorough analysis of his literary remains would probably yield far fewer fundamentally important insights about scientific practices and nineteenth-century society than the Darwin Industry has found. This is not to say, of course, that a less "industrial" historiography of Helmholz, employing standard published materials, could not produce new interpretations of merit.[14] Finally, if recent methodological trends in the discipline of history of science continue, i.e., if the shift toward systems theory, actor networks, deconstruction or other post-modern approaches gathers speed, we may well find diminished interest in any attempts to construct, at the detailed archival level, the intellectual or social trajectories of a single individual, even one as significant as Helmholtz. However despite such trends, rumors of the death of the individual in the history of science have probably been greatly exaggerated. In any case, when in the year 2021 scholars produce a volume to commemorate the two-hundredth anniversary of Helmholtz's birth, I shall be eager to see to what extent newly found manuscript materials, alone, will have made obsolete our commemorative papers from 1994.

14 For a recent example of an important new interpretation of a nineteenth-century scientific figure based solely on traditional sources, see Caneva 1993.

Acknowledgments

Research for this essay has been supported by grants from the National Science Foundation, the Alexander-von-Humboldt Foundation and Dartmouth College. For their helpful suggestions I thank the other authors in this volume, especially David Cahan, Dieter Hoffmann and Lorenz Krüger. And I thank Herbert Hörz and R. Steven Turner for criticizing an earlier draft of his essay.

Appendix

Helmholtz's Most Prolific Correspondents
(Five or More Letters Exchanged)

Correspondent	Number of Letters		Occupation
	To	From	
Acland, Henry	2	4	Medical professor
Althoff, Friedrich	19	4	State official
Barry, Anton de	0	5	Unknown (published on bacteriology)
Becker, Otto	0	7	Ophthalmology professor
Beetz, Wilhelm	2	10	Physics professor
Beltrami, Eugenio	0	5	Mathematics professor
Bence-Jones, Henry	0	14	Practicing physician (& chemist)
Beseler, Wilhelm Hartwig	2	5	State official (Bonn Curator)
Binz, Karl	2	11	Pharmacology professor
Boetticher, Karl Heinrich von	15	4	State official
Boll, Franz	9	5	Physiology professor
Boltzmann, Ludwig	0	12	Physics professor
Borchardt, Carl Wilhelm	2	12	Mathematics a.o. & academician
Bowman, William	0	5	Practicing opthalmologist and surgeon

Correspondent	Number of Letters To	From	Occupation
Braun, Karl Ferdinand	0	6	Physics professor
Brown, Alexander Crum	0	12	Chemistry professor
Brücke, Ernst von	0	54	Physiology professor
Bunsen, Robert Wilhelm	1	5	Chemistry professor
Busch, Carl David Wilhelm	0	19	Surgery professor
Caspary, Robert	0	6	Botany professor
Clausius, Rudolf	0	8	Physics professor
Dohrn, Felix Anton	0	5	Zoologist
Donders, Franciscus Cornelis	14	31	Physiology professor
Dorn, Friedrich Ernst	0	6	Physics professor
Du Bois-Reymond, Emil	92	80	Physiology professor
Eckhard, Conrad	0	7	Physiology professor
Ellis, Alexander John	0	14	Acoustician & phonologist
Exner, Sigmund von	0	7	Physiology professor
Foerster, Wilhelm	0	7	Astronomy professor
Friedreich, Nicolaus	0	6	Medical professor
Gueroult, Georges	0	5	Acoustician
Heintz, Wilhelm Friedrich	1	19	Chemistry professor
Helmholtz (Eltern)	19	2	Parents
Helmholtz, Anna	70	23	Wife
Helmholtz, Caroline	2	3	Mother
Helmholtz, Ferdinand Julius	48	18	Father
Hermite, Charles	0	5	Mathematics professor
Hertz, Heinrich Rudolf	10	16	Physics professor
Heynsius, Adriaan H.	0	7	Physiology professor
Jahn, Otto	0	5	Philology & music professor
Javal, Louis-Émile	0	10	Ophthalmology professor
Jolly, Julius August Isaak	5	5	State official
Karsten, Gustav	1	4	Physics professor

Correspondent	Number of Letters To	From	Occupation
Kayser, Heinrich	4	6	Physics professor
Kirchhoff, Gustav	1	14	Physics professor
Knapp, Herman	5	0	Practicing ophtalmologist & ophth. professor
Koenigsberger, Leo	3	2	Mathematics professor
König, Arthur	0	11	Physics professor
König, Karl Rudolph	0	5	Acoustic instrument maker
Kronecker, Hugo	11	6	Physiology professor
Kronecker, Leopold	0	32	Mathematics professor & academician
Krönig, August Karl	0	14	Teacher, editor, physicist
Kultusministerium	5	13	State officials
Kundt, August	1	12	Physics professor
Liebreich, Richard	0	6	Practicing ophthalmologist
Lipschitz, Rudolf	26	53	Mathematics professor
Lockyer, Joseph Norman	1	6	Astrophysicist
Ludwig, Carl	24	49	Physiology professor
Magnus, Heinrich Gustav	0	8	Physics professor
Mohl, Pauline	6	0	Sister-in-law
Mommsen, Theodor	14	4	Law professor, historian, politician
Muehler, Heinrich von	4	2	State official
Olshausen, Justus	2	7	State official
Overbeck, O.	5	0	Teacher
Pflüger, Eduard	1	18	Physiology professor
Quincke, Georg Hermann	0	8	Physics professor
Raumer, Karl Otto von	7	3	State official
Robertson, George Croom	5	3	Philosophy professor & editor (*Mind*)
Rodenberg, Julius	10	3	Journalist, novelist, poet, travel-writer

Correspondent	Number of Letters To	From	Occupation
Roscoe, Henry Enfield	11	20	Chemistry professor
Rowland, Henry Augustus	5	3	Physics professor
Schering, Ernst Christian Julius	3	5	Mathematics professor
Schultze, Max Johann Sigismund	0	7	Anatomy professor
Schulze, Johannes Karl Hartwig	4	5	State official
Sella, Quintino	0	6	Mathematics & mineralogy professor
Siemens, Ernst Werner von	0	10	Industrialist
Soret, Jacques-Louis	12	4	Physics professor
Spies, Gustav Adolf	0	9	Practicing physician
Stokes, George Gabriel	4	1	Physics professor
Strutt, John William (Rayleigh)	1	6	Physics professor
Tait, Peter Guthrie	4	20	Physics professor
Taylor, Sedley	4	8	Acoustician, private scholar
Thomson, Frau	3	4	Wife
Thomson, William (Baron)	25	60	Physics professor
Tiersch, Otto	0	5	Teacher & music theorist
Tommasi-Crudeli, Corrado	0	7	Pathology professor
Treitschke, Heinrich von	2	7	History professor
Tyndall, John	23	59	Physics professor at Royal Institution
Ungern-Sternberg, Johann von	3	4	State official
Velten, Olga von	53	2	Wife
Vierordt, Karl von	0	5	Physiology professor
Vieweg & Sohn	129	25	Publisher(s)
Virchow, Rudolf Ludwig Karl	0	5	Pathology professor
Volkmann, Alfred Wilhelm	0	22	Physiology professor

Correspondent	Number of Letters To	From	Occupation
Volkmann, Paul	1	4	Physics professor
Voss, Leopold	0	21	Publisher
Weber, Heinrich Friedrich	0	6	Mathematics professor
Weber, Karl Otto	0	28	Surgery professor
Wittich, Wilhelm Heinrich von	5	0	Physiology professor
Wüllner, Adolph	0	9	Physics professor
Zeller, Eduard Gottlob	1	9	Philosophy professor (Neo-Kantian)

Literature

Belloni, Luigi: Hermannn von Helmholtz und Franz Boll. *Medizinhistorisches Journal*, 17, 1982, 129-37.

Bevilacqua, Fabio. Helmholtz's *Ueber die Erhaltung der Kraft*: The Emergence of a Theoretical Physicist. In: Cahan 1993b, 291-333.

Bierhalter, Günter: Zu Hermann von Helmholtzens mechanischer Grundlegung der Wärmetheorie aus dem Jahre 1884. *Archive for History of Exact Sciences*, 25, 1981, 71-84.

Burkhardt, Friedrich/Smith, Sidney (eds.): *Calendar of the Correspondence of Charles Darwin, 1821-1882.* New York: Garland, 1985.

Cahan, David (ed.): *An Institute for an Empire: The Physikalisch-Technische Reichsanstalt, 1871-1918.* Cambridge: Cambridge University Press, 1989.

Cahan, David: *Letters of Hermann von Helmholtz to His Parents, 1837-1846.* Stuttgart: Steiner Verlag, 1993a.

Cahan, David (ed.): *Hermann von Helmholtz and the Foundations of Nineteenth-Century Science.* Berkeley: University of California Press, 1993b.

Cahan, David: Helmholtz and the Civilizing Power of Science. In: Cahan 1993b, 559-601.

Caneva, Kenneth L.: *Robert Mayer and the Conservation of Energy.* Princeton: Princeton University Press, 1993.

Churchill, Frederick B.: Darwin and the Historians. *Biological Journal of the Linnean Society*, 17, 1982, 45-68.

Desmond, Adrian/Moore, James: *Darwin.* New York: Viking Penguin, 1991.

Di Gregorio, Mario A. (ed.): *Charles Darwin's Marginalia.* New York: Garland, 1990.

Goldstein, E.: Helmholtz: Erinnerungen eines Laboratoriumspraktikanten. *Die Naturwissenschaften*, 9, 1921, 708-11.

Elkana, Yehuda: Helmholtz' 'Kraft': An Illustration of Concepts in Flux. *Historical Studies in the Physical Sciences*, 2, 1970, 263-98.

Heidelberger, Michael: Force, Law, and Experiment: The Evolution of Helmholtz's Philosophy of Science. In: Cahan 1993b, 461-97.

Heimann, P.M.: Helmholtz and Kant: The Metaphysical Foundations of *Über die Erhaltung der Kraft*. *Studies in the History and Philosophy of Science*, 5, 1974, 205-38.

Helmholtz, Hermann von: *Vorträge und Reden*. 4th ed. Braunschweig: Vieweg, 1896.

Helmholtz, Hermann von: *Über die Erhaltung der Kraft*. Facsimile reprint ed. Transcribed by Christa Kirsten; Introduction by H.-J. Treder. Weinheim: Physik-Verlag, 1983.

Herbert, Sandra (ed.): *The Red Notebook of Charles Darwin*. Ithaca: Cornell University Press, 1980.

Hermann von Helmholtz' philosophische und naturwissenschaftliche Leistungen. *Wissenschaftliche Zeitschrift der Humboldt-Universität zu Berlin, Math.-Nat. R.*, 22 (1973), 277-361.

Herneck, Friedrich: Die Stellung von Hermann von Helmholtz in der Wissenschaftsgeschichte. *Wissenschaftliche Zeitschrift der Humboldt-Universität zu Berlin, Math.-Nat. R.*, 22 (1973), 349-55.

Hörz, Herbert/Laaß, Andreas. *Ludwig Boltzmanns Wege nach Berlin*. Berlin: Akademie-Verlag, 1989.

Holmes, Frederic L.: *Lavoisier and the Chemistry of Life*. Madison: University of Wisconsin Press, 1985.

Kay, Lily E.: Constructing Histories of Twentieth-Century Experimental Life Science: The Promise and Perils of Archives. *Mendel Newsletter*, N.S. 2, 1992, 1-4.

Kelvin Papers: Index to the Manuscript Collection of William Thomson, Baron Kelvin in Glasgow University Library. Glasgow: Glasgow University Library, 1977.

Kirsten, Christa (ed.): *Dokumente einer Freundschaft: Briefwechsel zwischen Hermann von Helmholtz und Emil du Bois-Reymond, 1846-1894*. Berlin: Akademie-Verlag, 1986.

Klauß, Klaus: Ein neuentdecktes frühes Dokument zur Geschichte der Erfindung des Augenspiegels durch Hermann v. Helmholtz. *NTM-Zeitschrift für Geschichte der Naturwissenschaft, Technik und Medizin*, 18, 1981, 58-61.

Koenigsberger, Leo. *Hermann von Helmholtz*. Braunschweig: Vieweg, 1902-3.

Kohn, David (ed.): *The Darwinian Heritage*. Princeton: Princeton University Press, 1985.

Kremer, Richard L. (ed.): *Letters of Hermann von Helmholtz to His Wife, 1847-1859*. Stuttgart: Steiner Verlag, 1990.

Kremer, Richard L.: Innovation through Synthesis: Helmholtz and Color Research. In: Cahan 1993b, 205-58.

Kries, Johannes von: Helmholtz als Physiolog. *Die Naturwissenschaften*, 9, 1921, 673-93.

Kuchling, Heinz: Vorwart. *Wissenschaftliche Zeitschrift der Humboldt-Universität zu Berlin, Math.-Nat. R.*, 22 (1973), 277.

La Vergata, Antonello: Images of Darwin: A Historiographic Overview. In: Kohn 1985, 901-72.

Lenoir, Timothy: Essay Review: The Darwin Industry. *Journal of the History of Biology*, 20, 1987, 115-30.

Lenoir, Timothy: Laboratories, Medicine and Public Life in Germany, 1830-1849: Ideological Roots of the Institutional Revolution. In: Andrew Cunningham/Perry Williams (eds.): *The Laboratory Revolution in Medicine*. Cambridge: University Press, 1992, 14-71.

Lenoir, Timothy: The Eye as Mathematician: Clinical Practice, Instrumentation, and Helmholtz's Construction of an Empiricist Theory of Vision. In: Cahan 1993b, 109-53.

Lummer, Otto: Der Physiker und die Technik. *Vossische Zeitung*, 28 August 1921, Blg. 4.

Nernst, W.: Die elektrochemischen Arbeiten von Helmholtz. *Die Naturwissenschaften*, 9, 1921, 699-702.

Olesko, Kathryn M.: *Physics as a Calling*. Ithaca: Cornell University Press, 1991.

Olesko, Kathryn M./Holmes, Frederic L.: Experiment, Quantification, and Discovery: Helmholtz's Early Physiological Researches, 1843-50. In: Cahan 1993b, 50-108.

Richards, Joan L.: The Evolution of Empiricism: Hermann von Helmholtz and the Foundations of Geometry. *British Journal for the Philosophy of Science*, 28, 1977, 235-53.

Riehl, A.: Helmholtz als Erkenntnistheoretiker. *Die Naturwissenschaften*, 9, 1921, 702-8.

Scharlau, Winfried (ed.): *Rudolf Lipschitz: Briefwechsel mit Cantor, Dedekind, Helmholtz, Kronecker, Weierstrass und anderen*. Braunschweig: Vieweg, 1986.

Scheel, Heinrich (ed.): *Gedanken von Helmholtz über schöpferische Impulse und über das Zusammenwirken verschiedener Wissenschaftszweige*. Berlin: Akademie-Verlag, 1972.

Schiemann, Gregor: Die Hypothetisierung des Mechanismus bei Hermann von Helmholtz: Ein Beitrag zum Wandel der Wissenschafts- und Naturauffassung im 19. Jahrhundert. In: This volume, 1994.

Stumpf, Carl: Sinnespsychologie und Musikwissenschaft: Helmholtz' Grundlegungen. *Vossische Zeitung*, 28 August 1921, Beilage 4.

Tuchman, Arleen Marcia: *Science, Medicine, and the State in Germany: The Case of Baden, 1815-1871*. Oxford: University Press, 1993.

Turner, R. Steven: Hermann von Helmholtz and the Empiricist Vision. *Journal of the History of the Behavioral Sciences*, 13, 1977, 48-58.

Turner, R. Steven: Vision Studies in Germany: Helmholtz versus Hering. *Osiris*, 2d ser., 8, 1993, 80-103.

Turner, R. Steven: *In the Mind's Eye: Vision and the Helmholtz-Hering Controversy*. Princeton: Princeton University Press, 1994.

Vieweg: Voranzeige, January 1902. In: Generallandesarchiv, 76/9939, Bl. [74v], Karlsruhe.

Warburg, E./Rubner, M./Schlick, M.: *Helmholtz als Physiker, Physiologe und Philosoph: Drei Vorträge*. Karlsruhe: Müllersche Hofbuchhandlung, 1922.

Wassersug, Richard J./Rose, Michael R.: A Reader's Guide and Retrospective to the 1982 Darwin Centenial. *Quarterly Review of Biology*, 59, 1984, 417-37.

Wien, W.: Helmholtz als Physiker. *Die Naturwissenschaften*, 9, 1921, 694-99.

Wilson, David B. (ed.): *Catalogue of the Manuscript Collections of Sir George Gabriel Stokes and Sir William Thomson, Baron Kelvin of Largs in Cambridge University Library*. Cambridge: Cambridge University Library, 1976.

Wilson, David B.: *Kelvin and Stokes*. Bristol: Adam Hilger, 1987.

Sachverzeichnis

Personenverzeichnis

Autorenverzeichnis

Fabio Bevilacqua
Dipartimento di Fisica “A. Volta”
Università di Pavia
Via Bassi 6
27100 Pavia
Italy

Robert M. Brain
Department of History and Philosophy of Science
Frec School Lane
Cambridge, CBI 2NL
United Kingdom

Jed Z. Buchwald
Dibner Institute, MIT
E 56 - 100
Cambridge, MA 02139
USA

David Cahan
Department of History
University of Nebraska
610 Oldfather Hall
Lincoln, NE 68588 - 0327
USA

Martin Carrier
Zentrum Philosophie und Wissenschaftstheorie
Universität Konstanz
Postfach 55 60 024
78434 Konstanz
Deutschland

Olivier Darrigol
83 rue Broca
75013 Paris
France

Hartmut Hecht
Europa-Universität Viadrina
Fakultät für Kulturwissenschaften
Große Scharrnstraße 59
15230 Frankfurt an der Oder
Deutschland

Michael Heidelberger
z.Zt.: Humboldt-Universität zu Berlin
Institut für Philosophie
Unter den Linden 6
10099 Berlin
Deutschland

Elfrieda Hiebert
Erwin Hiebert
Harvard University
Widener Library 172
Cambridge, MA 02138
USA

Frederic L. Holmes
Section of History of Medicine
Yale University School of Medicine
P.O. Box 3333
New Haven CT 06510
USA

Walter Kaiser
Rheinisch-Westfälische Technische Hochschule Aachen
Lehrstuhl für Geschichte der Technik
Kopernikusstraße 16
52074 Aachen
Deutschland

Horst Kant
Forschungsschwerpunkt Wissenschaftsgeschichte und -theorie
der Förderungsgesellschaft Wissenschaftliche Neuvorhaben mbH
Jägerstraße 10/11
10117 Berlin
Deutschland

Richard L. Kremer
Department of History
Dartmouth College
6107 Reed Hall
Hanover, NH 03755
USA

Lorenz Krüger
Georg-August Universität Göttingen
Philosophisches Seminar
Humboldtallee 19
37073 Göttingen

Wolfgang Küttler
Forschungsschwerpunkt Wissenschaftsgeschichte und -theorie
der Förderungsgesellschaft Wissenschaftliche Neuvorhaben mbH
Jägerstraße 10/11
10117 Berlin
Deutschland

Kathryn M. Olesko
Department of History
Georgetown University
Washington/DC 20057-1058
USA

Ulrich Röseberg
Forschungsschwerpunkt Wissenschaftsgeschichte und -theorie
der Förderungsgesellschaft Wissenschaftliche Neuvorhaben mbH
Jägerstraße 10/11
10117 Berlin
Deutschland

Gregor Schiemann
Technische Hochschule Darmstadt
Institut für Philosophie
Schloß
64283 Darmstadt
Deutschland

Volkmar Schüller
Forschungsschwerpunkt Wissenschaftsgeschichte und -theorie
der Förderungsgesellschaft Wissenschaftliche Neuvorhaben mbH
Jägerstraße 10/11
10117 Berlin
Deutschland

Annette Vogt
Forschungsschwerpunkt Wissenschaftsgeschichte und -theorie
der Förderungsgesellschaft Wissenschaftliche Neuvorhaben mbH
Jägerstraße 10/11
10117 Berlin
Deutschland

Renate Wahsner
Forschungsschwerpunkt Wissenschaftsgeschichte und -theorie
der Förderungsgesellschaft Wissenschaftliche Neuvorhaben mbH
Jägerstraße 10/11
10117 Berlin
Deutschland

M. Norton Wise
Department of History
Princeton University
129 Dickinson Hall
NJ 08544 - 1017
USA

www.ingramcontent.com/pod-product-compliance
Lightning Source LLC
LaVergne TN
LVHW081358011026
849161LV00013BA/2846